Technology, World Politics
& American Policy

INSTITUTE OF WAR AND PEACE STUDIES
of the School of International Affairs of Columbia University

Technology, World Politics, and American Policy is one of a series of studies sponsored by the Institute of War and Peace Studies of Columbia University. Among those Institute studies also dealing with war, peace, national security, and future world order are *Defense and Diplomacy* by Alfred Vagts; *Man, the State and War* by Kenneth N. Waltz; *The Common Defense* by Samuel P. Huntington; *Changing Patterns of Military Politics* edited by Samuel P. Huntington; *Strategy, Politics, and Defense Budgets* by Warner R. Schilling, Paul Y. Hammond, and Glenn H. Snyder; *Stockpiling Strategic Materials* by Glenn H. Snyder; *The Politics of Military Unification* by Demetrios Caraley; *NATO and the Range of American Choice* by William T. R. Fox and Annette Baker Fox; *The Politics of Weapons Innovation: The Thor-Jupiter Controversy* by Michael H. Armacost; *The Politics of Policy Making in Defense and Foreign Affairs* by Roger Hilsman; *Inspection for Disarmament* edited by Seymour Melman; *To Move a Nation* by Roger Hilsman, jointly sponsored with the Washington Center of Foreign Policy Research, Johns Hopkins University, *Planning, Prediction and Policy-Making in Foreign Affairs* by Robert L. Rothstein; *The Origins of Peace* by Robert F. Randle; *European Security and the Atlantic System* edited by William T. R. Fox and Warner R. Schilling; *American Arms and a Changing Europe: Dilemmas of Deterrence and Disarmament* by Warner R. Schilling, William T. R. Fox, Catherine M. Kelleher, and Donald J. Puchala; *The Cold War Begins: Soviet-American Conflict Over Eastern Europe* by Lynn E. Davis; *The Crouching Future: International Politics and U.S. Foreign Policy—a Forecast* by Roger Hilsman; *Germany and the Politics of Nuclear Weapons* by Catherine M. Kelleher; and *Developing the ICBM: A Study in Bureaucratic Politics* by Edmund Beard.

Technology, World Politics & American Policy

victor basiuk

COLUMBIA UNIVERSITY PRESS   NEW YORK 1977

Figure 2-1 is reprinted from the author's *Technology and World Power* (1970), which appeared as an issue of the "Headline Series" of the Foreign Policy Association. Used by permission.

Figures 14-1 and 14-2 are adapted from William C. Gough and Bernard J. Eastlund, "The Prospects of Fusion Power," *The Scientific American*, February 1971, p. 61. Copyright © 1971 by Scientific American, Inc. All Rights Reserved.

Quotations from Daniel Bell's *The Coming of Post-Industrial Society* (New York: Basic Books, 1973; rev. ed., 1976) are used by permission of the author and Basic Books, Inc.

Library of Congress Cataloging in Publication Data
Basiuk, Victor
  Technology, world politics, and American policy.
  Includes bibliographical references and index.
  1. International relations. 2. World politics—1945–
3. United States—Foreign relations—1945– 4. Technology and state.
5. Technology and civilization. 6. Twentieth century—Forecasts.
7. Twenty-first century—Forecasts. I. Title.
JX1255.B37   327   76–51841   ISBN 0–231–04154–1

Columbia University Press New York Guildford, Surrey

Copyright © 1977 by Columbia University Press
All Rights Reserved
Printed in the United States of America

To MY MOTHER AND THE MEMORY OF MY FATHER

Contents

Foreword   xi

Preface   xiii

1. Introduction   1

Part 1. The Sociopolitical Impact of Technological Trends

2. The Levels of Technological Impact   11

   The Nuclear Umbrella 11; "Subnuclear Umbrella" Military Technology 12; Nonmilitary Technology 15

3. Technological Trends and Prospects   17

   Decline in the Deterministic Effect of Resource Location 17; The Integrative Effect of Technology 19; Global Projection of Influence Through Technology 24; The Growing Scale and Costs of Technology 27; The Rapidity of Technological Change 29; Technological Impact and Social Discontinuities 33

Part 2. The Developed Regions in a Technological Future

4. The United States   41

   America's Strong Points 41; Weaknesses 43; Policy in Flux 46; Problems and Issues to Be Resolved 53; The Prospect 58

5. The Soviet Union   60

   Postwar Progress 60; Emerging Problems 61; Soviet Strengths 64; Soviet Weaknesses 69; National Resources and the Military Sector 75; Potential Models of Soviet Society 78; The Prospect 83

6. Western Europe   89

   Post-World War II Resurgence of Europe's Vitality 89; The "Technology Gap"—Illusion or Reality? 90; Western Europe's Weaknesses 92; The Trends of Change 100; Europe at a Crossroad 103; Model One: A Technological-Trends-Resistant Western Europe 103; Model Two: A Functionally Oriented Western Europe 107; Model Three: A Unified Europe 109; Technological-Trends-Resistant or a Modified Functionally Oriented Western Europe? 110; A Functionally Oriented Western Europe: Directions of Evolution 111

viii   Contents

7. Japan   114

Favorable Conditions and Factors 114; Organization and Will 116; Japan's Technological Capability 120; The Benefits and Challenges of the Future 122; The Prospect 130

Part 3. Technological Change and the World Arena

8. Technology and Distribution of Power   135

Change in the World Distribution of Power through Military Technology 135; Nonmilitary Technology and Power 137; Less Developed Countries and International Power 139; New Candidates for International Power 144; The Sociopolitical Impact of Technology on Advanced Societies 146; The Changing Importance and Nature of Power 150

9. Technology and International Stability   154

Strategic Nuclear Weapons, the Superpowers, and World Stability 155; Nuclear Weapons and Third States 157; The "Conventional" Military Sector and World Stability 159; Transnational Violence 160; Nonmilitary Technology and World Stability 161

10. Technology and World Society   167

World Distribution of Power and the Nature of International Society 167; World Distribution of Power and Some Possible Scenarios of International System 169; Technology and Regional Integration 174; Technology, Functionalism, and World Society 175; Global Functional Institutions and World Society 178; The Multinational Corporation 181; Marine Resources Development 183; The International System: Directions of Evolution 186

Part 4. American Policy and the Impact of Technological Realities

11. Concepts, Institutions, and Values   195

Control of Technology and National Strategy 195; The Rational and the Constituency Method of Decision Making 197; Institutions and Characteristics of Decision Making 200; Institutions and Values 207; Toward a New Balance between the Rational and the Constituency Decision Making 210

12. The Home Front   215

Technological Strategy 215; Nonmilitary Technology and the Question of Priorities 219; Modification of the Technological Superstructure 225; Government and Business 228; Technological Impact and Sociopolitical Discontinuities 231

13. The Outside World   235

U.S. Policy and the Soviet Union 236; Western Europe 245; Japan 250; The United States and a Future International System 252

Contents ix

Part 5. Over the Horizon: Technology and Directions of Societal Development

14. The Changing Nature of Advanced Societies   259

The Postindustrial Society: Where Is It Taking Us? 259; The Postindustrial Society, Political Authority, and Collective Human Will 266; The Postindustrial Society: Whither Power? 268; The Semistationary Society: A New Direction of Societal Evolution? 275; Sociopolitical Implications of the Advent of the Semistationary Society 280; International Implications of the Advent of the Semistationary Society 287

15. The National Purpose in an International Context: Whither America? 293

America's Purposes and the Extent of Their Achievement 294; Some Perceived National Goals Reexamined 298; The Crisis of National Purpose 303; Criteria for a National Purpose and Goals 304; Toward an Articulated National Purpose and Societal Goals 306; A Road to Implementation 310; The National Purpose in a Global Context 315; Conclusions: America and a Changing International System 326

Notes   331

Index   385

List of Illustrations

1-1. Technological Advance and Its Impact on Societies	3
2-1. Scientific-Technological Spectrum and Its Effect on Distribution of Power	16
10-1. Changes in the Relative Incidence of Power and Authority	191
12-1. Desirable Direction of Re-allocation of Resources	219
14-1. Present-day "Linear" Economy	277
14-2. "Closed-materials" Economy	277

Foreword

THE WIT WHO DESCRIBED the Pentagon as like a log floating downstream on which there were twenty thousand ants all imagining that they were steering might equally well have used his metaphor to describe the United States and the way in which Americans long exercised their asserted technological leadership in a world of bewildering change. By now, however, it is banal to say that the technological variables are among the most difficult to control in the equations of world politics. Can one therefore simply treat them as independent variables, to be studied and accepted as given but not themselves the proper concern of policy makers at all? Victor Basiuk thinks not.

Before one can know what opportunities exist to make the technological variables work to support one's policy preferences one must understand some of the major technology-induced transformations in world politics which have already occurred in this century and develop a sensitivity to those changes that lie "over the horizon." *Technology, World Politics, and American Policy* is meant to provide that understanding and promote that sensitivity.

Some of Dr. Basiuk's readers may see technological change as a great beast in headlong motion from whose back humanity dares not jump. For others it may be a tractable and ethically neutral genie ready to serve whoever may control the levers of power. Many will see it as some Jekyll-and-Hyde combination of bad beast and good genie. All must, however, agree that in thinking about world politics scholars and policy makers alike must take more explicit account of that change. They need to understand better the complex ways in which changes in technology widen or narrow the range of policy choice and complicate or facilitate the making of choices. How, they may ask, can the processes of technological change be made to add to, or kept from substracting from, the public good (however that may be defined)?

This volume is both more and less specialized than its title alone may suggest. In our pluralistic world system, advanced Western technology is not the only engine of seemingly irresistible change. The now largely completed process of dismantling overseas empires opens a whole new era in North–South relations. There is a revolution in human dignity as well as a revolution in technology. A book by a Third World scholar on "Technology, World Politics and the Third World" addressed to the needs of Third World policy makers would be very different from the present volume. There is need for both kinds of volumes.

Victor Basiuk is a political scientist who has specialized in international relations. He is a pioneering and largely self-taught student of advanced technological processes as they bear on changing power patterns in world politics. Finally, he is a participant-observer in Washington decision processes with respect to science and technology in both the civilian and military spheres. He has brought all of these perspectives to bear in this ambitious and wide-ranging study.

WILLIAM T. R. FOX

Preface

GENERAL OMAR BRADLEY once remarked that in world affairs the United States should be steered by the stars, and not by the lights of each passing ship. It is difficult to challenge the wisdom of this statement, and yet there is always the danger of the opposite extreme: neglect of the lights of passing ships can lead to collision courses and thus jeopardize the ability of the American ship of state to reach her destination.

At the expense of some oversimplification, one could say that there is a tendency in Washington to be preoccupied by the lights of each passing ship, whereas the academic world tends to display its own partiality for pursuing the stars. A dichotomy between the two thus exists. This dichotomy has been of concern to me for some time, largely because I have had the benefit of being involved in both the academic and the governmental worlds for a number of years.

In part, this book attempts to bridge the existing gap. First, it tries to identify some of "the stars," since they alone can provide us with a sense of direction. But it does not neglect the obstacles—in the governmental process and in the external world—which may prevent us from getting to our chosen destination. The analysis is conducted with particular reference to the impact of technology and the kind of constraints and opportunities this impact has created or is likely to create for us in coming years.

A book that covers the world and attempts to look 75 years into the future has to lean on a good number of individuals for information, ideas, critical comments, and other forms of assistance. Not all of them can be given the credit which is justly theirs, but particular indebtedness to some must be singled out.

To William T. R. Fox, who was for many years Director of Columbia University's Institute of War and Peace Studies, my indebtedness is multiple and includes my initial interest in technology and international relations. When, as a graduate student at Columbia years ago, I registered for his seminar on this subject, little did I know that it was the beginning of my career in what has eventually become an area in its own right. The Institute of War

xiv   Preface

and Peace Studies, which sponsored this book, provided a stimulating and most pleasant environment for research and writing. Professor Fox's comments and judicious advice were especially valuable in that, in a book leaning on a number of disciplines, they kept me from losing sight of the fundamentals of international relations.

Caryl P. Haskins' pioneering book on *Scientific Revolution and World Politics* bears a resemblance to mine in scope and approach that is more than coincidental. As both a "hard" and a political scientist, Dr. Haskins provided the kind of perceptive advice and support to my work on the book which probably no other single individual could.

Christopher Wright, then Director of Columbia's Institute for the Study of Science in Human Affairs, commented on an early draft of the book, and his remarks led to significant improvements. Robert L. Rothstein, then at Johns Hopkins University, read the manuscript with care and made a number of thoughtful and useful suggestions. Donald Fink, then at the Environmental Protection Agency, not only contributed to the book in his professional capacity as an economist and environmentalist, but provided editorial assistance as well.

Anna Hohri, Administrative Assistant of the Institute of War and Peace Studies, was most helpful in many ways, by making various arrangements as necessary and solving problems with her customary efficiency, tact, and cheer. Linda Wangsness in New York and Jean M. Flanagan and Maude Claiborne in Arlington provided particularly competent typing assistance.

I wish to thank Columbia University Press—especially Bernard Gronert, the Executive Editor, and Leslie Bialler, my manuscript editor—for their responsible and efficient handling of the complex process which results in a finished book.

Particular thanks are due to the following individuals: Arthur G. Ashbrook, Jr., Washington, D.C.; Peter F. Drucker, Claremont, California; Stephen E. Doyle, National Aeronautics and Space Administration; Bertrand M. Fainberg, ABC Television Network; P. J. Fellon, formerly at The British Embassy, Washington; Wreatham E. Gathright, Department of State; Robert Gilpin, Princeton University; T. Keith Glennan, Department of State; Leland M. Goodrich, Columbia University; John P. Hardt, Congressional Research Service, Library of Congress; Robert S. Hayes, National Defense University; Louis Henkin, Columbia University; John D. Holmfeld, Staff, Committee on Science and Technology, U.S. House of Representatives; David A. Kay, American Society of International Law; Oliver J. Lissitzyn, Columbia University; T. Dixon Long, Case Western Reserve University;

James W. Morley, Columbia University; E. Raymond Platig, Department of State; Zdenek J. Slouka, Lehigh University; Istvan Szent-Miklosy, Queens College; Jiro Tokuyama, Nomura Research Institute of Technology and Economics; Kei Wakaizumi, Kyoto Sangyo University; Stanislaw Wasowski, Georgetown University; Howard M. Wiedemann, Department of State; Thomas W. Wolfe, RAND Corporation.

It is usually understood that individuals whose assistance is acknowledged are not responsible for the book's faults, whether of commission or omission. However, since this book includes normative considerations as well as an analysis based on empirical research, I should like to make my sole responsibility for its content explicit.

Arlington, Va. VICTOR BASIUK
January 2, 1977

Technology, World Politics
& American Policy

Introduction

1

THIS BOOK ANALYZES the future impact of technology on international relations and societies, focusing on the developed regions. It raises this question: What are the implications of this impact for American policy at home and abroad? The time frame is approximately 75 years, until about the year 2050. Parts 1 through 4 of the book deal with the first 25 years (until about the year 2000); they examine technologies, their potential impacts, and the policy options they create. Part 5 deals with broader issues, such as the effect of technology on directions of societal development, and looks beyond the year 2000.

That more distant future is not left here to coming generations to cope with nor is it treated merely as an entertaining subject to the present generation (although, no doubt, some aspects of the future are quite entertaining). Part 5 examines the meaning of likely developments in the early twenty-first century to us now and what we can do to meet their challenges and, in so doing, solve some of the critical problems facing us today. In particular it analyzes requirements of the future in the context of America's national purpose, past and present, and suggests how the national purpose can be an instrumentality for attaining a better future in the near as well as in the long run.

This is a study in international relations.[1] In recent years the scope and concerns of this field have been broadening. International relations began to emphasize the utility of interdisciplinary approaches, to examine more thoroughly internal national developments and decision making as a clue to understanding forces and group behavior on the international scene, and to raise questions about how the field can be made more useful to the policymaker, especially in regard to the future.[2] I attempt to follow these new trends and, when it seems necessary, to broaden the concerns of the field.

Since the analysis is focused on technology and its impact—a pervasive, multifaceted, and dynamic force, with a high propensity to introduce change in social entities from the smallest and most local to the global—opportunities to detect areas of analysis that might become of concern to international relations arose. These new concerns might include the body of knowledge that the philosophers and historians of civilizations have created. When subjected to a more systematic scrutiny with the help of some of the existing tools of international relations, it might significantly expand and enrich the field. Another example is that of a societal or "national" purpose that, paradoxically, might be an appropriate area of attention precisely when pluralization within societies and integration among them make their boundaries and definitions obscure. Attention to these two areas—civilizations and societal purposes—might help us elucidate normative criteria, so important for policy research.

International relations has not yet developed a comprehensive, generally accepted theoretical framework. In fact the field at present encompasses several different schools of thought challenging one another's approaches.[3] This book is a policy-relevant, not a theoretical study; it therefore makes no attempt to remedy the situation by building a comprehensive theory or paradigm. However, there is a definite need in policy related to the impact of technology to formulate concepts and intellectual tools of analysis that would have more than an *ad hoc* value. To that extent an effort to conceptualize is made here. Perhaps some of the tools of analysis or the concepts suggested may serve as components of a broader theory or paradigm, if and when developed. However, considerations of policy prevail. If a particular categorization, a tool of analysis, or a concept is useful in terms of policy, it is employed here on the strength of its usefulness, even though it may as yet lack the finesse or precision of a theoretical instrument.

The various forms of technological impact are discussed at length in the main body of the book, but one form continuously reemerges. Technological advance introduces a high degree of complexity into societies; it produces an increasing structural and role differentiation, and this leads to articulation of interests and the formation of interest groups. The multiplication of interests increases the scope for symmetrical (as distinguished from asymmetrical or hierarchical) relations among them [4] and tends to result in two divergent phenomena in the resolution of issues: either a deadlock among the interests concerned or a continuous, amorphous interplay and (relatively minor) adjustments among them to avoid a sharp confrontation or a deadlock. Actually, it is usually a combination involving both, in varied orders and proportions.

Technology thus creates something of a paradox. On the one hand it immensely broadens horizons for societies to change their environments and to mold their own futures by making available the necessary technological instruments for that purpose. On the other hand it greatly handicaps decision

Introduction 3

making and mobilization of human will to steer societies effectively into better futures. To the extent that interests are deadlocked, societal will is stymied. Insofar as a highly pluralistic, amorphous situation involving interplay of interests exists, adequate leverages are not available for society to steer itself into clearly defined directions or to achieve rationally other objectives important for society as a whole.[5]

Yet another and closely related effect of technology is beginning to emerge: Technological advance complicates societies and their problems so much that a rational understanding of the problem and the finding of an intellectual solution become difficult or nearly impossible. It thus tends to handicap human intellect as well as societal will. But so far, between the intellect and the will, the latter is handicapped much more. The impact of technological advance on societies can be visualized conveniently when presented in the form of a diagram, as in figure 1-1.

The foregoing discussion should make it clear by now that this book is concerned with analysis of the future and finding of solutions for its problems, as well as with the ability of society to implement them. The latter aspect leads inevitably to the nature of decision making, organizations, and institutional change, which are examined in chapters 11, 13, and 15.

Since technology will bring both great benefits and serious perils to society, one of my central concerns is *controlling* technology and its impacts; the alternative is to be controlled by them. Control of technology requires a sys-

Figure 1-1. Technological Advance and Its Impact on Societies

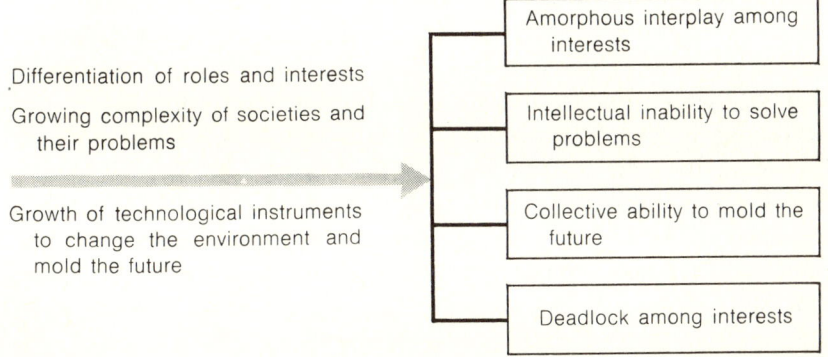

Technological advance over time.

tematic effort by society to maximize technology's beneficial possibilities and minimize its harmful effects in the context of social goals and purposes.

Another closely related central concern is the viability and vitality of the United States as a society in the coming decades. This concern stems from the propensity of future technology to have far-reaching political ramifications affecting stability and survival of societies and nations. The criterion of vitality and viability in the future is also applied to other advanced regions discussed in this book, namely, Western Europe, the Soviet Union, and Japan.

Two concepts were found particularly convenient. One is the distinction between technology as a *dependent* variable and as an *independent* variable; it is especially helpful in dealing with one of the principal concerns of the study—the need to control technology rather than allow it control us. In the last analysis technology is always a dependent variable—inasmuch as it takes human beings to create and apply it. However, from the point of view of a particular society, technology can be a dependent, a partially dependent, or an independent variable.

Let us take the case of a particular major technology—nuclear, outer space, or marine—that appears to be within reach, promises definite advantages, but is not yet developed. If the government decides to undertake a program to develop and apply a given technology for the benefit of the nation, we have a reasonably straightforward case of technology as a dependent variable.

But technology produces an impact on society as an independent variable for at least three reasons. In the long run no single nation can monopolize a particular technology. When other nations acquire it, for whatever purpose, the given technology produces an impact as an independent variable on the first nation. An exception to this would be the case in which the first nation deliberately transfers a technology to another nation. Under these circumstances the technology may remain, with respect to the nation of origin, a dependent variable, or, perhaps, as a "semidependent" variable, since a degree of control over it will thus be lost. In short, the existence of the nation-state system with its decentralized power and decision making is one reason why technology tends to produce an impact as an independent variable.

Secondly, even if technology is developed and used within a particular nation, it may produce an impact on that nation as an independent variable. This occurs, for example, when private individuals or organizations develop or use technology to achieve their own objectives—be they economic gain, pleasure, or power. The effect of such technology on the nation itself may not be in its best interests. Thus, the decentralized nature of a given society with respect to the development and application of technology is another reason technology may produce an impact as an independent variable on society as a whole.

Finally, technology may produce an impact as an independent variable if its impact cannot be fully foreseen. A nation may decide to develop and use a particular technology—say, a form of weather modification. But eventually it may become apparent that the conversion of Nevada into a blooming region changed the pattern of precipitation in California and Oregon, making some parts of these two states less fertile and others more so. Accordingly the impact of our hypothetical improvement in the climate of Nevada turned out to be: (1) changes in land values in the neighboring two states, with resultant political and legal (damage suits) repercussions; (2) migration of population from California to Nevada and consequent reduction of the significance of California as the most populous state and the pacesetter for American culture; (3) an initial jump, and eventual collapse, in prices of all other desert lands in the United States outside of Nevada; (4) a wave of antitechnology demonstrations in various parts of the country. Not all of these developments were likely to have been foreseen when the decision to improve Nevada's climate was made.

In summary, technology can be a completely dependent variable only under conditions of world government, centralized decision making, and perfect foresight of its impact. If these conditions are met, some technologies will begin to manifest signs of acting as an independent variable with regard to our unified world society should contact be made with intelligent life on another planet.

From the point of view of a particular society—say, the United States—"control" of technology and of its impact means making it a dependent variable. Realistically viewed, and for the reasons already explained, technology cannot be a completely dependent variable; its function will cover the entire spectrum from dependent to independent variable. "Controlling" technology—or making it a dependent variable—does not necessarily mean that a given society must centralize all decision-making processes with regard to the development and application of technology. Actually all the development and application of technology can stay in private hands, be decentralized, and still largely fulfill the condition of "control" from the societal point of view as long as society observes technological development, curtails its growth in undesirable directions, and encourages progress in the direction beneficial to society as a whole.

A second concept frequently used here is the *differential* impact of a particular technology, that is, its ability to affect individuals, groups, and nations in unequal ways. This phenomenon can be illustrated by a discussion of the differential impact of technology on nations in terms of power. There are at least two reasons why some nations gain more from a particular technology than others do.

A given technology may benefit some states more than others because, in conjunction with certain attributes possessed by the "favored" states, the technology can be more effectively used by them. The introduction of steam propulsion for marine use in the nineteenth century was bound to benefit Great Britain, with her extensive sea power, more than Russia, with her limited access to warm seas. Conversely, the introduction of railroads in the same century was bound to benefit the United States, Russia, and Germany more than Great Britain because these three great powers had a larger land base suitable for effective integration by railroads than Great Britain did. In terms of national power Great Britain was thus disadvantageously affected by the introduction of railroads, which not only helped to develop inland resources of the continental giants but also gave them overland mobility rivaling the seaborne mobility then dominated by the British. In this regard the effect of technological change on the distribution of power is almost "deterministic."

A second reason for differential impact of technology stems from what may be called—for lack of a more suitable word—its "volitional" effect. A particular technology may not inherently favor a given state—it may potentially favor other states much more—but the state in question realizes the technology's long-term value and promotes it vigorously and thus eventually augments its power in spite of the handicaps. A case in point is the present promotion of technologies for exploitation of the oceans by the Soviet Union. Similarly a state may, *a priori*, be favored by a particular technological development (a case of "favorable determinism"); it realizes the technology's potential benefits and, through a concerted and farsighted effort, exploits it well beyond what would normally be expected. Thus, in view of her global interests, Great Britain was bound to benefit from technological developments in telecommunications (cable and radio) in the nineteenth and the early decades of the twentieth century. In fact, through farsighted policies, she achieved a very effective worldwide ascendancy in this field that was not undermined until the requirements of World War II forced the United States to allocate extensive resources to global telecommunications. Another example is the effective development and use of airpower and the tank by Hitler in the early phases of World War II. Moreover, through a timely volitional effort some nations may strengthen their relative power position by forestalling an injurious impact of technology on society. Such a situation may arise when rapid technological change tends to produce insecurity and political unrest because of occupational and social dislocations or similar developments.

The "deterministic" and "volitional" elements of the impact of technology are closely interdependent, albeit antagonistic. If a national effort is aimed against an unfavorable technological impact or trend, then the gain in power is made in direct opposition to the deterministic impact of technology. If the deterministic impact is favorable to a nation but the nation makes a strong volitional effort to capitalize on it, then volitional elements take over. The deterministic impact of technology implies that technology acts as an independent variable. "Volition," in our context, involves the use of conscious societal will; whether it runs against the existing trend in technology—the trend's continuance based on the provision that events are allowed to take their course without interference—or capitalizes on a favorable technological trend through a conscious and purposeful policy, it views and uses technology as a controlled (dependent) variable.

From a system characterized by a zero-sum game, in which each state competes with other states and where one state's gain is (or is thought to be) another's loss, international society is gradually developing characteristics of a positive-sum game, in which—through international cooperation—every participant gains. Advanced technologies, because of their huge costs, large scale, and, in the case of nuclear weapons, immense destructive power, provide an important impetus to international cooperation. So does the effect of technology on the environment. However, in the midrange future, the stimulus for cooperation will be only partially on a global scale; many of what may be called "imperatives" of technology and ecology can be met on a level short of global—regional and, in some cases, bilateral. Accordingly, whereas forces for certain forms of international cooperation on a global scale will grow, a great deal of room will be left for competition—between regions, among nations, and among other actors in world politics.

Competitive elements will still be strong in this century and beyond, although in interstate relations (but not necessarily in all relations in the international arena), they will be gradually declining. Competition will be especially felt in nonmilitary technology on which this book is focused. But a technologically dependent world can no longer be fully understood in terms of traditional adversaries and competitive zero-sum games any more than policy can be classified as domestic or foreign.

Finally, the question of definition of technology should be addressed. Technology can be narrowly defined as the totality of physical man-made tools, or, more broadly, as all man-made tools, physical and nonphysical (e.g., management techniques), intended to modify man's environment. My concern here is with physical tools, although at times I discuss nonphysical tools. An even broader definition of technology includes the tools, material results of their application, and their social impact. As used here, the term *technology* frequently includes its impact.[6] This broader definition permits the use of phrases such as "technology does" this or that. This is a useful shortcut, which spares the reader from the verbiage and the tedium of numerous qualifications. Unless this is clearly understood, many statements in subsequent chapters may appear to be gross oversimplifications, with technology viewed as almost reified, with a will of its own. Technology does not have such a will, but the impact "it"—or its particular forms—produces may create imperatives of its own.

The Sociopolitical Impact of Technological Trends

part 1

The Levels of Technological Impact

2

For our analysis it is convenient to distinguish three levels of scientific–technological impact on international relations, viability and vitality of societies, and national power (the latter term being used in its broadest sense to connote a spectrum ranging from mild forms of influence to the ability to wreak devastating destruction). The first is the nuclear-umbrella level, which mainly consists of the strategic nuclear weapons of the United States and the Soviet Union. It does not include tactical nuclear weapons insofar as they are used—or intended to be used—short of an all-out war. The second is the subnuclear-umbrella military level. It encompasses general-purpose forces (including tactical nuclear weapons) and those military–technological instruments employed or employable for political purposes short of war, such as a naval demonstration, "quarantine," blockade, and the like.[1] The third is the nonmilitary level. It primarily includes technology relevant to industrial and agricultural development. Obviously technology on this level is relevant to the other two to the extent that it provides a base for military production and employment of military instruments of power.

The Nuclear Umbrella

The existence of a nuclear umbrella is a recent phenomenon—it rose to a spectacular degree of importance in the 1950s. Politically it has proved over time to be limited in its power rewards. To be sure, the initial acquisition of nuclear weapons gave the United States a potentially vast military superiority. But the great destructiveness of these weapons and the unwillingness to use them have shown that the opportunities to transfer overwhelming nuclear superiority into political power are limited. The picture has become further complicated by the success of the Soviet Union in developing its own nuclear technology, which accounted in large measure for the rise of Soviet military power in the last 30 years. However, after a certain point in weapons development and accumulation of hardware had been reached, the achievements of the two superpowers merely tended to neutralize each other, without a meaningful change in the world distribution of power.[2] Realization that a point of sharply diminishing marginal returns on the strategic nuclear level had been reached resulted in the decline of expenditures for strategic nuclear weapons in the United States over the past several years and contributed to the 1972 agreement by the Soviet Union and the United States on a strategic arms limitation treaty.

What would be the significance of the acquisition of strategic nuclear weapons by third powers in the next decade or two? If such weapons are acquired by a responsible industrial power (e.g., Japan), the nation's power would be enhanced via greater prestige, potentially greater independence in the global arena, and, probably, greater voice in international councils. But, short of an all-out nuclear war, the nation's nuclear capability would be absorbed in the overall nuclear umbrella, with the corresponding effect of rapidly diminishing marginal returns for any additions to the nuclear arsenal.

The situation would be somewhat different in the case of the lesser states whose nuclear weapons capability would have rudimentary means of delivery, with strictly regional effectiveness. An early acquisition of nuclear weapons by a state such as Israel or Argentina would significantly change the regional balance of power, but as other states in the same region developed nuclear weapons, their marginal utility would also rapidly decline. Thus regional nuclear umbrellas would be established; the actual use of the weapons would be discouraged by their destructiveness and by the likely efforts of the superpowers concerned about the possibility that a regional nuclear action might trigger off global nuclear war.[3]

"Subnuclear-Umbrella" Military Technology

Historically what is now "subnuclear-umbrella" military technology (or "general-purpose forces" technology) has been an important means of changing the world balance of power. Thanks to the sailing ship and to gunpowder European states had initially succeeded in gaining an ascendancy over the rest of the world that lasted for several centuries.[4] The employment of military technology in wars between major powers was critical in the rise of some states and the decline of others.

After its early preoccupation with the nuclear umbrella in the 1950s, the United States increasingly turned her attention to this level of military technology. It was thought that individual countries in Southeast Asia, Africa, and Latin America might be gained or lost to the Free World, depending on our success in developing, producing, and applying technological instruments on the subnuclear level. A number of political trends—the rise of the less developed countries (LDCs), the disappearing bipolarity, the instability and decentralization of power in the world arena—were significant in bringing technologies pertaining to "limited wars" back into fashion.[5]

However, there are signs that the stalemate dominating the nuclear level is beginning to spread into the conventional military sector as well and that opportunities for changing the world balance of power on this level are decreasing. The reasons for this are political, military, and technological. Concerned that a local conflict may escalate and trigger a nuclear war, the superpowers appear eager to restrict or smother it in an early stage and thus forestall the stronger party from effecting significant change in the local balance of power through force of arms (e.g., the Middle East). The stalemate over Vietnam suggested that substitution of technology for political solutions in limited conflict did not necessarily provide a solution. Vietnam also indicated that modern conventional military technology, pitted against itself, has a tendency to deadlock: American technology was fighting not only the Vietnamese but also the products of Soviet and Chinese factories. Our aircraft, though superior, were practically stalemated by the effectiveness of Soviet-built surface-to-air missile sites, assisted by gun antiaircraft batteries.[6] The lessons of Vietnam, including its costliness, are not likely to be lost on either the Americans or the Soviets.

Lastly, the increasing global projection of Soviet power on the subnuclear military level tends to narrow the opportunity for the United States to intervene effectively in local situations and neutralizes our military power. Given the Soviet naval presence in the Mediterranean, the U.S. government would think twice before undertaking another military landing in Lebanon. During the *Pueblo* incident (January 1968), the U.S. Navy moved a number of ships toward Korean waters; the presence of Soviet vessels in the area neutralized the intended effect of our action.[7]

Future technology is likely to reinforce the growing stalemate. Greatly improved and nearly instantaneous satellite surveillance will help the United States and the Soviet Union to be well and swiftly informed on the location of each other's forces, particularly at sea. Mobility, both in the air and at sea, will significantly improve, permitting each to block the other's movements. In coming years we shall witness a growing introduction of electro-optical weapons, a development of major significance in the history of warfare. Because of their importance it might be useful to review briefly some potential implications of electro-optical weapons at this point.

Electro-optical weapons employ guidance or control based on precise information about target location, information obtained from television, infrared, or laser sensors. The essential feature of such weapon guidance is its accuracy: Almost any target that can be seen—by TV, a night vision mechanism, or simply a human eye—can be hit with a precision ranging from several inches to several feet. As a result almost every weapon becomes lethal. This development, taking place now, is historically unprecedented; of the many millions of conventional weapons—unguided projectiles, bombs, and missiles—expended in modern warfare, only relatively few hit the target. However, these extraordinary electro-optical weapons are expensive and are suitable for use only against high-value targets (aircraft, bridges, ships, etc.), and not against individual soldiers.

Although an early acquisition of electro-optical weapons would probably give the offensive side a decisive advantage, in the long run these weapons overwhelmingly favor the defense. The following developments—possible, perhaps, by 1990—are particularly relevant in this regard: (1) electro-optically guided defensive weapons sufficiently precise to intercept individual missiles, (2) laser antisensor weapons designed to blind or confuse the electro-optical guidance on strike weapons, and (3) high-energy laser weapons capable of burning through an approaching missile or aircraft. If most or all of these developments materialize, they would greatly inhibit tactical strike warfare and thus significantly strengthen stalemate among militarily advanced nations.

There would be two principal means of breaking the stalemate: (1) A saturation attack—one so massive in numbers of missiles that the highly computerized and automated defense, equipped with laser weapons, cannot meet it in adequate numbers. Such saturation tactics would be extremely expensive. (2) A resort to tactical nuclear weapons. Since tactical nuclear weapons do not have to hit the target to be devastatingly effective, short-range electro-optical defensive weapons provide a generally poor defense against this kind of attack. But a resort to tactical nuclear weapons threatens escalation to all-out nuclear war and thus discourages this approach to the breaking of the stalemate. What would remain then is the old-fashioned, bloody war of the foot soldier, without sophisticated weaponry involving high-value targets. Although such a war is conceivable, the difficulty of separating its conduct from the available highly lethal and devastating advanced weapons in itself might discourage warfare on the subnuclear level.[8]

The foregoing discussion is not intended to suggest that the subnuclear-umbrella military level is likely to be stalemated in the way the nuclear umbrella is. It will never be completely stalemated. There may be situations in which the superpowers may not choose to stalemate each other (e.g., the USSR may be willing to let the United States defeat China in a localized conflict); there might be cases in which the superpowers will not stalemate each other, because of certain characteristics of the particular situation; furthermore the superpowers may not choose or be able to participate in a particular localized conflict. The important point is that the subnuclear-umbrella military level is being partially stalemated, and the stalemate will increase in coming years.

Nonmilitary Technology

Over the centuries nonmilitary technology accounted for at least as much change in the distribution of world power as military technology. In recent times nations have grown powerful thanks in large part to peacetime technologies that made them great industrial centers.[9] Military power usually followed in the wake of industrialization. Unlike the other two sectors there is no sign of the emergence of a stalemate to prevent nonmilitary technology from effecting changes in the distribution of power of nations and regions. To mention but a few examples: Computers can provide a major impetus to the growth of other areas of a nation's technology; facilitate economic planning, investment, and marketing strategy of corporations; and thus provide differential advantage to the countries or regions leading in this area of technology. Technological capability for the development of marine resources opens up new vistas in the three quarters of the globe that is virtually unexplored, unexploited, and unsettled, with concomitant geopolitical implications. An efficient internal system of telecommunications and transportation stimulates the economic pulse of a country and enhances its vigor and viability in the international arena. Those advanced nations that solve their energy problems and reduce the cost of energy sooner than others will gain important advantages in world markets and strengthen their industrial capability and leverages of influence in the less developed world. The ability to manipulate hurricanes, covertly remove moisture from the atmosphere over another nation, to interfere with the layer of ozone surrounding the earth (thus subjecting the ground underneath to severe burns by ultraviolet rays of the sun) [10]—for all of which the technologies could be developed by the mid-1990s—would provide a more direct and dramatic means of changing the power distribution.

The impact of nonmilitary technology, when viewed in the context of general scientific and technological trends, suggests at least one important con-

16 Technological Trends

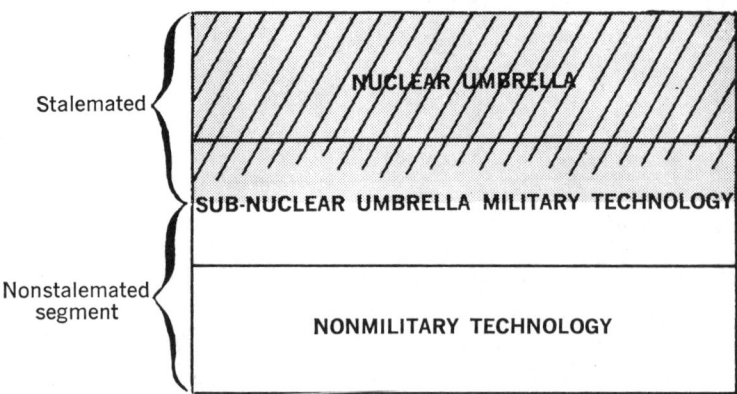

Figure 2-1. Scientific-Technological Spectrum and Its Effect on Distribution of Power*

*This diagram is intended for visual illustration only and is not designed to quantify the power impact of technology. In this respect, the division of the spectrum into three equal blocks does not necessarily indicate that each is equal in its power effect.

clusion. Because nonmilitary technology now occupies a larger part of the spectrum of technologies capable of changing the world distribution of power, because its share of the nonstalemated area of the spectrum is growing, and because the pace of technological change is more rapid now than in the past, the importance of nonmilitary technology in changing the distribution of power and affecting national security will be correspondingly greater than it was heretofore. Figure 2-1 illustrates the relationship of nonmilitary science and technology to the other segments of the spectrum. Historically the entire spectrum of sciences and technologies played a role in changing the distribution of power; the nuclear umbrella and the concomitant stalemate are barely 25 years old.

Whereas military technology is primarily focused on considerations of power, the purpose of nonmilitary technology is broader, and its impact is much broader than its purpose. The significance of nonmilitary technology for viability and vitality of societies will be increasingly critical in coming decades. Trends in the nonmilitary sector of technology and their impact are discussed at greater length in the next chapter.

Technological Trends and Prospects

3

ONE MAY DISTINGUISH five major trends in the technological revolution transforming the modern world.

Decline in the Deterministic Effect of Resource Location

Technology has historically played a major role in creating new resources or helping to exploit traditional ones, but in most instances it has tended to be tied to the location of its resources base. In recent history major industrial centers were built around coal–iron ore complexes, such as those in the Durham–Cleveland (Britain), the Great Lakes–Pennsylvania, the Ruhr, and the Krivoi Rog–Donbas regions. The development of railroads was closely tied to the availability of resources—either mineral or agricultural—on the land that the railroads crossed. Electricity made water power a significant source of energy, but hydroelectric stations and the economic activity stimulated by them were narrowly confined to rivers and their vicinity.

What is noteworthy today is the decline in importance of the location of raw materials and sources of energy as a determining factor in economic development and the rise of new power centers. No single invention was responsible for this decline, but the most significant technologies initially responsible were chemistry and transportation. By combining and recombining elements and converting them into more useful products, chemistry undertook to make "resources" out of the most ubiquitous materials, such as air, water, and sand. Substitution of raw materials became relatively easy and prevalent; for example, if natural rubber was not locally available, synthetic rubber could be produced from either coal, oil, or foodstuffs.

Technological advance in transportation, resulting in the reduction of costs, greatly increased the mobility of raw materials. Perhaps the most striking example of this is the post–World War II development of Japan's steel industry, which by 1964 was the third largest in the world (after the United States and the Soviet Union). In 1965 Japan imported 88 percent of her iron ore and 64 percent of her coking coal over an average distance of 5500 miles.[1] The magnitude of Japan's achievement (aided by, besides transportation, other technological improvements) was driven home in April 1969, when the merger of the two largest Japanese steel companies into a single enterprise, Shin Nippon Steel Company, generated serious concern on the part of American and European steel manufacturers about their ability to compete with Japan. In 1970 Shin Nippon Steel Company produced more steel than the U.S. Steel Corporation and thus became the largest steel company in the world.[2]

The decline in the deterministic effect of resource location has many implications. The vistas for organized human will backed by advanced technology have broadened immensely. On the other hand the inherited technological superstructure [3] of advanced societies creates rigidity and deterministic effects of its own. Thus what will increasingly matter in producing differential benefits from science and technology among the advanced nations will be, not so much the extent of present advances, but rather farsightedness in planning, determination in pursuit if selected goals, and willingness to reshape and restructure existing social institutions and technological superstructure to exploit opportunities presented by future technology. In a potentially less constraining future the margin between an effective application of human rationality and will and one that is less effective will be considerably wider than has been the case heretofore.

Although the foregoing analysis is addressed primarily to advanced countries, it is also applicable, with some qualifications, to the less developed countries (LDCs). One might argue that, between the advanced nations and the LDCs, a new "determinism" has emerged—that of the aggregate of the technological superstructure and the availability of skills in the advanced nations that tends to perpetuate itself and to which the flow of raw materials gravitates. The LDCs are thus condemned to an inferior position, and a significant change in the distribution of industrial power between the advanced and the developing regions cannot take place. It is true that the preponderance of technologies and capital in the advanced nations does impose constraints on the ability of the LDCs to change the global pattern of industrial activity in their favor; however, these constraints are not nearly so great as those of the location of resources in the past. This is so for at least the following reasons: (1) Technologies, skills, and capital are, in most cases, much more transferable than resources were in the past. (2) In recent decades a certain obligation has been recognized by the advanced nations to help the LDCs to industrialize. (3) As a matter of self-interest the advanced nations find it advantageous to invest capital in the LDCs and build up their technological superstructure; this is being done not only to extract resources (as was usually the case in the past), but also to build highly advanced manufacturing facilities to take advantage of abundant labor and avoid pollution and overcrowding at home (e.g., Japanese investments in Southeast Asia and Latin America; Western European and American investments in Hong Kong). Aided by these favorable factors, organized human will and rationality are hardly less important for the LDCs in being able to mold their future than they are for the advanced nations.

The Integrative Effect of Technology

A second trend consists of two interdependent but mutually supporting tendencies: the integrative effect of technology on society and the integrative process within technology itself. Much has been written about this trend in general and in mostly loose terms: It has been pointed out how progress in transportation and telecommunications has "shrunk" the world and how interdependent—because of modern and highly complex technology—the peoples in the world or a given region have become, and so on. In fact the integrative trend of technology is a highly complex and multifaceted phenomenon that has not been fully understood and probably will not be for some time. It operates on different levels with impacts on society that frequently run at crosscurrents.

One level is horizontal and physical—it involves integration of human activity and spatial units into one global geotechnical system. Emanating from industrial and political centers this integration has been proceeding for many years through a multiplicity of instruments—transportation, communications, new means for exploration and exploitation of natural resources, and improved methods of manufacturing. Three areas that have not been fully—or at all—integrated into the global system are the inhabited but underdeveloped regions, the uninhabited or scarcely inhabited regions (the Amazon, the Arctic), and the ocean floor. Technological developments in the next decade or so will contribute significantly toward their integration.

Thanks largely to satellites, costs of telecommunications are rapidly decreasing. Inflation being allowed for, direct-dial costs between New York and London in off-business hours decreased by more than 50 percent between 1967 and 1975.[4] The trend is likely to continue. The drop in telecommunications costs will be a major factor in integrating the LDCs internally and with the world system. A satellite-based television network for LDCs as large as India could be developed, built, and made operational in about four years.

To construct a comparable terrestrial network would take perhaps as long as two or three decades and the cost would be about twice as great.[5] Provided the government of the receiving state cooperates, such a network would make possible direct transmission of programs from an advanced nation into community receivers of the LDCs. If the presently available receivers are appropriately augmented (at a cost as low as $100 each) a television network, providing programs on a global basis and involving direct transmission through satellites into home receivers, could be operational no later than 1985. Without such augmentation an advanced country could develop, at substantial cost, the ability to transmit programs directly into home receivers of foreign nations by 1985.[6]

Public preferences in advanced countries for local programs, differences in time zones, and political considerations will probably hinder establishment of a global television network in this century. It is likely that a greater interchangeability of programs among the advanced nations will develop in the near future; selected programs from abroad will be increasingly transmitted, with the aid of a satellite, through local television stations. Since the LDCs have only limited resources for their own programs, they will rely to a greater extent on programs from abroad, beamed into community receivers or perhaps directly into home sets. New audiovisual techniques of instruction, combined with satellite-aided television, promise not only a much more rapid integration of the LDCs with the rest of the world but also a social transformation whose effects are not entirely clear.

The techniques of controlling rainfall are expected to improve significantly by the early 1980s [7] and thus help to increase the number of geographic areas capable of sustaining agricultural productivity and expand the areas of human habitat in general. The Japan Desert Development Institute appears to be successful in producing similar results by injecting a 3-mm-thick, 3-m-wide layer of hot asphalt by a 40-ton bulldozer in the sand at a depth between 60 and 90 cm. This is accomplished at a speed of 1 m per sec. The asphalt prevents rapid evaporation of irrigation water; as a result a crop of vegetables can be produced in desert lands. An experimental 60-acre farm is successfully operating in Japan, and in 1975 the Japan Desert Development Institute concluded an agreement with the United Arab Emirates to build a 2-acre pilot farm near Abu Dhabi.[8]

In July 1972 the first Landsat (Land Satellite, formerly called Earth Resources Technology Satellite, ERTS), designed to monitor the biosphere through remote sensing, was placed into orbit. The data that Landsats provide have multiple functions, one of which is to facilitate the discovery of mineral resources. When the transmitted data are processed and further augmented by conventional aircraft employing sophisticated scanners, the discovery of resources on a large scale is significantly expedited. This may eventually lead to the generation of economic activity in the geographic regions hitherto not considered to be economically attractive and perhaps not even inhabited.[9]

Exploitation of oil and gas from the seabed has moved rapidly in recent years. In 1975, 16.4 percent of total U.S. production of oil came from the seabed, and the annual revenue of the U.S. offshore oil and gas industry runs into billions of dollars.[10] Extensive as the exploitation of offshore oil is, present efforts barely scratch the surface of its potential. Oil companies drill deeper and deeper into the ocean. In mid-1976, they were exploiting offshore oil resources at a water depth of about 656 ft (200 m), and exploratory drilling for oil was conducted to depths of up to 2150 ft (655 m).[11] By the end of 1971 ocean mining had come of age. Plants were at work extracting magnesium metal, magnesium compounds, sodium chloride, bromide, or deuterium oxide from sea water. Dredging operations were conducted for diamonds, gold, tin, heavy mineral sands of titanium and thorium, ironsand, calcerous sands and shells, and ordinary sands and gravels.[12]

Man's ability to operate in deep water has been continuously expanding. On June 6, 1975, the U.S. Navy conducted a joint experiment with the United Kingdom in which U.S. Navy divers reached a depth of 350 m (1148 ft) off Panama City, Florida, under conditions of saturated diving (i.e., unprotected against the pressure of the water) and swam in the natural environment. In the same month a French company, COMEX, reached a depth of 326 m (1069 ft) off Labrador in a commercial dive, also under conditions of saturated diving.[13] Limited working capability under conditions of saturated diving at a depth of 2000 ft (610 m) is expected to be achieved by 1980, and prolonged commercial operations at 2000 ft, or perhaps at 3000 ft (915 m), are expected by 1990. Within this time frame extensive underwater commercial operations by men in an environment protected against water pressure will take place at considerably greater depths.[14]

Looking further into the future, scientists envision the emergence of global electrical power transmission grids. The likely shortages of electric power in the next decades will create incentives for full exploitation of time zone differences and seasonal diversities to equalize the overall global daily demand for electricity. Power transmission over thousands of miles is expected to be accomplished by a combination of superconductive cables, satellites, microwave, and laser beams.[15]

Another level of the integrative effect of technology is in-depth integration, proceeding mainly through continuous improvements in transportation, communications, and data processing. Modern media of transportation and communications made possible huge, multimillion-inhabitant cities whose viability is supported by sprawling suburbs. The trend now is in the direction of large metropolitan regions, extending for hundreds of miles. Increasing use of helicopters and short takeoff and landing (STOL) aircraft, improvements in high-speed surface transportation, introduction of innovations such as trains operating on the air cushion principle or the magnetic levitation (suspension) concept and, eventually, the development of urban gravity–vacuum transit systems [16] will facilitate mobility within metropolitan regions. The introduction of videophones in the United States in 1970 and their expected spread will reduce the need for personal contact and thus alleviate potential problems created by the large scale and rapidity of the integrative process. Expanded transmission of documents via telecommunication media will produce a similar effect. The greatly expanded needs for communications within and between metropolitan regions will be met by innovations such as the millimeter waveguide and optical fibers, later combined with lasers.[17]

In the United States computing has become a public utility. In 1974 there were about 168,800 computers and more than 150 commercial network services available to anyone.[18] Thus an integrated research project, leaning on dozens of data banks, could be conducted, at moderate costs, by several researchers scattered in various parts of the country. And the potential of computers is far from exhausted. The present trends are toward vast increases in storage capability (memory) of computers; greater compactness in storage of data and, hence, smaller computers with high capacity; and a significant lowering of costs.

Present advanced computers based on large-scale integrated (LSI) microcircuits offer a storage density of 1 million bits [19] per sq. in. and a typical capacity of 8–16 million bits per computer. Bubble memories, introduced in recent years,[20] will offer capacities of from 10 million to 10 billion bits and a projected density of 1 billion bits per sq. in. Computers based on laser technology, expected to be operational in the 1980s, will offer memories of from 100 billion to 10 trillion bits but a relatively moderate density of 10 million bits per sq. in. The most remarkable development, however, is the decrease in costs. From about one bit per one cent in the high-speed LSI technology and ten bits per one cent in slower LSI computers (both of which are technologies of today), the cost is expected to drop to 33,000 bits per one cent in the case of laser technology; moreover, it is likely to drop to 1 million bits per one cent in the follow-on generation of computers based on holography now under development.[21] In short, the use of computers will become extremely cheap and available to broad masses of people.

A third type of integration, closely related to integration in depth, is that of individual instruments for the accomplishment of particular social, economic, military, or technological goals. This trend is not new but has become pronounced in the post-World War II decades. The compartmentalization between land, sea, and air power is a thing of the past; what reigns supreme now is an effective application of force, with the instruments closely intertwined and tailored to the task.[22] In a sense the National Aeronautics and Space Administration (NASA) presents a marvel of this type of integration. Several years of complex scientific, technological, and managerial activity in many laboratories, plants, and offices, supported by a budget of nearly $25 billion, were integrated to achieve a single result: placing a capsule containing two men on the moon and bringing them back.

Finally, integration proceeds within science and technology itself. The combination of computers, satellites, atmospheric sciences, and oceanography opens up broad vistas for weather prediction and modification. Transplantation of body organs and their substitution by synthetic items—which is revolutionizing medicine—represents the meeting of medicine, mechanics, and chemistry. The combination of electronics and chemistry, electronics and physics, and biological sciences and chemistry are further examples of the growing trend. It appears that the synergistic effect resulting from the meeting and integrating of two or more sciences and technologies produces the most rapid progress.

What are the implications of this integrative trend? From the instrumental viewpoint, the capability of understanding and controlling the integrative trend of technology is important and will be increasingly crucial in the future. There is a growing need for conceptual instruments on how to integrate technologies to achieve particular social goals. Systems analysis is but one such instrument, and it has not yet been adequately developed. The ability to control the various forms and phases of the integrative trend—particularly on the "horizontal" level—is an increasingly important leverage of international power and influence. A better understanding and fostering of integration between the various sciences and technologies will accelerate technological innovation and progress.

The integrative trend of technology creates an appreciable measure of global, regional, and local interdependence but does not necessarily create unity. Indeed in a number of its forms the integrative trend of technology can create or exacerbate conflict. A growing settlement and integration of northern China and eastern Siberia is more likely to increase rather than diminish conflict between the USSR and China, and the better integrated instruments of military force are likely to intensify a clash. There is no conflict for the seabed in the deep ocean areas at this time, but such a conflict looms as a possibility in the future as the ocean floor begins to be integrated into global economic activity.

These examples are not intended to confirm the somewhat simplistic statement to the effect that "technology unites, while humans divide." Technology itself *as an independent variable,* may divide. Computers, by integrating human and material resources of educational institutionsn make it possible to create a megauniversity, but these same computers, by facilitating impersonality, may split asunder the megauniversity and even the entire society.[23] In terms of political division or unification or political impact in general, it is not always clear what the integrative trend signifies for the future. The rise of multinational corporations and the technological integration of the ocean floor—which may involve companies of various nationalities working, and eventually multinational personnel living, side by side on the bottom of the sea—are truly major developments. They might modify the present nation-state system in the next decades.[24] But it will take some time before the exact nature of their political impact becomes clearly discernible.

Global Projection of Influence through Technology

A third trend—in some aspects closely related to the one previously discussed—is the increasing, and increasingly versatile, global projection of national influence and power through technology. In the nineteenth century this phenomenon was manifested by Great Britain's global projection of power, first through the Royal Navy and later through telecommunications (cables and radio). In the post-World War II period the United States eclipsed all other powers in the magnitude of its global projection and the multiplicity of its instruments. After the massive projection of U.S. military power to Europe and Asia during World War II, the United States made an effort to withdraw, but was compelled to stay. The forms of its military power grew in variety and size—from bombers to ICBMs and Poseidon IRBMs trained on various targets in Eurasia. Global military operations have been improved—and are continuously improving—through nuclear propulsion in naval vessels, better communications systems, more effective logistics support, worldwide surveillance through satellites and other means, and long-range airlift.

On the nonmilitary level the Voice of America provides one form of projection of global influence. Physically the lunar landings were aimed at the moon, but politically they were aimed at Earth. Foreign aid is, in part, a means of extending America's influence through more advanced technology into the less developed world. The multinational corporation (MNC)—largely a subsidiary of the technologically and managerially superior American companies—is a projection mainly into industrialized nations.[25]

The Soviet Union imitated American global strategic nuclear projection, and now the Soviets are expanding their global capability in subnuclear military technology. The USSR is roughly equal to the United States in global projection of its influence through outer space technology (but not communications satellites), broadcasting, and oceanology. In other respects its efforts at projecting Soviet power are less significant than those of the United States.[26]

In comparison with the United States and the Soviet Union, Western Europe has not scored conspicuously in the global projection of its power and influence. It takes a unified political will for an effective worldwide projection of one's power and influence, and Western Europe does not have this. Europe participates in Intelsat, the Europeans have an outer space program of their own, and Great Britain and France have a nuclear striking capability; but on the whole the present capability of Western Europe to project its power and influence globally is small compared with United States or Soviet capabilities. There are two exceptions, however. One is the official foreign aid program of the European Community (EC) nations, which now exceeds that of the United States.[27] Another exception is the large merchant marine of the EC nations.[28] Under peacetime conditions, however, a nation's merchant marine is not one of the most important instruments of power and influence. By shipping goods (rather than people) and having low visibility, the merchant marine has relatively limited impact on human psychology and values and is basically at the disposal of the customer, of whatever nationality he might be.

Technological instruments for projecting one's power and influence globally will be significantly improved in the next decade or two, and entirely new instruments will appear. Global voice communications will become cheap and readily used within the next 10 years. By the early 1980s some highly advanced nations will likely be able to beam television programs, with the consent of the nations concerned, directly into the receivers of the less developed world. Computers, global communications, and high-speed transportation make it possible to control a scattered international enterprise from a central point; further progress in all these areas will enhance the role of foreign investment in extending a nation's influence. The era of global mass travel has just begun, and we shall witness further growth as the use of "air buses" spreads and as supersonic aircraft become more generally used in civil aviation.[29] Control of the means of telecommunications and of mass travel—through ownership of manufacturing facilities, the media themselves, or both—is a lever of power. Quite apart from this leverage, proliferation of global mass travel and mass communications tends to favor those who are technologically more powerful, affluent, and numerous: They have much more to project and can afford to project more. The influence exerted may include an important modification in the value systems of those who are the primary objects of the projection.

The 1980s are likely to be highly important in clarification of physical processes involved in large-scale climate and weather modification, and the 1990s will witness the transformation of fundamental understanding into technological capabilities.[30] For nations leading in this field, climate and weather modification provides an important instrument for projecting influence externally. The cost of energy is likely to rise for a while, but then, by the late 1980s, as intensified research and development (R & D) in energy begins to bring in results, it will start to decline. Major technological breakthroughs might occur that could significantly lower the cost of energy. The nations that capitalize on technologies where major breakthroughs are likely—which would include areas such as fusion energy, superconductivity, and magnetohydrodynamics (MHD)—and that possess the capability of using them on a sufficiently large scale will be in a position to gain major advantages in the world market through superior and cheaper products.

Development of marine resources is also important. It will permit the projection of national substance into hitherto unoccupied and unexploited parts of the globe, and its geopolitical impact will grow in the next decades.

We live in a period in which attributes of power and influence become increasingly diffuse; unlike attributes in the pre-World War II decades, they do not necessarily accrue narrowly to individual nations. In an integrating world, where regional differences are perhaps more important than national, it would be an oversimplification to equate each manifestation by a nation of ability for global projection with the actual attainment of power and influence. The effect of some external projections of national substance will be neutralized by other developments, and still other projections are likely to be counterproductive in terms of power and influence. This may pertain to the potential impact of global projections on the values and motivations of those peoples who become their targets. However, in the coming decades we are not likely to have a homogeneous world, and on balance the ability of global projection through technology will be an important attribute of vitality and influence of nations, regions, and those international institutions capable of commanding the necessary resources for it.

The Growing Scale and Costs of Technology

A fourth trend in technology is its growing economies of scale, which require enormous resources for its use. Even in such "conventional" items as transformers for power networks, the minimum economic size of a plant needed to produce the huge transformers necessary to take advantage of the economies of scale is pushing producers into cooperation or mergers. Most European companies are incapable of producing the huge transformers needed for the continuously increasing voltage in the networks. This development has forced even such giants as Siemens and Telefunken to join forces in the heavy electrical engineering field. In the United States 5 producers supply the entire market, whereas in Europe there are more than 20.

The scale and cost of the present advanced technology are of gigantic proportions and impose a heavy burden, even on a country of America's wealth and resources. Future technologies will become even more demanding in this respect.

Present Energy Research and Development Administration (ERDA) program goals call for construction of a demonstration fusion reactor, based on toroidal magnetic confinement, between 1995 and 1997. Estimates of the costs of the magnetic confinement program (including the demonstration fusion reactor) are between $8 and $10 billion.[31] Whereas the demonstration fusion reactor is expected to be relatively small (500–1000 megawatts), commercial fusion reactors (possible around the year 2000) are expected to have a capacity of 2000–10,000 megawatts (electric). Reactors of 20,000 megawatts are conceivable and would be even more economical. A reactor of this size would have a generating capacity more than twice as large as that of Poland in 1967. Particularly responsive to economies of scale is MHD power. If expected progress in the technology of MHD power materializes, the economies of scale might push the size of generating plants even higher than these figures suggest; furthermore huge MHD plants are likely to become operational before commercial fusion reactors become a reality.[32]

The cost requirements for R & D are complicated by technological uncertainties. A successful breakthrough in a particular technology may result in great savings, but there is no certainty which technology will be successful; this leads to further expenditures to hedge against potential failures. This can be illustrated by the case of commercial fusion power. Until quite recently laser fusion—a method of producing electric power from fusion technology different from magnetic confinement [33]—was not viewed as especially promising. But the picture has changed, and at present ERDA is allocating substantial funds to laser fusion: $98.9 million, as compared with $214 million for magnetic fusion (requested budget, fiscal year [FY] 1977). Governmental projections for laser fusion are about equal to those for magnetic fusion—a demonstration reactor producing economical electric power from fusion around 1995. However, it is possible that fusion may become an economical source of energy sooner than that by using still different technological approaches, *viz.*: (1) a fusion-fission hybrid reactor which need not meet the high technological requirements of a pure fusion reactor; and (2) a dual-purpose fusion reactor, producing both electric power and hydrogen, which might make fusion economical in the late 1980s.[34]

Other technologies on the horizon will require major outlays. It is possible that nuclear-propelled air-cushion vehicles (some types are known as "hovercraft"; others, as "surface-effect ships"), perhaps as large as an aircraft carrier, will be built around 1990. These vessels, moving at speeds up to 100 knots (about five times the average speed of the present conventional ships), might have mobility on both land and sea. The amphibious variety of surface-effect ships is likely to be especially attractive commercially after the initial R & D costs are met, since there will be no need to construct dock facilities and roads; these ships will move directly from a point inland, cross the sea, and reach another point inland. They will also cross marshes, rivers, and minor solid obstacles.[35]

Various schemes for large-scale climate modifications will demand immense expenditures. The present exploitation of ocean resources is limited to relatively shallow depths. To move to greater depths and to widen the scope of exploitable resources, the necessary exploration and R & D will require billions of dollars.[36]

In a somewhat different category—but nonetheless real—are the costs generated by undesirable byproducts of technological progress. The present advanced nations have been developing and applying science and technology with little heed to how this activity might affect the environment. It will require vast expenditures to compensate for years of neglect and to effect the necessary planning and readjustment for the future. Preliminary estimates of the U.S. Council on Environmental Quality indicate that the costs of water and air pollution alone for the United States between 1973 and 1982 will amount to $266.9 billion in public and private funds.[37]

The foregoing discussion suggests at least three points. First, rising costs are such that even the superpowers will not be able individually to take advantage of the full potential of future technology. This will generate increasing pressure for international cooperation in technological development. Second, huge scales of effort will be needed to use advanced technology. This will put pressure on middle-rank powers and smaller nations to cooperate in its use—a step that would involve restructuring institutions and the existing technological superstructure—or to forego the advantages of the technology altogether. Third, some forms of future technology—such as large-scale climate modification—will require international cooperation, not so much because of the costs involved, but because more than one geographic region will be affected, and the participation of those concerned will be essential. All three considerations have significant implications for functionalism, either as in independent variable affecting the future international system or as an instrument of national policy for the attainment of specific objectives in international affairs.

The Rapidity of Technological Change

The continuous acceleration of technological innovation and change constitutes another major trend. More than half of the products manufactured by American industry today did not exist 20 years ago. When the Communications Satellite Corporation was established, it was viewed as an investment in a technology several years away; its operational capability and earnings potential proved to be much faster and greater than originally anticipated. The technological growth and the spread of computers have considerably exceeded early expectations. Although there are indications that, in some areas, the speed of technological advance appears to be tapering off,[38] the base and momentum of modern technology are huge and, in their aggregate, continue to generate extensive socially significant change.

Rapid strides are taking place in biology; indeed the entire field, especially in its application to medical science, is on the threshold of significant breakthroughs. A few examples can be given.

Significant progress is likely in decreasing the rate of aging. In this century the average life expectancy in advanced countries has been continuously increasing; in the United States it grew from 47.3 years in 1900 to 72.4 years in 1975.[39] This was primarily achieved by eradicating various diseases that caused premature death and by improving living conditions. Nothing has been done with the so-called biological clock, the finite lifespan of human beings that varies in different individuals but appears to be about 100 years.[40] Gerontologists have relatively recently begun to apply themselves to the aging process itself, an effort that could increase human vitality and vigor within an existing life expectancy, expand life expectancy within the framework of the biological clock, or expand the limits of the biological clock itself.

At present the United States is allocating massive resources—nearly $1 billion annually—to eradicate cardiovascular diseases and cancer.[41] If this effort is successful, 14.3 years would be added to the average life expectancy from birth.[42] But partial or complete elimination of these diseases would merely add to the number of infirm, increasingly dependent, aged population. Awareness of the full implications of this development [43] will likely combine with the pressure of the growing number of voters over 65 to result in allocation of large resources to research of the aging process itself, an area that has very low funding at present. In view of recent advances in genetics and molecular biology, and provided gerontology receives strong support in the near future, it is conceivable that, by the year 2005, vigorous, potentially productive life could be increased by 15 years or more in advanced societies.[44]

Rapid advance is taking place in unconventional means of reproduction, such as artificial insemination and test-tube fertilization. One of the most extraordinary developments in this area is cloning—production of genetically identical copies of organisms (eventually including human beings) from single body cells.

By the method of DNA hybridization, it was experimentally demonstrated several years ago that a single specialized cell of an organism contains all the genetic information necessary to re-create the entire organism. It has also been experimentally shown that a single cell, extracted from virtually any part of the body of certain plants and animals and placed in a suitable environment, is capable of growing into a complete organism, genetically identical to the one from which the cell was derived. In the 1960s carrots and tobacco plants were successfully cloned. This was followed by the growing of whole frogs from intestinal cells; the cloned frogs were essentially carbon copies of the original body. In 1970 skin cells were successfully used for cloning frogs, a development that further confirmed experimentally the potential of duplicating organisms from specialized cells.[45]

The cloning of higher animals—which is expected to be achieved in this century—is likely to have major implications for agriculture. Cloning might make it possible to produce hundreds or thousands of identical copies of a particular type of horse, cow, or sheep, resulting in a genetically pure strain. The cloning of people will probably have to wait until after the year 2000, although some projections place it in the 1990s. For our purposes the differences in forecasts may not be important. What is important is the awesome potential—in our lifetime—of a science that ranges from the ability to synthesize life itself (one further step in this direction was made in June 1970 when the first complete synthesis of a gene was reported), and to manipulate human intelligence, to the duplication of genetically identical human beings by an asexual method.[46]

The total body of existing science and technology is so large and versatile that a socially significant impact need not depend on progress in one area of technology. It may come from an advance or a breakthrough in one area or another, or it may come from all of them at about the same time, in which case the impact is likely to be revolutionary. This can be illustrated by the potential progress in reducing the cost of energy.

Reduction in the cost of power could be achieved through progress in superconductivity, that is, the property of certain metals or alloys (pure unstrained lead, tin, niobium–tin, niobium–zirconium, and niobium–titanium) to provide no resistance to the transmission of electric current when their temperature is lowered to or near the absolute zero. So far the greatest advance in application of superconductivity has been in the construction of magnets and the generation of power. Operational and experimental superconducting generators have been built in the United States and the United Kingdom. In large generators, weight is reduced from 3 to 10 times. Volume is reduced by a factor of three. Their efficiency and stability are greater. With size and weight reduced, generators of higher power can be built. Because of reduced weight, manufacturing, shipping, and construction costs are lower. Conservative estimates for 1990–2000 indicate that, if expected additions to the U.S. electric utility industry in this period are made in superconducting generators, savings to the industry within the decade will amount to $4.3 billion.[47]

These projections assume that there will be no change in the superconducting temperature used. If, however, the temperature at which metals become superconducting could be increased from the present upper limit of 21° Kelvin (degrees Celsius above absolute zero) to 25–30° Kelvin—and there is a distinct possibility that this could be achieved in the near future—

then the impact could be revolutionary.[48] Given these conditions the relatively inexpensive hydrogen (instead of costly helium) could be used for the refrigeration of superconductors. More efficient refrigeration technology could be employed at these higher temperatures that would also permit far higher magnetic fields for superconducting magnets and thus make possible considerably cheaper technologies and higher fields and currents where these magnets would be applied. At this point the cost of refrigerating the lines for transmission (as distinguished from generation) of power is too high for commercial use, but the increase of superconducting temperatures would go far toward making superconductivity applicable to transmission. This would make it possible to transmit electric power economically over thousands of miles.[49]

Cheaper energy could be achieved through progress in fusion, especially since the source of power for fusion—water—would be abundant and inexpensive. More important in the midterm future (7-15 years), developments in MHD power could significantly reduce energy costs. In a conventional fossil-fuel power-generating station, thermal energy produced by coal, oil, or gas is first converted into mechanical energy by means of an engine or a turbine, and then it is further converted into electric energy by a generator. The efficiency of such a station is, at best, about 40 percent which means that, for a given amount of thermal energy, only 40 percent is obtained in electric power while 60 percent of energy is lost in the conversion process. MHD plants would, in part, use direct conversion from thermal energy into electric energy and thus increase efficiency to 50-60 percent, and eventually perhaps even higher. This would result in significant savings, since, for a given amount of thermal energy, from 25-50 percent more electric power would be obtained.[50]

Whereas each of these areas—superconductivity, fusion, and MHD—offers important promise for reducing energy costs, such costs would be substantially reduced if significant advance is achieved in all of these, since the three technologies are complementary and can be applied in a single power station. Quite apart from any major technological innovation that may take place in a particular field, synergistic effects involving innovations from various fields that reinforce each other provide a major dynamic force in modern technological change.

Rapid and accelerating technological change has two effects. First, countries in the forefront of technological development reap disproportionately large benefits from it, economically and politically. Countries only a little behind frequently find themselves beginning to manufacture products already on the threshold of obsolescence; unless they have a commercial advantage in another respect (e.g., cheap labor or an unusually effective marketing organization and techniques), these nations are significantly handicapped in the world market. Second, the rapid technological change produces a fundamental impact on societies, as discussed below.

Technological Impact and Social Discontinuities

In the post-World War II years, the locus of the technological impact on society was the military field: Through nuclear weapons and missiles military planning and warfare were revolutionized. It is only relatively recently that advanced technology has begun to affect the daily lives of individuals and the work of institutions. It maybe that the impact of technology on society will produce some important discontinuities in social and political trends in the not-so-distant future. One form of such discontinuities was suggested by the turbulance on U.S. campuses in the 1960s. It found its roots in a change in values of American youth.

Two major *Fortune* magazine surveys of college and noncollege youth conducted in 1968 and 1969 shed some light on this subject.[51] According to the surveys three-fifths of American college students were "practical-minded"; for them, college was a natural road to career, status, and financial success. About two-fifths of college students, however, rejected these values, and since they then appeared likely to become a majority in the future, the surveys called them the "forerunner group." The forerunners were not concerned about money. They were more likely than the practical students to have privileged backgrounds; they took affluence for granted, were disdainful of "careerist" values, and vague about their own career expectations. The forerunner group was most likely to major in humanities and seemed to be interested in finding work that was intellectually challenging and related to its social concerns.

About half of the forerunners believed that the United States was a "sick society," and nearly a fifth had a sense of "solidarity and identification" with the New Left. They tended to scorn partriotism. However, only a small group (between 3 and 4 percent) of college students were classified as "revolutionary," and less than 10 percent as "radical dissidents." The vast majority (76.4 percent) fell into the categories of "reformers" and "moderates," and a little more than 10 percent were "conservatives." Noncollege youth took nearly as critical a view of society as college youth.

Not all aspects of this development could be attributed to technological impact, but its important role (direct and indirect) is unmistakable. The "affluent society" and the values and attitudes stemming from it are largely a result of the high productivity made possible by modern technology—a new factor that has created the previously unthinkable option of eradicating poverty. The penetrating influence of technology makes the irrelevance of old attitudes obvious: What matters now is whether one possesses the skill to operate computers or to conduct research in laboratories, not whether one is black or white, Protestant or Catholic. The rise of what Kenneth E. Boulding calls the "superculture"—"the culture of airports, throughways, skyscrapers, hybrid corn and artificial fertilizers, birth control and universities" [52]— worldwide in scope, with science as its common ideology, induces questions in the minds of young people about the relevance of nationalism in the present world system.

The rapid pace of technological change (practically all of the socially conscious life of the college generation of the late sixties was within the previous 10 years marked by the impact of innovations such as the transistor radio, worldwide jet transportation, the Pill, the landing on the moon, and other near-miracles) created among the young a different perspective on how long it takes to get things done. When this perspective was applied to social problems like poverty, racial prejudice, and wars, it not only widened the generation gap but also made a large percentage of young people action oriented. Society's affluence provided money, transportation, bullhorns, and duplicating machines to translate the propensity to action into organizationally and politically meaningful terms. Nearly instantaneous television coverage also fostered organizational efforts: It showed radically inclined students all over the nation where the action was and made rapid response possible. Perhaps more important, television compelled emotional involvement in the coverage of violence (such as clashes of demonstrators or the torments of wounded soldiers in Vietnam) but was much less effective with complex, abstract issues. It thus encouraged "commitment" and discouraged rational resolution of social problems.[53]

Other factors contributing to the trend in value orientation exemplified by the forerunner group included the high complexity of modern technological society. A young man must go through many years of schooling and then years of work in routine or subordinate capacity before he reaches a stage where he can do something important or be influential. The impersonality of the existing institutions, their huge size and complexity producing rigidity and poor adaptability to change, induces a feeling of helplessness and alienation from society as presently governed. Furthermore, as young people observed the rapidity of change, they began to wonder whether the skills they obtained through long years of schooling would be applicable to the society they will live in. As they tried to look into the postindustrial society of coming years [54]—with its emphasis on superior knowledge, meritocracy, and highly complex, even more impersonal, computerized decision making by increasingly larger organizations—many young people found the picture not particularly reassuring. Hence the shift to social and moral concerns, the humanitarian aspects of life, to "feeling" and "simple justice," as distinguished from the highly rational, complex, impersonal, and at times quite arbitrary realities induced by technological advance.[55]

The campuses of the seventies are no longer turbulent: The tight job market makes students more security conscious and more willing to conform; the state of the economy—inflation, shortages, and tight governmental and educational budgets—results in less money from home and less scholarship assistance and frequently induces students to work. Students have won a measure of influence in academic institutions—they now sit on many academic and disciplinary bodies in colleges—and appear to be willing to accept the existing educational system rather than to change it at its roots.

Many social and humanitarian concerns remain, but students seem to have learned that certain objectives cannot be achieved overnight, and they are more willing to act on them constructively through appropriate career pursuits. Some explosive issues—such as Vietnam—no longer exist. Resentments against, and intolerance of, large, impersonal institutions remain, but they are no longer expressed in violence. The explosive growth of the youth segment of the population has ended with the sixties, and the student generation finds itself less focal in society's attention and not so influential.[56]

Although gone, the turbulence of the sixties remains instructive. It indicates that technological impact, when it coincides and interacts with other factors—demographic, external military (Vietnam), and so on—can sharply change the direction of social and political trends. We do not know yet the full depth of the impact of the sixties on society, but they suggest a question: What sort of social discontinuities will result when institutions and daily lives of individuals are much more strongly affected by future technology? The 1980s, in particular, seem to be important in this respect. At that time a number of technological impacts will converge and may produce major, perhaps radical, changes in society. Among the relevant factors can be mentioned the pervasive impact of computers; truly global mass communications and travel; development of ocean resources; the multiple problems of the postindustrial society then in full swing; and the imponderable effects of biomedicine. Moreover it will be a time when nations will have to begin adjusting their technological superstructure and institutions for the even larger scale of the technologies of the 1990s: MHD power; thermonuclear plants; large-size, high speed, surface-effect ships; and large-scale climate modification.

The potential impact of the likely increase in human longevity discussed in the previous section deserves elaboration. This impact will probably begin to be felt in the late 1990s, but its full effect will be felt early in the next century. If accompanied by increase in the vitality and vigor of a large percentage of the population, longer life is likely to produce serious conflict and instability for two reasons.

It is probable that, by 1990, all compulsory retirement will be eliminated.[57] Therefore, it is entirely possible that many mentally and physically vigorous senior citizens would stay in top positions in business, government, and the professions and thus block for many years the promotion of younger people. The great majority—those less successful in their careers and the working classes—will probably choose to retire. Through sheer numbers, retired senior citizens may be a sufficiently powerful force in politics to consistently implement legislation—such as that dealing with retirement income or the extent of welfare support for higher age brackets—that acts in their favor. As a result serious cleavages may develop in society because of vested interests associated with differences in age. Age-related riots in the future may thus overshadow in magnitude the U.S. racial strife of the 1960s.

The situation could be more acute—and more complex—if, in addition to the vested interests involved, the various age groups were characterized by serious value or ideological cleavages. An expansion of the average age from 72 to 90 years or more adds at least one more generation to the population living at a given time, and the lowering of the voting age to 18 further expands the age of the politically involved population. These developments might seriously increase potential differences among generations. Whether or not this will be so depends on whether generations tend to preserve the values, attitudes, and ideology of their socially formative years (say, late teens and early twenties) or change them as young people assume family and job responsibilities.

There is no definitive empirical evidence one way or the other, and disagreement on this subject persists. Karl Mannheim, basing his analysis largely on historical observations, has developed the concept of "generation units"—groupings that share common experiences in a given time and that tend to preserve their attitudinal and ideological leanings throughout their lives.[58] Ronald Inglehart conducted an empirical study of values in Western Europe in 1970. Although his study needs to be corroborated by longitudinal data over a period of time, it nevertheless provides strong evidence in support of intergenerational changes similar to those advanced by Karl Mannheim. In particular Inglehart notes the persistence of "acquisitive" (related to economic security) values in the generations that experienced the wars and scarcities of the era preceding the West European prosperity of the last 20 years, whereas the younger generation, reared in the more recent affluent years, appears to have developed "post-bourgeois" (or "postindustrial") value priorities relating to the need for belonging and to esthetic and intellectual needs.[59]

Seymour Martin Lipset and Everett Carll Ladd examined college generations in the United States from the 1930s to the 1960s and concluded that each generation tends to become more moderate as it grows older but nevertheless is more liberal than the preceding generation.[60] If correct, this conclusion would exclude the possibility of discontinuities in value formation.

Peter Drucker did not explicitly address himself to the theories of generational changes, but, by implication, he strongly embraced the life-cycle interpretation. Drucker pointed out that from 1964 until 1971 the 17-year-old group was the single largest age group in the United States and that this accounted for social instability. He predicted that, as in the 1970s the population gravity in the United States shifts from teens to youth in their twenties, and as large numbers of young people begin to pursue scarce jobs, economic concerns would prevail, and the 1970s would be very different from the 1960s.[61]

On the evidence available it is reasonable to conclude that reality embraces a blend of the generational and the life-cycle influences. The effect of economic, demographic, and other factors on the value pattern and concerns of a particular age group later in life cannot be lightly dismissed as irrelevant. However, it would be a gross oversimplification to assume that there is an inevitable "logical progression" in the values and attitudes of a particular age group, although some mellowing of youthful exuberance and a greater degree of responsibility can be expected with advancing age. Strong influences in later years can profoundly change the value system of a nation and create major discontinuities in the direction of its development. The use of nuclear weapons against Japan in World War II and the impact it produced on the attitudes of the Japanese population, young and old, is but one example in this regard.

On the other hand there is substantial evidence to the effect that, such strong influences being barred, the value pattern formed by a young generation in the years of its socialization has a strong tendency to persist; it is not likely to fade or disappear as the young people get absorbed into the process of making a living. This alone may account for heightened intergenerational conflict as longevity of population increases. But if there is a prevalent tendency on the part of older age groups to control the top positions in society for a long time, not only the young people but the middle aged as well will not find "the system" so attractive as before in terms of rewards; in fact, jobs will be frustrating in terms of their intrinsic interest, promotion potential, power, and so forth. Hence the propensity of those actually working (as distinguished from students) to challenge "the system" would be considerably greater under these circumstances, providing an impetus for discontinuity in social and political trends.

In summary, multiple and multicausal social discontinuities are likely to arise in the future. The nature and extent of such discontinuities will depend on a timely appreciation of the emerging problems and on effective policies to cope with them. Their nature and extent will also depend on recognition of the degree to which the change being introduced by technological impact is inevitable and legitimate and on the adjustment of institutions and the creation of new ones to meet the requirements of change. Not all societies will be equally successful in coping with the impact of technology. The price for failure could, however, be large: It may involve extensive social dislocation, widespread personal insecurity, severe social instability, and a serious weakening, if not collapse, of governmental authority and capability for defense.

Although the exact consequences of the accelerating impact of future technologies are impossible to foresee, it is clear that the traditional view of technology as producing "long-range impact" no longer applies. Social, political, and institutional changes induced by technology may be rapid, either as a result of a pent-up response to the already existing technology that has not yet come to the surface or as a result of a swift advance in technology itself inducing concurrent sociopolitical change.

The Developed Regions in a Technological Future

part 2

The United States

4

ANALYSIS OF America's ability to deal with problems arising from future developments in science and technology reveals strengths and weaknesses. Which will prevail—and the extent to which they will prevail—depends on our understanding of the stakes and choices involved and on our willingness to adjust our institutions to the making of rational decisions on a national scale.

America's Strong Points

The United States has an impressive record in science and technology. During the past 25 years or so, the conceptual, institutional, and physical capacity of America for exploiting its scientific and technological potential has increased significantly. From World War II emerged the concept of the federal research and development (R & D) contract, which joined government, industry, and universities in pursuit of new scientific and technological goals. This cooperative approach has remained one of the principal pillars of American technological growth.[1]

Systems analysis, originally applied by the military during the war to cope with complex problems created by expanding technology and large-scale projection of military power overseas, was applied to industry in the late 1950s and the civilian branches of government in the mid-1960s. Its application, still evolving, facilitates both large-scale use of technology and its integration with other elements for the realization of given social goals. Cost-effectiveness analysis, introduced in the Department of Defense in the early 1960s, provides a useful means of determining priorities in the development of weapons for a given mission. A related instrument—cost-benefit analysis—began to be applied to the appraisal of utility to the nation of existing or prospective scientific and technological programs in a broader sense, encompassing political, economic, and military considerations.

Planning for a technological future, albeit mainly on a quite compartmentalized and specialized basis, has become an accepted activity on the part of the various government agencies and private corporations. "Futurism," with strong technological overtones, has gained in interest among society at large.

The two major federally sponsored scientific–technological endeavors—the nuclear missile program and the exploration of outer space—provided the dynamism and the patterns of large-scale operations that have perhaps become the most important single components of the "spin-off" into the economy. Largely because of the tie established between the granting of graduate degrees and research (for which government funds were available), the United States developed a large body of scientific and engineering manpower. A number of sectors of technology, assisted by the government or partially developed in conjunction with military programs, took off on their own and thus injected new momentum into the economy.

America's traditional assets—large territory, energetic and skilled population, richness of natural resources—were further strengthened by the central geopolitical position of the United States in the post-World War II years; by the requirements of sustained global operations, military and civilian; and by the drastic decline of prewar industrial and political centers in Western Europe and Japan. The airlines, sustained by subsidies in the prewar years, became much less dependent on government assistance in the postwar period; their efficiency was boosted by adoption of the jet engine developed by the military. The spectacular rise of the television industry from humble beginnings in the late 1940s was somewhat indebted to militarily developed electronics but rapidly proved its own economic viability. So did the multibillion-dollar American computer industry, now preeminent in the free world.[2] A different pattern is presented by the chemical industry inasmuch as its growth (about 7 percent annually since World War II) was a result of internal dynamics of its own research and development, with little government aid.[3]

The strength of American science and technology, aided by the energetic and competitive spirit of the private enterprise system and the availability of capital, led to the setting up abroad of subsidiaries of American companies, in some respects further strengthening the nation's economic viability.[4] The private sector of the economy was further benefited by an influx of skilled, technically oriented personnel into the managerial ranks of large corporations.

Weaknesses

However encouraging the picture painted above may be, the record of the post-World War II years is not uniformly bright. Furthermore there are some weaknesses in the American system dealing with science and technology that raise questions about its adequacy for the present, let alone the decades ahead.

Some U.S. industries showed rigidity in adapting to technological change. The decay of the shipping industry—which is undergoing change only now, while a number of foreign shipbuilders had made major technological progress in the last 15 years—reflects both lack of initiative in the industry and hyperprotective and inflexible policies of the U.S. government.[5] Even the automotive industry—traditionally a proud example of American inventive genius applied to mass production—began to show some glaring symptoms of sluggishness, as the early case of the rotary engine and the tardiness in introducing antipollution devices indicated. A number of other industries—like construction, fishing, railroads, and metal casting—are technologically behind their counterparts in Western Europe or Japan.[6]

The United States is behind Western European nations and the Soviet Union in certain individual technologies. The French and British are ahead of the United States in the technology of air-cushion trains. The Soviet Union leads in development of magnetohydrodynamic (MHD) power. The Soviets may be ahead in magnetic and laser fusion, and, perhaps, in weather and climate modification.

More serious reservations arise with respect to the national machinery for decision making in science and technology. It was not very effective and, in recent years, it has been plagued by changes that have not necessarily provided improvement.

The old machinery for science and technology policy operating in the White House was not very effective. It responded mainly to the pressures coming from the agencies of the U.S. government, the Congress, and society at large and was not capable of providing true leadership and a strong sense of direction for policy. Its relative ineffectiveness, plus President Nixon's personal dissatisfaction with his science and technology advisory machinery in general and with its members' political activity in particular, resulted in the Reorganization Act No. 1 of January 1973.[7]

The reorganization weakened rather than strengthened the national science and technology advisory machinery. The Director of the National Science Foundation was given an additional function as Science Adviser to the President, but his access to the President was extremely limited. Military science and technology were taken from under his jurisdiction and placed under the National Security Council (NSC). A new NSC Military Technical Consultant Mechanism (MTCM), assisted by a single staff member, became principally concerned with this sector. A pool of outside consultants from which *ad hoc* panels were created to examine particular technologies, systems, or programs, it became more of an auxiliary to the Department of Defense than a source of independent review of military science and technology. The Office of Science and Technology (OST) and the President's Science Advisory Committee, two principal organizations dealing with science and technology policy on the White House level and headed by the Science Adviser to the President, were abolished.[8]

Largely in response to pressure from the scientific and engineering community, the Administration decided, in May 1975, to restore the science and technology advisory machinery in the White House. This decision was implemented in May 1976, but the new machinery is not especially strong, and the question of the overall organization of the federal science and technology is still not settled (see chapter 11).

A 1968 Organization for Economic Cooperation and Development (OECD) report provides useful insights into the nature of U.S. science and technology. This examination by West Europeans was accompanied by the hope that some of the American experience would be transferable across the Atlantic, but the exercise did not prove to be especially rewarding in this regard. Although the system—both private and public—was found vital and dynamic and although its pluralistic nature was acknowledged as a source of its strength, it was considered to be weak in terms of an overall purpose, too rich, too amorphous, and too wasteful in its frequent duplication or overlap of effort to provide a useful pattern for imitation, however admirable it may have been in parts.[9] In the 1970s, U.S. science and technology is relatively no longer quite so rich, but most other features remain the same.

From a historical perspective it becomes apparent that the fundamental forces underlying America's technological growth have not significantly changed in the post-World War II era as compared with earlier periods. Those forces have always been essentially quasi-deterministic and included some occasional volitional overtones. One such force was described in chapter 1 as "favorable determinism," the aggregate of America's resources—locational, territorial, demographic, and similar assets—favoring development of particular technologies in the process of economic growth. This favorable determinism has been assisted by mild volitional action—a mere

first time since 1893—the U.S. balance of trade became unfavorable, the deficit amounting to $1.5 billion.[11]

Along with the decline in the growth of labor and capital productivity, an important factor in this development was the deterioration in the U.S. balance of trade in technology-intensive products in which the United States historically excelled. As the United States had taken its technological preeminence and industrial strength for granted and placed emphasis in its R & D on outer space, defense, and atomic energy, our trade in commercial technology-intensive products became significantly threatened by foreign manufacturers. These, to a large extent, used American technology and, along with exporters, enjoyed strong support of their governments.[12]

The problem of mobilizing technology in support of the economy is truly fundamental, and its solution will play a critical role in the nation's viability and vitality in coming years. The response of the United States to this problem illustrates the shortcomings of the U.S. government in general and the old White House science and advisory machinery in particular in harnessing technology for national needs.

OST had been slow in appreciating the importance of this development and in acting on it. As a result the President appointed, in September 1971, William M. Magruder, an energetic engineer, as a Special Consultant to the President to mobilize technology in support of the economy. Operating on a crash basis with a number of task forces and study groups throughout the U.S. government, Magruder produced in a couple of months what came to be known as the New Technological Opportunities (NTO) program. It consisted of programs intended to improve the American position in international trade, to apply high-technology know-how to major domestic problems, and to reduce the unemployment rate among American scientists and engineers. The list produced would have cost $1.5 billion in FY 1973 and $11 billion through FY 1977 in federal, state, and local government funds and industrial cost-sharing programs.

The NTO program was, however, drastically pared by the White House in December 1971, to the point of annihilating it as a distinct entity. Accounting for the program's demise were its costs in an inflationary period, the resistance of several agencies to certain programs, and the difficulties of their implementation within the time proposed. Above all, the Administration realized that the harnessing of technology in support of the economy is a highly complex undertaking that would require additional study and time and could not be solved on a crash basis by a list of programs.[13]

A second response of the Administration to the challenge of the economy and technology—the message of the President on science and technology of March 16, 1972, heralded as the first message on this subject ever to be presented by a U.S. President to the Congress—proved to be equally abortive. It did call for the strengthening of science and technology in support of the economy and provided some broad guidelines for policy. It further proposed a "new Federal strategy" in science and technology, basically consisting of reorientation of scientific and technological programs toward society's domestic needs.[14] However, neither the Nixon Administration nor the Ford Administration backed the message by political, intellectual, or fiscal commitment. With a few exceptions initial steps made in this direction were not pushed vigorously, have bogged down, or have been shelved.[15]

One tangible effect of the message was the continuation of a trend, begun in the late 1960s, of gradual reorientation in the national R & D budget toward civilian programs, as distinguished from the earlier predominance of defense and outer space. However, except for the field of energy, the allocation of resources to R & D in support of the economy is still quite modest.[16]

To sum up: Our shortfalls in technology and productivity, as revealed by the changing pattern of the balance of trade in the late 1960s and early 1970s, were principally coped with by short-term measures—such as devaluation of the dollar—while the more fundamental longer range correctives have been relatively neglected. The onslaught of the energy crisis in the winter of 1973–74 provides a partial explanation, if not justification, of this neglect. The near-term effect of the energy crisis on the balance of trade was much more severe than that of the declining exports of technology-intensive products.[17] Moreover, its impact on some of our principal competitors—Western Europe and Japan—was more damaging than on the United States, and thus, at least for a while, the commerical pressure coming from these regions was alleviated.

Although the attention of the Administration began to be absorbed by the energy crisis, the results were short of impressive. Effective action would have required a strategic approach in the scientific and technological effort closely fitted into an overall energy policy, involving a competent orchestration of such elements as conservation, emergency stockpiling, imports, taxes, and investment. The initial U.S. reaction to the energy crisis consisted of a number of *ad hoc* measures and unrealistic expectations that added to the existing uncertainty and confusion in science and technology.

Nongovernmental studies of Project Independence, announced by President Nixon in June 1973 and professed to make the United States free of foreign supplies of petroleum by 1980 with the assistance of $10 billion in energy R & D, indicated that its objective was unrealistic. At best the nation could become independent of foreign supplies of energy by 1985. This could be achieved principally not through energy R & D—which, because of long lead times, could not produce effective results until after 1985—but through a vigorous policy in a number of areas, including reduction of demand, and at a very substantial cost to the economy.[18]

Money began to flow generously into energy R & D, at a rate well in excess of the $10 billion for five years envisioned by Project Independence, although an overall strategy for energy R & D had not yet been developed. In June 1975 the Energy Research and Development Administration (ERDA) presented a National Plan for Energy Research, Development, and Demonstration. The Plan placed emphasis on R & D related to abundant domestic resources; it stressed the need for a broad approach involving the development of several technologies, as distinguished from concentration on only a few; and it elevated the development of solar electric energy to a priority comparable to fusion and the breeder reactor and thus indicated that the earlier heavy and single-minded support of the breeder reactor may not hold for the future.[19]

Whereas ERDA thus undertook to reorient its program toward a strategic approach in energy R & D, the picture in an overall energy policy is less encouraging. The President's energy program, intended to free the nation by 1985 from the threat of an oil embargo, finally passed the Congress late in December 1975, but it had aroused serious doubts about its feasibility as originally designed, quite apart from the diluting effect of compromises generated in the legislative process.[20]

America's response to the fundamentals of the energy crisis was predominantly unilateral. When, however, as an aftermath of the Washington Energy Conference of February 1974, the energy-consuming nations established an International Energy Agency (IEA) in November 1974, a promise for a broader, multinational activity emerged. This promise still awaits fulfillment. The IEA appears to have been reasonably successful in providing for various emergency measures in the event of interruption of the supply of oil, but the member nations agreed on little more than token programs in energy research and development. And it is precisely on these programs that the long-range future and the ultimate effectiveness of the IEA depend.[21]

Just as the energy crisis tended to overshadow and obscure the more fundamental need for mobilizing technology in support of the economy on a broad front, there is also a potential danger that other developments, capable of ameliorating the situation, might be mistaken for fundamental solutions. They include the potential of agricultural exports and reimbursable technical assistance for buttressing the U.S. balance of payments position.

In recent decades U.S. agriculture has greatly benefited from scientific and technological advances and has, in effect, become a high-technology sector. Our exports of agricultural products have grown from $5.7 billion in FY 1969 to $21.5 billion in FY 1975 and are expected to rise to between $27 billion and $33 billion (in 1974 dollars) by 1985.[22] These projections may not materialize: Protectionism overseas (e.g., in Western Europe), humanitarian exports of food in response to starvation, and increases in the production of food abroad as a result of technological advance may significantly change the picture.

The importance and full potential of reimbursable technical assistance (including industrial equipment, skilled personnel, turnkey plants, etc.) have not yet been adequately appreciated in the United States, but the awareness is spreading. The immediate impetus is being provided by oil-producing countries, but in the longer run the more successful less-developed countries (LDCs), such as Brazil, will claim a large share of reimbursable technical assistance from the United States and other advanced nations.

Arab countries are energetically seeking a more stable and durable foundation for their growth than the bonanza of oil revenues can provide. Industrialization is a logical means to this end. According to a Brookings Institution study, exports of goods and services from the United States to oil-producing countries will rise nearly threefold between 1973 and 1980—from $4.4 billion to $12.4 billion (1973 dollars).[23] Exports of industrial equipment and technical assistance could be substantially higher than these estimates suggest if other LDCs, which eventually obtain loans and grants from oil-producing countries, are included and if the United States creates an effective internal organization and develops a farsighted policy promoting technical assistance on a reimbursable basis. As things stand, the United States is still a long way from tapping the potential of the excess reserves of petrodollars, estimated at from $250 billion to $300 billion for 1980 in current prices.[24] And in coming years petrodollars will not be the only source of potential demand for reimbursable technical assistance, especially when one considers the possibilities in some of the more rapidly growing LDCs.

The energy crisis is not an unmitigated evil in still another sense. In view of the vast resources of various forms of energy available in the United States—coal, thermal energy, and oil shale plus the energy that could be obtained by advanced technology (e.g., solar power and fusion)—it is conceivable that some 30 years from now the United States might become a major exporter of energy. Such a development, however, if possible, would take a careful estimate of costs, early planning, and a farsighted strategy.

In summary there are several means by which the United States could offset the unfavorable balance of payments situation threatening the nation's economy now and in the future. But these short-term means, if not handled as parts of a long-range design, constitute a grave danger to the nation. Ultimately the nation's future depends on its ability to mobilize science and technology in support of its economy across the board. Only in this way can America avoid the fate of Great Britain with regard to the well-being of her population and her vitality and viability in world affairs. And it is precisely in this broader area that the U.S. government so far is not ready, and does not know how, to produce effective results. It has chosen to tackle less fundamental problems, where an appearance of progress could be demonstrated and some secondary problems ameliorated.

The pitfall of this approach lies in that what appear to be solutions may produce serious damage to the nation. Reimbursable technical assistance provides such an example. If organized "correctly," backed by sustained U.S. leadership in key areas of science and technology, and conducted with due regard to the laws of comparative advantage, reimbursable technical assistance could significantly strengthen the U.S. economy and open new frontiers for U.S. international science and technology policy. If these prerequisites are not met, it can only contribute to America's eventual decay as other nations capitalize on technology transferred from the United States.

Our discussion so far has focused on problems and shortcomings. In some areas, however, promising developments have taken place. After some false starts in institution building in the area of the environment, by mid-1971 a degree of stability appeared to have been achieved. In January 1970 a three-member Council on Environmental Quality responsible for policy in that area had been set up in the White House. In December 1970 an Environmental Protection Agency (EPA) responsible for operations was established. The Agency's budget authority, including construction grant obligations to states and local communities, had grown to $7 billion by FY 1977. The National Oceanic and Atmospheric Administration (NOAA), established within the Department of Commerce in 1970, extended environmental activities into marine areas.[25]

Ecological and environmental concerns have been projected to the international forum; U.S. policy in this field has mostly been addressed to the support of initiation of global or regional efforts. The United States gave broad support to the 1972 Stockholm Conference on the Human Environment and the institutional arrangements (including the U.N. Environment Program) stemming from that conference. The United States provided the initiative for the Ocean Dumping Convention, agreed upon in London in December 1972. Among regional endeavors U.S. activities in the Environment Committee of OECD and the North Atlantic Treaty Organization's (NATO's) Committee on the Challenges of Modern Society are noteworthy.

Useful as this activity has been, it was primarily *ad hoc.* U.S. environmental policy still has quite a way to go in developing environmental standards in the context of U.S. economic realities and requirements, projecting a similar analysis to other nations, and developing a comprehensive policy capable of global application. Such a policy would need leverages for its implementation that would go beyond a mere verbal persuasion and might, for example, include development and production of environmental control equipment available for export at acceptable costs.

Finally, important developments for national science and technology policy took place in the legislative branch of the government. The Office of Technology Assessment (OTA) was established in October 1972. Since OTA largely responds to the requests of individual congressional committees for technology assessment studies, its activity has a virtually built-in bias for an *ad hoc* approach to technological impact and thus imposes a limitation on its contribution to the development of a comprehensive national science and technology policy. On the other hand OTA does provide the Congress with a potentially influential instrument, not only for passing judgment on programs and priorities in science and technology submitted by the Executive Branch, but also for generating new initiatives in the development of technology and control of its impacts.[26]

Perhaps even more important has been the strengthening of the committee structure in the U.S. House of Representatives with regard to science and technology. In 1974 the Committee on Science and Astronautics was renamed the Committee on Science and Technology, and its jurisdiction was expanded to an oversight of the entire federal R & D program. The reform did not quite abolish extensive compartmentalization of science and technology policy in the U.S. Congress; there is still a separate Committee on Energy and Environment in the House, and there is no single committee or subcommittee in the U.S. Congress that provides a systematic oversight of the Bureau of Oceans and International Environmental and Scientific Affairs (a major source of the Bureau's weakness) in the State Department. But the establishment of the House Committee on Science and Technology and, in particular, its cautious extension into the international sphere through creation of the Subcommittee on Domestic and International Scientific Planning and Analysis are steps in the right direction.[27]

Problems and Issues to Be Resolved

As one looks at actual accomplishments in recent years, one must be less than enthusiastic about the extent of progress achieved. However, the problem of control of technology and of its impact—which, in fact, is the heart of an effective science and technology policy—is much broader and much more complex than a mere sum total of institution building and policies at a given time. To have a realistic and forward-looking science and technology policy for the United States, one must place its formulation in the context of global developments and over a period of decades. Our policy-making will require a continuous process of modifying values, institutions, and policies to meet the requirements of the time.

Thus what ultimately matters is the success of the process, and not necessarily the occasional tardiness in individual accomplishments. There is, however, a definite relationship between the timeliness of action and the success of the enterprise as a whole: Undue delays can be dangerous inasmuch as they can aggravate the situation beyond the available means for a successful resolution. Accordingly, although we must view the problem of controlling technology as an evolutionary process, both the speed of technological developments and the inheritance of the need for action overdue from previous years suggest that there is little room for complacency.

Our success in controlling technology and its impact and in meeting the requirements of the future will depend largely on a satisfactory resolution of a number of issues. One central issue—that of a greater degree of consensus in society on values and goals—lies outside the sphere of technology. We did not hear many voices criticizing U.S. science and technology policy in the 1950s and the early 1960s, not because that policy was ideal (which was far from the case), but because there was a fundamental consensus in society supporting that policy and because, funds being generously available, the scientists themselves saw no reason to disagree with the existing state of affairs.[28]

Ironically, technology itself played a major role in breaking up that consensus, and it thus exacerbated difficulties in achieving a coherent science and technology policy characterized by clear guidelines for priorities in the allocation of resources. Technology was instrumental in producing this effect by contributing to the change in values among the young, by enhancing pluralization of society, by generating undesirable side effects that turned a number of people against technological progress, and by producing insecurity, instability, and alienation of the individual through the rapid pace of technological change.

Accordingly a forward-looking science and technology policy—used in the broad sense of the term—can be more than just a passive recipient of whatever consensus of values happens to come from society. Insofar as a timely science and technology policy effectively tackles the problems generated by technological impact that are distinctly harmful to society and its stability, it may go far in building a foundation for a greater consensus on values. It is thus important to recognize that there is a reciprocal relationship between such consensus and an effective policy with respect to science and technology.[29]

Another central issue is the need for a healthy balance between centralization and decentralization in science and technology policy that would be most appropriate to a given time. At this point pluralism in science and technology is very extensive and the entire field is amorphous; even if we had a clear-cut consensus on values and priorities, it would take the United States many years to reorient the science and technology complex in the direction of that consensus under the existing system. The solution appears to lie in changing the system to allow a healthy measure of pluralism so essential for dynamism and, at the same time, to introduce an adequate degree of centralization at the top, to make science and technology more responsive to politicosocial goals.[30]

A closely related issue is the abovementioned need for a better organized will and purposiveness in our society. The forces that historically fueled America's dynamism in technological growth are no longer adequate. Dependent on foreign energy and raw materials sources, the United States can no longer rely on a strong natural resource base to provide a powerful stimulus to the growth of its technology. The highly competitive nature of the present world and the rapid pace of technological change require a high degree of national initiative in technology and related pursuits. I do not suggest that the United States should imitate Japan by becoming "America, Inc.," but greater cohesiveness among the various components of the nation—political, economic, and technological—will be necessary to meet the challenges of the future.[31]

An important subissue is the need for closer cooperation between government and business. Unlike the case in other nations (Germany, Japan), business in the United States is suspicious of government, and a degree of antagonism exists between them. One can detect some change in both the attitudes and the policies. For example, it has been suggested in a prominent business magazine that the loan guarantees extended by the federal government to Penn Central and Lockheed in 1971 are the start of a trend, and a desirable one at that.[32] It is not important whether this is so or not. The real question for the next decades is this: Will a closer cooperation between government and business come in a timely fashion and will its form meet the requirements of the times?[33]

An appropriate balance between involvement in the world and withdrawal from it is another issue: Overcommitment of resources to external pursuits can significantly sap the energy of the nation and produce a multitude of undesirable repercussions. On the other hand the nation can hardly afford the isolationism of the old days: The world is now simply too highly interdependent—politically, economically, ecologically, and technologically.

In trying to find some guidelines for allocation of resources in this connection, it might be helpful to lean on the perceptive analysis by Harold and Margaret Sprout,[34] who distinguished between "internal" and "external" costs. The former encompass mainly various social programs—like education, health services, welfare, and housing—and the latter include those expenditures pertaining to the nation's position internationally (the military budget and other costs related to national involvement abroad and to international power).

The Sprouts point out that Britain's ability to maintain her leading position was due largely to her keeping the great majority of her population poor, while reinvesting the capital saved and maintaining her naval power. When, in response to the growing demands of her population, consumption and allocation of resources to social services rose significantly in the twentieth century (a phenomenon distinctly related to the pluralizing effect of technology and the creation of multiple interests, discussed in chapter 1), she was no longer able to meet the growing external costs, and her position declined. The Sprouts further extrapolate from the earlier experience of Britain and raise the pointed question of the viability of the superpowers—especially that of the United States in the late 1960s—at a time when both internal and external costs were mounting.

As we look into the future, three considerations affect the concept of viability and vitality of nations when placed in the context of internal and external costs.

First, the concept of international power has been changing in recent decades; it is certainly not the same as it was in the last century. While in absolute terms military power has increased tremendously, its relative influence internationally has declined. New elements of international power have emerged; among them is the image of a nation as a model for emulation—in terms of its ability to solve internal problems, to achieve social justice, and to grow.

Second, with global economic interdependence growing, the option for a superpower to place its resources primarily into the "domestic" sector is likely to decrease further. A certain minimum (not necessarily small) of international involvement will be essential to maintain the viability of the home front, because the latter heavily depends on a particular international order and the ability of the United States to protect and foster such an international order.

Lastly, the impact of technological advance creates imperatives of its own for developed societies, especially after a certain threshold of growth has been reached. A large allocation of resources to "external" pursuits (of whatever nature) at the expense of the social infrastructure (such as education, health care, housing, public transport, government services, and the preservation of the natural environment) may create dislocations in the economy and impose stresses on the human environment and the social structure itself.[35] Such conditions can hardly sustain the nation's vitality and viability in the global arena.

In short, America's viability and vitality in the future will depend on finding among those "internal" and "external" pursuits a balanced synthesis that meets the requirements of a given period and on supporting it with the total resources available to the nation.

Another issue is that of striking a healthy balance between environmental concerns and technological progress. A central issue here is this: How much of the hitherto "external" or "social" cost (i.e., the cost borne by society as a whole in the form of pollution) is to be converted into the economic cost borne by manufacturers and, ultimately, by consumers of products?[36] In the last analysis, it is the function of a rational and effective leadership to strike an appropriate balance between the two in such a way as to ensure both the environmental and economic viability of the nation, while keeping in mind that this issue is ultimately manageable only in an international context.

One major issue was implicitly referred to above—that of priorities in the allocation of resources to science and technology. Such priorities must be based on society's value pattern (or hierarchy of values) at a given time. The preceding discussion, in broad terms, contained value connotations; this can be detected in the concept of viability and vitality of the United States, in the definition of this concept as requiring a synthesis of internal and external pursuits, and so forth. Whereas the foregoing issues are yet to be resolved in specific terms, these concepts themselves provide a certain broad guidance for priorities. The next step would be to analyze both the trends in the impact of technology with the view of determining where the payoffs lie and the areas where resources can be obtained, since priorities (and values, on which they are based) are to a considerable extent conditioned by the total extent of resources available to the nation, that is, by what it is possible to achieve.

In short, we need what can be conveniently called a "scientific–technological strategy," a continuous endeavor at rationalization of resource allocation in the context of national goals and pursuits. In turn rationalization would help us to crystallize values and priorities inasmuch as they depend on a realistic prospect for their fulfillment.

Another problem the nation faces is the shortage of resources. Capital resources are essential, not only for technological innovation (R & D), but also for its use in the form of actual technological superstructure. Without resources results of R & D are useless. Considering the mounting demands on our resources for technological and other social needs, do we have any potential resource reserves to meet present and future needs? Indeed we do, although it would take time and effort to make them available.

One reserve of resources is the unnecessary luxuries in the economy, such as the remodeling of cars every year, whose cost runs into billions of dollars annually. There is evidence that manufacturers are beginning to trim this fat under the pressure of costs and public demand for economy and more value rather than frills.[37] Considerable waste in our society can be eliminated. In recent years defense has been a prominent target of the press and Congress for eliminating waste; voices have begun to be raised in other areas, such as welfare, that promise opportunities for more effective results and savings in resources.[38] Raising productivity in the nation as a whole—and particularly in such laggard industries as construction, railroad equipment, and a good part of shipbuilding—can result in savings of many billions of dollars annually. Research and development for this purpose and an improved application of science and technology to industrial production in general can go far in achieving this result. According to the U.S. advertising industry, $28.3 billion was spent on advertising in 1975, and total expenditures are growing at a rapid rate. Advertising is essential for the functioning of a free economy, but could not the economy function just as effectively at a smaller expenditure? Still another source of capital is cooperation with other nations in scientific and technological research and development.

Lastly, under the impact of technology, the nation is facing the problem of adjusting its institutions and policies to minimize the alienation of the individual and reduce personal insecurity and social instability. Indeed, the problem of stability, this time external, will gain in importance for American foreign policy as well. In coming decades the United States may or may not have a Department of Domestic Stability and a Bureau of International Stability in the Department of State, but if such institutions are found necessary, this will only indicate a failure of policy to resolve the problems of instability by other means (see chapters 9 and 12).

The Prospect

Will the United States successfully resolve the issues and problems discussed above? It is not an easy question to answer, and under the best of circumstances it cannot be answered definitively. There are some reasons for cautious optimism as well as for serious concern.

As compared with other nations or regions—for instance, Western Europe—the United States is better equipped to solve its problems. At least it has a central government and impressive resources that the government can mobilize. Awareness of a good number of problems exists, and efforts are being made, both nationally and locally, to solve them. Some of the problems we face begin to appear less formidable than initially anticipated. For example it has been pointed out that environmental regulations are not hurting the economy and may actually be benefiting it; moreover, they are likely to create a big industry of their own and thus further invigorate the U.S. economy.[39] The United States has a strong asset in the pluralism of its institutions. Aside from being the ultimate safeguard of our system, pluralism manifests what might be conveniently called "anonymous rationality." Somehow, by somebody, a particular solution is suggested. It may lie dormant for some time; but then, under more favorable circumstances, it begins working its way to the top. On the other hand, largely because of that very pluralism, there is no guarantee that the solution will reach the top or, if it does, that top leaders will be successful in implementing it. It may therefore take a serious crisis or two to shake the nation and set it on a course toward solution of its problems.

Our single most important problem is decision making. The United States is not likely to solve its problems and meet the challenges of the future unless it modifies its institutions and improves its decision-making processes. This subject is discussed at greater length in chapter 11.

The Soviet Union

5

THE TECHNOLOGICAL HISTORY of the USSR since World War II has been a mixture of striking progress, backwardness, and emerging problems. Some new developments still underway represent important departures from Soviet prewar practice. If they are successfully carried out, the USSR can be very effective in the use of technology for national viability and vitality in the coming decades. Success in the present Soviet efforts is especially important, for the Soviet Union—like the United States—is facing pressing shortage of resources in relation to mounting needs.

Postwar Progress

Despite its destructive effects, in certain respects World War II contributed to the strengthening of Soviet technological capability. Limitations of resources and facilities forced the Soviets to concentrate in selected areas on high quality—as distinguished from sheer quantity—of output and thus upgraded the technological level of defense-relevant industry.[1] Lend-lease contributed to the transfer of advanced technology, both military and nonmilitary, from the West. The Soviet leadership deliberately adopted a policy of requesting machinery and industrial equipment under lend-lease to facilitate postwar reconstruction.[2] Apparently a part of this equipment was allowed to rust—as were a number of East German factories dismantled and shipped to the Soviet Union—but it nevertheless helped increase Soviet technological strength.[3] These factors, plus the importation of German scientists and the sheer drive on the part of the Soviets, go a long way in explaining the rapidity of the country's postwar recovery and its impressive technological performance in the late 1940s and the 1950s.

With the outbreak of the Korean War in 1950 the Soviet Union stepped up its military and outer space efforts.[4] However, along with the emphasis on the nuclear-missile and outer space technologies, the first postwar decade also witnessed a considerable degree of neglect of technological progress in other sectors of the economy. Beginning with 1955 the Soviet leadership addressed its efforts against "stagnation in science and technology" in such sectors and called for "a struggle against [technological] conservatism."[5] The changes that followed involved more than simply updating technological hardware in selected industries. A shift of emphasis took place, away from the "traditional" technologies fostered by the Soviet leadership in previous decades, largely based on coal and iron ore, to more modern ones dependent on electric power, oil, and gas.

Significant changes took place in the energy base of the USSR. In particular, exploitation of natural gas—little developed before World War II and growing but slowly after the war—began to be pursued in earnest after 1955. By 1966 the USSR was producing more than a quarter of the U.S. output and, by 1972, the Soviet production of gas had increased to more than a third of that of the United States.[6]

New emphasis was placed on production of petroleum and on trunk pipelines for its distribution, which eliminated the long hauls of coal that were taxing the transportation system. By 1967 Soviet output of petroleum had reached two-thirds of that of the United States, and by 1972 it had increased to 87 percent.[7] A major change also took place in electric power generation. The policy of constructing small stations based on local fuel was abandoned in favor of large central stations (mostly thermal) of 1 million kilowatts or more and construction of major transmission lines and distribution grids to meet the rising demand for power.[8]

The revolution in the Soviet energy base helped bring about other important technological transformations. Between 1955 and 1965, Soviet railroads largely changed from steam to electric and diesel-electric traction. In 1965 the railroads carried more than three times as much traffic as in 1950.[9] In 1958 the Soviet Union launched a rapid expansion of its lagging chemical industry; by 1965 chemical output increased almost two and a half times.[10] Increasingly, using oil and gas as raw materials, the industry shifted to modern products such as petrochemicals, synthetic rubber, synthetic fibers, and plastics. Greater emphasis was placed on nonferrous metals, especially those, like aluminum, using electric power for processing.

Emerging Problems

Stimulated by these developments, the Soviet economy maintained a fairly vigorous growth throughout the 1950s. However, by the early 1960s it began to experience difficulties, some of which came from the very remedies introduced previously. For a comprehensive picture of the problems that began to emerge in those years, they must be placed in a broader context of the Soviet economic growth policy.

Aside from the emphasis on certain "power-rewarding" sectors of technology (military and nonmilitary) and sheer drive, the Soviets were historically successful in producing a fairly high rate of growth in the economy (somewhere between 6 and 9 percent annually) mainly because of two factors: (1) A very high rate of reinvestment of capital, some 20 to 30 percent of gross national product (GNP), as compared with 15 to 20 percent in the United States. This high rate of investment was achieved predominantly at the expense of consumers, deprived of consumer goods.[11] (2) Transfer of labor from low productivity areas into areas of high productivity. This mainly involved the transfer of surplus labor from agriculture into industry and the employment of women, previously engaged in household work and hence statistically nonproductive. Although the total input of capital and labor continued to be high in the first half of the 1960s, the rate of growth of the Soviet economy between 1961 and 1965 declined from about 6 percent annually in 1956–1960 to about 4 percent in 1961–1965.

There were several reasons for this decline, the single most important of which appeared to be the lower rate of growth in productivity, which found its roots in the changing economy and some of its endemic problems. As priorities shifted from single-minded emphasis on a few industries and as the technological base of industry broadened, the economy became too complex to be run effectively from the center by traditional methods involving close and minute control of individual enterprises. In some sectors of the economy new technology was introduced faster than it could be effectively assimilated. The result was to aggravate further the problems of inefficiency and inferior quality of output that have plagued the economy for decades. Many sectors of the economy were patently backward technologically. In spite of the shift of emphasis to some new areas, the economy was still essentially out of balance. Protracted neglect of such areas as agriculture, residential construction, and consumer goods had reached a point at which major remedial action was necessary. The satisfaction of the needs of consumers became especially important because incentives to produce suffered badly in the absence of worker rewards in tangible material forms. The economy thus needed major changes.

Although Khrushchev introduced some reforms in 1957–1958, they were not adequate to avoid the problems of the 1960s.[12]

In the debate that ensued among the Soviet economists in the late 1950s and continued into the 1960s, two remedial trends—not necessarily mutually exclusive—began to crystallize. One relied on some of the traditional tenets of Western economics, such as incentives and partial introduction of the market mechanism. The other placed its faith in cybernetics, which—by using computers and mathematical models—was viewed as capable both of providing more effective and rational control from the center and allowing a measure of decentralization. Cybernetics revived faith in the Soviet Union in rational management of a planned society. Although no firm commitment to planning based on cybernetics was made in those early years, cybernetics received the Party's blessing in 1961.[13]

In September 1965 the Soviet leadership announced a series of economic reforms. In some respects, they were vague. In general they preserved—and, in some respects, strengthened—central control of the economy, but at the same time they were intended to introduce greater flexibility into the economy and to permit economic levers to operate automatically or semiautomatically and thus to promote efficiency.[14] As we shall see later, their implementation left something to be desired, but they nevertheless represented an important break with the past and a willingness on the part of the leadership to adapt to new conditions.

For our purposes perhaps even more important changes have taken place in the policy approach to science and technology. By 1960 Soviet science (but not necessarily technology) had developed to the point where it could no longer look to the West for guidance; it had to break its own paths for growth. The new maturity was reflected in organizational changes in science and technology and in the new emphasis placed on this field. The Soviet Academy of Sciences was reorganized in 1961 and 1963. Whereas in the past the Academy had been assisting the Soviet leadership in achieving industrialization goals, its function has now become largely concerned with obtaining worldwide preeminence in basic research. While still subject to the industrial goals of the Soviet state (which, in part, include due regard to the applicational value of basic research and the conduct of some applied projects), scientists of the Academy have been given a considerable degree of autonomy in defining directions of basic research.[15]

Concurrently with these developments a new institution—the State Committee for Science and Technology [16]—was created. It has become, in effect, a "general staff" for formulating the nation's nonmilitary applied science and technology policy and directing its implementation. It is a powerful body headed by a Deputy Chairman of the Council of Ministers and directly responsible to the Council. It appears to work closely with the State Planning Commission (Gosplan) and the Soviet Academy of Sciences. The Committee's full-time staff is believed to total more than 3,000, in addition to about 20,000 part-time personnel located throughout the USSR.[17]

In the late sixties and the early seventies, Soviet leaders began to stress the significance of science and technology as a decisive element in the competition between the "socialist" and the "capitalist" world systems.[18] In particular the Soviet leadership has placed heavy emphasis on two aspects of science and technology: improvement in its development and its more effective introduction into, and utilization in, the economy. This relatively recent drive is aimed at increasing productivity of the Soviet economy and its growth. As a part of this effort, the Soviets have been fostering the development of "science of science," which, in their context, means a more effective development of science for a more effective growth of technology and the economy.[19] Soviet writers maintain that the USSR has certain important advantages over capitalist countries in using science and technology in competition with the West and that these advantages—namely, a planned development of the economy and a society purposefully organized to pursue rational goals—must be capitalized upon.[20]

In the early 1970s, as a part of détente, the Soviet Union has embarked on a massive drive to upgrade its technological capability by infusion of technology and technological know-how from advanced noncommunist nations, mainly the United States. The USSR employs a variety of techniques for transfer of technology; among them are intergovernmental agreements on economic, scientific, and technological cooperation with Western countries, such as licensing, joint ventures, and construction of turnkey plants.[21] The scope of agreements and operations is huge, involving many billions of dollars.[22]

What are the implications of these developments for the future? Do they suggest that the Soviet Union is likely to cope effectively with its problems and, leaning on technology coming from outside its borders, eventually become more successful than the West in using science and technology for its viability and vitality? For an answer to these questions, Soviet successes and assets have to be examined along with failures and constraints on the nation's resources.

Soviet Strengths

In the postwar decades the Soviet Union demonstrated a remarkable degree of adaptability to new conditions, in particular to the requirements of an effective use of science and technology. The Soviets are applying themselves in earnest to developing new conceptual instruments for this purpose. In a relatively short period of time—from the mid-1950s to the mid-1960s—the Soviets made major strides in restructuring the technological composition of their economy. In recent years the Soviet leadership has appeared to exhibit less dogmatism and to seek the solution of society's problems by listening to various views coming from below. A considerable degree of public discussion of policy questions has been permitted, a development inconceivable under Stalin.[23]

The leadership is aware of the problems of the economy and, in a number of cases, seems to be pragmatic in its approach, even if this pragmatism conflicts with ideology or with traditional policy. The Soviet leadership appears to be sensitive to the needs of consumers, both as a means of improving productivity and as an area of potential danger, as exemplified by consumer riots in Poland in 1970. In the Ninth Five-Year Plan (1971–1975), for the first time since the USSR had embarked on its industrialization drive in 1929, the "B" sector of industry (consumer goods and light industry) was scheduled to grow more rapidly than the "A" sector (heavy industry). Consideration for the consumer was displayed in the decision to respond to the extremely poor harvest of 1972 by importing more than 29 million metric tons of food and feed grains from the West at a cost of $2 billion, an act that was painful in economic, ideological, and political terms.

In their efforts to upgrade productivity the Soviets suffer from two types of inefficiency—economic and technological—that are, at times, so closely intertwined as to be virtually indistinguishable. Economic inefficiency stems mainly from inadequate incentives to produce and rigidities in the structure and the functioning of the economy. Technological inefficiency stems mainly from the inadequacy of procedures for introduction and dissemination of technology in the Soviet economy, from inadequate assimilation of available technology outside of the few high-priority sectors (such as defense, outer space, and aircraft industries), and from sheer technological backwardness in many sectors not enjoying high priority. Technological inefficiency is also caused by the general rigidity of the economy, which inhibits the flexible interplay between technological instruments, know-how, and economic needs that usually generates high productivity.

In the Ninth Five-Year Plan the Soviet Union placed a strong emphasis on improving labor productivity. In industry it was expected to rise by 36–40 percent, a rise that, in turn, was to account for 87–90 percent of the total planned increase of industrial production. The 1965 reforms—which were intended to increase economic efficiency—failed to be effectively implemented. Although efforts to improve economic efficiency have not been abandoned, the Soviet leadership appears to have placed a greater relative emphasis on raising technological efficiency and thus stimulating economic productivity.[24] In 1971 the productivity of Soviet labor in industry was only 41 percent of that of the United States, and labor productivity in agriculture was only 11 percent of American.[25] Thus the elimination (or, at least, reduction) of inefficiency, both economic and technological, represents a very important reservoir for Soviet economic growth and a critical potential fountainhead of resources for the development of future technology and for other needs.[26] The Soviets are using several means in their earnest attempts to raise productivity of the economy.

The direct infusion of technology from the West has already been noted. Another rising instrumentality which may strengthen economic productivity of the USSR is Soviet multinational corporations (MNCs) or joint ventures with Western nations. These corporations (mainly subsidiaries of Soviet home-based firms) cover areas such as automotive vehicles, aircraft, petroleum, steel smelting, banking, chemicals, and computers. They are expanding principally in Western Europe, Latin America, and the Middle East.[27]

The growth of Soviet MNCs is expected to accomplish several objectives. They constitute an additional vehicle for the transfer of advanced technology to the USSR. The MNCs also provide a convenient means of learning Western product design, marketing techniques, and management practices. The necessity to meet foreign competition by the Soviet MNCs is likely to stimulate the efficiency of domestic production. Furthermore, Soviet MNCs earn hard currency that can be used for further imports of technology from the West.

The Soviet Union relies heavily on computerization of the economy and on employment of mathematical models as a means of revitalizing planning and its implementation and of increasing efficiency of the economy. A decision has been made to develop an All-Union Automated System for Planning and Management (OGAS), a vast network of regional computer centers linked with the central computer system in Moscow. Regional centers will come into operation as construction permits, and initial stages of the entire system are intended to begin functioning in 1990. American and British companies are assisting in the development of this system, in which both Soviet and foreign computers will be employed.[28]

The Soviet Union places a heavy emphasis on research and development (R & D) and on scientific and engineering manpower. During 1951–1970 the Soviet R & D force was growing at 9.2 percent annually, as compared with

6.3 percent for the United States.[29] By evaluating those types of Soviet R & D manpower that correspond to their American counterparts, Western analysts estimated that the Soviet R & D establishment in 1971 would have cost the United States about $40 billion annually to support. U.S. support for R & D effort in 1972 was $28.9 billion.[30]

In April 1973 the Soviet leadership initiated a major organizational change that, in part, is intended to remove an endemic problem in the Soviet economy: the separation between R & D and production activities and a resultant tardiness in the introduction of technological innovations into the economy.[31] In accordance with this change, the basic unit in Soviet industry will no longer be the individual enterprise but the "production association," a group of related enterprises usually led by the largest and the most modern one. Production associations would also include their own design bureaus and research institutions. With R & D directly linked with industrial requirements, a new vigor is expected in technological innovations.[32]

In its potential importance this reorganization of industry goes considerably beyond its impact on technological innovations in industry. In the establishment of the production association Soviet planners borrowed heavily from Western experience and, in effect, created an organization patterned on the American corporation. In the Soviet setting such an organization can help solve a number of problems that have been plaguing the economy for some time.

It would reduce the complexity of central planning, which will have to deal with some 6,000 organizations instead of 50,000. It can also contribute to the efficiency of production. Since bonuses and other rewards are awarded to the production association as a whole and no longer to individual enterprises, the more advanced and efficient enterprises within each association have an incentive to assist and modernize the more backward ones. By the same token smaller enterprises that were habitually discriminated against in the allocation of resources and therefore tended to compensate for this by hoarding spare parts and other materials well in excess of their needs no longer have the incentive to do so. A greater degree of specialization is likely to be encouraged within the production association, and the association itself is likely to develop greater self-sufficiency than individual enterprises could. In short, if effectively implemented, the new organization may significantly raise the productivity of the Soviet economy.[33]

Whereas Soviet leadership views organization of the economy and incentives for the introduction of technology as important means of achieving technological progress, it also regards the structuring of R & D and investment priorities in such a way as to favor the more advanced, future-oriented technological sectors—for example, oil and gas in preference to coal, light metals and chemicals in preference to steel, fusion and breeder reactors in R & D as distinguished from conventional nuclear generation—as equally important. In other words the Soviets resort to a planned orchestration of the technological composition of the economy as a way of spurring technological progress and stimulating the vitality of the nation. In recent years the Soviet Union placed heavy emphasis on long-range planning. For the first time in its history the USSR has introduced a 15-year plan (1976–1990), of which the Tenth Five-Year Plan (1976–1980) is viewed as a critical phase.[34] In both the orchestration of the composition of the economy and in long-range planning the Soviet thinking and activity are distinctly ahead of those of the United States, where long-range planning in particular is almost nonexistent.

In terms of both institutional organization for science and technology and prevalent attitudes of the Soviet elite concerned with this field the USSR appears to be well prepared for taking advantage of future technology. In some respects the Soviet claim that their system has advantages over the "capitalist" West in using the potential of science and technology has considerable validity. By controlling the commanding heights of their economy and having an institutional mechanism for central administration of priorities, the Soviet leadership is in a position to allocate large resources to a promising technology of the future and to be much less handicapped in implementing their decisions by internal political or legal considerations than Western leaders are.[35]

The Soviet Union demonstrated impressive capability in key areas of advanced technology to which it assigned high priority, and it is in the forefront of development in the majority of advanced technologies that carry high promise for the future. In the following areas the Soviet Union appears to be ahead of the United States: the engineering of magnetohydrodynamic (MHD) power, peaceful uses of atomic explosives, breeder reactors, thermionic converters for direct conversion of nuclear power into electricity, magnetic and laser fusion, and, possibly, research and experimentation in weather and climate modification. In still other areas Soviet capability is roughly equal to that of the United States: military R & D, outer space (exceeding the United States in the magnitude of the effort but not in its quality), and the theory of superconductivity.[36]

As reflected in Soviet policy-oriented publications and Western reports, Soviet scientists and planners are thinking in terms of truly large-scale projects and are thus responding to one of the basic trends of present advanced and future technology. Academician I. Gerasimov, Director of the Institute of Geography of the Academy of Sciences of the USSR, writing in the leading Soviet periodical on policy and ideology, *Kommunist*, called for no less than a "general plan for the transformation of our country's natural environment." [37] The Soviet press has been discussing the feasibility of damming the Bering Strait and, through the pumping of the water into the Pacific, changing the direction of the Gulf Stream and thus ameliorating the climates of Eastern Siberia, Alaska, and Northern Canada. Soviet planners have been working on a scheme whereby the direction of flow of the great Siberian rivers will be reversed from north to south. In this way the vast amounts of water that empty into polar regions would be used for irrigation and raising the declining level of the Caspian Sea. If approved, the project, spread over decades, would cost more than $100 billion.[38] The Soviets have been using peaceful nuclear explosives to stimulate the production of oil and gas and to construct canals.

Top Soviet leaders are keenly attuned to the potential of science and technology for the future of the USSR. Unlike top leadership in the United States a number of them (including Brezhnev and Kosygin) have engineering backgrounds. The Soviet Union being an autocratic state, scientists as a group do not seem to be particularly influential. However, individual prominent scientists appear to have ready access to key Party and governmental officials and, insofar as their ideas serve the interests of the Soviet state, seem to receive attention.

Soviet Weaknesses

Although the foregoing account shows the Soviet Union is well equipped for facing the future, there is also another side to the coin. One important weakness is the uneven level of Soviet technology; advanced and backward technologies exist side by side. It will take a considerable effort to upgrade technologically backward sectors of the Soviet economy. Although, as a rule, the areas that enjoy high priority are technologically very advanced and efficient, the USSR is considerably behind the West (and especially behind the United States), even in some "power-relevant" areas of concern to the Soviet leadership.

The Soviet Union lags behind the United States and the leading non-Communist countries in computers. Most Soviet computers currently in use are second-generation models, using transistors. In the spring of 1973, the USSR introduced a series of sophisticated machines, built jointly with other Comecon nations. Called the Ryad system, it is comparable to the IBM 360, which appeared in the mid-sixties. The goal of the Soviet Union's Ninth Five-Year Plan (reportedly, not achieved)—to increase its stock of high-speed computers to 22,000 by 1975—amounted to only one-sixth of U.S. computers in 1973.[39]

Soviet industry is weak in a number of areas in electronics. Whatever strides have been made in the Soviet chemical industry, it still does not measure up to the United States in some areas of chemical technology, let alone in overall output. The Soviets are behind the West in microminiaturization. Soviet R & D in other than high-priority areas is poorer in quality and less in quantity than in the United States.

A part of the technological backwardness of the USSR—and a reflection of it—is the poor quality of Soviet products. If the qualitatively poor products are machinery, the nation's ability to produce suffers. Soviet machinery is cruder than its American counterpart and consumes more metal. The Soviet leadership is aware of the problem and considers reduction of material expenditures per unit of product "an enormous reserve" in the economy.

Another major weakness of the Soviet Union is the rigidity, the high degree of centralization, and the inertia of its bureaucratic system. Soviet bureaucracy does not seem to be capable of remedying some of its shortcomings, and it frequently manages to defeat or neutralize changes initiated by the top leadership. Thus, whatever the shortcomings of the 1965 reforms, they did attempt to introduce some novel features into the Soviet economic practice that had the potential of stimulating the economy. These features included a degree of decentralization of the economy at its lower echelons and a substitution of "economic levers"—prices, profits, capital charges, and incentive funds—for "administrative methods" (the central issuance of directives) to regulate and control the economy. In implementing reforms, the bureaucracy defeated the "liberalizing" aspects through procrastination, further complications, and more regulations. The upshot was that the system has ended up, not with fewer, but with more administrative methods of control and with the number of state employees increased by 38 percent, or more than half a million, during 1964–1970.[40] There have also been reports of the resistance of bureaucracy to the introduction of production associations, in spite of Party pressure for their implementation.

In short, whatever the intent of the Soviet leadership to eliminate some of the weaknesses of the system and increase its efficiency, there is no guarantee that it will be effectively carried out. This is especially so when the task of implementing a change is left to the bureaucracy itself, as was the case with the 1965 reforms.

Another limitation of the USSR is a pervasive degree of conservatism among the top Soviet leaders. Although they have allowed some airing of various views, they have been cautious in introducing change. The reason is partly ideological (change may lead to further change, get out of control, and undermine the regime) and partly a reflection of adherence to the past and the proven formulas that did work. The Soviet Union rose to superpower status by developing its heavy industry and military power. There is no such thing as a complete unity among the top Soviet leaders, but by and large they still seem to be reluctant to deviate significantly from the proven path.

Although the Soviet press continues to emphasize the goal of raising the material standard of the Soviet people, Soviet leadership is not unequivocal in its commitment to this objective. When, in 1972, the fulfillment of the Ninth Five-Year Plan faltered, the brunt of the cut was carried by the consumer sector, and not heavy industry or defense.[41]

A major problem facing the Soviet Union is the increasing pressure on the nation's resources. In spite of a GNP only about 58 percent of that of the United States, the USSR is allocating more resources to the military sector and exploration of outer space than the United States. The Soviet oceanographic effort is comparable to—if not greater than—that of the United States. Future technologies are expensive and will require large initial outlays. In the meantime the Soviet Union is facing some relatively new claimants on its resources.

The deterioration of the Soviet–Chinese relationship and growing Chinese military power impelled the Soviets to deploy 57 combat-ready divisions (about 570,000 men) to protect the 4500-mile Soviet–Chinese border and the 2500-mile border between Outer Mongolia and China by 1973.[42] In the long run true security in that area can be provided only by building up the Siberian hinterland to minimize the need for expensive logistic support of military activity and, perhaps, to give a touch of finality to the Soviet occupation of the area against Chinese claims. But Siberia presents both a promise and a problem to the USSR.

The mineral and energy resources of Siberia complement the human, industrial, and agricultural resources of the Western USSR;[43] in the long run (15–25 years), their development would be very beneficial to the growth of the Soviet Union. In the near term West Siberia, with its extensive deposits of oil and gas in the Tyumen region and reportedly the largest deposits of natural gas in the world at Urengoy, is of particular importance to the USSR. Soviet planning places heavy reliance on this region to increase the country's output of petroleum and gas. But the extent of the deposits in West Siberia is still uncertain; their exploitation—because of harsh climatic conditions—

may require imports of American rotary-drill technology; and current projects are complex, difficult to carry out, and based on rather optimistic assumptions.[44]

The spectacular rise of prices for oil and gas in recent years has enabled the Soviets to capitalize on the current production and to avoid potential problems through a sizable inflow of foreign currency.[45] Also, the outlook for exploitation of West Siberian resources brightened up considerably in 1976 as the governments of the United Kingdom, France, and West Germany indicated their interest in providing most of the financing for the North Star (natural gas) project at Urengoy.[46] However, if the expected deals with the West do not materialize, the Soviets may experience shortages of energy for the domestic economy and thus impair its efficiency; they may have to curtail exports of oil and gas to the Comecon countries, which would be highly undesirable politically; and they might have to import oil from the Middle East at the cost of the scarce foreign exchange. Alternatively the Soviet Union could divert to Siberia major resources from other sectors currently enjoying high priority, such as defense.[47]

As things stand, the leadership appears to be committed to the development of Siberian resources, both east and west. A further indication of Soviet intentions with regard to Siberia has been the resumption of the construction, in July 1974, of the Baikal–Amur Mainline (BAM), a 2000-mile railroad that will cut across Siberia some 150–300 miles above the present Trans-Siberian railroad.[48]

Agriculture is another area requiring a large infusion of resources. The Soviet Union has practically exhausted the potential for extending the area of its arable land; even now a large percentage of its agriculture is found in climatically unfavorable regions.[49] The *kolkhoz* system of agricultural management is highly inefficient; its abolition could perhaps provide a solution to the agricultural problem. Another possible solution would be to import agricultural products from abroad. Both, however, appear to be politically unacceptable to the Soviet leadership, although, under emergency conditions, the USSR has been known to resort to imports. Thus the Soviets are pouring vast resources into agriculture, mainly to obtain larger yields from the existing arable land. In the Ninth Five-Year Plan the USSR has been investing about $34 billion annually in agriculture, compared to about $8.5 billion a year in the United States.[50] The tremendous cost of the Soviet drive to achieve agricultural self-sufficiency is further highlighted by the fact that Soviet agriculture uses nine times as much labor and 50 percent more land to produce roughly three quarters of what American farms produce. But the determination of the Soviet leadership to be self-sufficient in agriculture persists. The USSR is undertaking a vast agricultural development project, to cost $44 billion in the Tenth Five-Year Plan. The project will develop the non-black-soil zone, a dry steppe consisting of sandy, clayish brown marginal soils, running northward from Leningrad and Moscow across European Russia into Siberia.[51]

The Soviet Union will also have to spend massive sums of money for the environment. To be sure, the USSR is relatively less developed than the most advanced Western nations and, for the size of its population, it enjoys the benefits of a large territory capable of absorbing the effect of pollution. Moreover, as environmental concerns soared into prominence in advanced nations of the West in the late 1960s, the Soviets claimed that the environment was not a serious problem for the USSR since its planned system was much better equipped than Western "capitalist" systems for producing economic growth with minimal environmental disruptions. In fact, however, the single-minded stress on production in the USSR is perhaps even more damaging to the environment than industrial activity in the West. The meeting of production goals has been the traditional and foremost pursuit of the Soviet plant managers and ministries. The effect on the environment was paid little attention to, and, unlike in the United States, organized private activity to safeguard the environment was either nonexistent or carried little weight.

As a result, water pollution is serious in the USSR; in some cases, it has reached crisis proportions. Air pollution is a less pressing problem, but the situation is likely to deteriorate as the Soviets enter the automobile age.[52]

It appears that, by the second half of the 1970s, the magnitude of the problem had compelled the Soviets to become more open on the subject and more direct in facing up to it. According to Soviet sources, for 1976 1.7 billion rubles ($2.2 billion) were allocated for capital expenditures alone to combat pollution, while operational costs for safeguarding the environment were said to be mounting.[53] Quite unprecedentedly for Soviet professional literature, where the military is treated very cautiously, at least one analyst pointed out that détente and the resultant reduction of expenditures for defense may allow a considerably larger allocation of resources for the environment.[54]

The consumer sector (including residential housing) is another area likely to require an increasing share of resources. The Soviet leadership is committed to the improvement of the lot of the consumers, who have a long way to go to approach the level of their counterparts in the West; consumer standards in the USSR are quite low even if compared with Eastern European countries. The unprecedented goal of the Ninth Five-Year Plan for the consumer sector (Sector "B") to grow faster than heavy industry did not actually materialize and in the Tenth Five-Year Plan the Soviet planners returned to what—to them—comes naturally: a higher rate of growth for heavy industry (38–42 percent vs. 30–32 percent for the consumer sector).[55]

The willingness of the Soviet government to allocate greater resources to the consumer sector in the future is speculative at best, but it is possible to envision at least three factors—in addition to the consideration of work incentives—that may impel the government to allocate increasing resources to consumer goods:

1. Widespread consumer unrest, which may lead not only to a greater allocation of resources to this sector but also to major economic reforms.

2. As the lot of the consumers is gradually improved, a kind of steady but rising pressure of consumer expectations may emerge and acquire a momentum of its own. In the light of the internal dynamics within the Soviet ruling group, the leadership may consider it prudent to give in gradually to this momentum rather than oppose it.

3. The supply of certain consumer goods inevitably stimulates demand for certain products and services that may require major resources. This is well illustrated by the case of automobiles, to whose provision the Soviet government is strongly committed. Automobiles require reasonably well-kept roads, and yet the Soviet Union has a total of only 139,000 miles of paved roads, which amounts to only 11 percent of corresponding roads in the United States and to only 16 percent of the entire mileage of Soviet roads.[56] The USSR is extremely limited in its automotive service and repair facilities. All these shortcomings will require huge investments.

Another problem closely related to the availability of resources is that of manpower. We have noted earlier that the excess manpower in the agricultural sector and in households (unemployed women) has virtually dried up and thus eliminated the possibility of increasing labor productivity through a mere transfer of manpower from these areas of low productivity into the industrial sector. As we look further into the future, the problem of manpower becomes more serious. In the 1970s, the situation with manpower is being significantly alleviated by an influx of more than 2 million 16-year-old youths each year (2.1 million annually in 1976–1980). The situation changes drastically, however, for the 1980s: The annual influx of workers drops to about 600,000. Moreover, most of these increases come from the central Asian republics, where population is much less mobile and where the level of urbanization and rate of labor force participation are considerably lower. Thus, unless the USSR significantly raises productivity of labor in the 1970s, it faces a potentially disastrous situation in the 1980s.[57]

Aside from the more general means of increasing labor productivity, such as the introduction of technology, use of financial incentives, and increases of consumer goods, an important instrumentality to this end has been the so-called Shchekino system. First introduced at the Shchekino Chemical Combine in 1967, the system allows individual enterprises to cut their labor force and to use wages saved as incentives for both management and the remaining workers. Although there is evidence that the Shchekino system reduces redundancy in labor, increases productivity, and releases marginal labor for other employment, by the beginning of 1976 it had been extended only to 6,000 out of 52,000 enterprises.[58]

In the first four years of the Ninth Five-Year Plan "over 80 percent" of growth in the industrial sector was attributed to increases in labor productivity.[59] Although this outcome fell short of the 87–90 percent envisaged in the original target of the Plan, it was a creditable performance nonetheless. In the future, however, even larger increases in productivity will be required to maintain the present rate of economic growth. The influx of labor into industry in the Tenth Five-Year Plan is expected to decrease to 4 percent, as compared with 7 percent in the Ninth,[60] and things will get even worse in the 1980s. In view of the likelihood of diminishing marginal returns in the rate of growth of labor productivity in coming years, Soviet economic growth may be severely curtailed. There is, however, an important potential source of mature and, in a number of cases, highly skilled manpower—the large Soviet military establishment. This brings us to a broader consideration of the role of the military sector in Soviet economic growth.

National Resources and the Military Sector

As we have seen, the Soviet Union requires very large resources to sustain its economic growth and national vitality. Soviet needs can be briefly summarized. The USSR needs large resources for the development of Siberia in general and the region's energy resources in particular. The Soviets need multibillion-ruble capital investments to modernize and expand their agricultural production. The neglected consumer sector requires extensive investments. Environmental needs will require increasing resources. The USSR needs resources for obtaining technology from abroad to increase the productivity of Soviet labor. And, especially in the 1980s, the USSR will require manpower resources to enable its economy to grow.

This spectrum of resource requirements is interdependent. Successes in one area may alleviate pressures elsewhere. Tradeoffs between and among the various areas are possible. All this being allowed for, two salient conclusions emerge:

1. The course of action the Soviet leadership chooses in solving the problem of shortages of resources will be a highly important factor in influencing the nature of the Soviet system and the direction of its evolution.

2. Despite whatever tradeoffs among the areas discussed above are possible and whatever amelioration in some areas can be achieved because of a successful solution of problems in others, the basic fact remains that the pressure on Soviet resources is real and compelling. It is therefore appropriate to take a look at the military sector and its actual and potential role in the Soviet economic growth.

Apparently influenced by the Cuban missile crisis of 1962, the present Soviet leadership has embarked on an intensive buildup of its military forces. Strategic nuclear weapons have received special emphasis, but other sectors—especially naval power—have also been substantially expanded and modernized.

According to U.S. intelligence estimates, the USSR is allocating up to 15 percent of its GNP to defense, as compared with 5.5–6.0 percent for the United States. The CIA estimated that, in 1975, Soviet military programs exceeded those of the United States by 50 percent, if military retirement pay was excluded on both sides. The USSR maintains about 4.8 million men under arms, more than twice as many as the United States does. However, nearly half of Soviet military forces are engaged in work considered civilian in the United States or are assigned to missions that do not directly threaten the United States.[61]

The cost to the Soviet economy of the heavy emphasis on the militarily relevant sectors has been considerable. A large share of the best scientific, engineering, and managerial talent has been diverted toward defense. This has been a major factor in the low productivity in the civilian economy. In analyzing the relationship between defense and the civilian sector, Stanley Cohn concluded that a shift of resources (including highly skilled manpower) from defense to the civilian economy, if decided upon by the Soviet leadership, would contribute significantly to productivity and the growth of GNP, considerably more so than a direct increase in investment at the expense of the military.[62] The question arises, however, whether the Soviet leadership would be willing—or, for that matter, able—to change the pattern of resource allocation of recent years that has tended to favor the military.

Western students of the USSR point out the emergence of a Soviet military–industrial complex that penetrates deeply into the economy.[63] Its head, Dmitri F. Ustinov, became a full member of the Politburo in 1976 and, in the same year, succeeded the late Marshal Grechko as the Minister of Defense. The military has an important leverage in the Soviet decision-making process through the Defense Council, a highly influential advisory body of the Politburo that brings together the leading political and military officials.[64]

I do not intend by my discussion to suggest that the influence of the military in the USSR is all-powerful and that it will necessarily preclude reallocation of resources from the military sector to other sectors of the economy. In fact there are indications that the debate on the direction of resource allocation has been going on in the USSR and that nonmilitary influences are strongly felt. For example the announcement of the Ninth Five-Year Plan was delayed more than once, and, as late as January 1972, a number of aspects of the plan had not been made public, suggesting that a ferment with regard to allocation of resources exists in the USSR. Furthermore Western observers note that other groups—in particular, the economic planners/enterprise managers—are emerging in the USSR and gaining in strength.[65] They may provide a countervailing force to military influence. Soviet participation in disarmament negotiations (both SALT and Mutual and Balanced Force Reductions [MBFR]) is indicative of a consciousness of the economic burden of military weapons.

The important point in appraising the outlook for a major diversion of resources from the military sector into the civilian economy is not just that the military is powerful and may therefore prevent such a diversion but that it has systemic implications for the USSR. In a Soviet Union that shows distinct signs of an evolving institutional pluralism the military is influential, not only—and perhaps not primarily—because it enjoys support of certain top leaders, but also because it occupies a powerful position in the existing institutional configuration of power. A successful withdrawal of major resources from the military would change the internal balance of power, strengthen countervailing institutional forces, and possibly trigger fundamental long-range changes. Furthermore, a move to withdraw major resources from the military sector is likely to have serious implications for the fortunes of individual members of the ruling group, and they might feel that this subject, at least for the time being, should best be left alone.[66]

Moreover the diversion of resources from the military sector is not the only alternative to the solution of Soviet economic problems. Another alternative would be more extensive economic reforms, but the Soviet leadership is apparently concerned about their impact on the system. The requisite nonmilitary technology, potentially capable of boosting Soviet productivity and thus generating extensive internal resources, can also be obtained from the West. The latter is precisely the course of action the Soviet leadership has chosen. It has the advantage in that many types of technology imports are not only self-liquidating but also directly profitable. As a result of the rise in prices of raw materials, the materials pay for Western technology brought in to extract them and leave a substantial surplus of foreign currency. To be sure, the choice made by the Soviet leadership also has very important systemic implications, perhaps much more important in the long run than diversion of resources from the military sector. But it seems that, at this point, the Soviet leadership has found cooperation with the West to be the preferred solution to its economic problems. Whether it alone will suffice remains to be seen.

However, insofar as the military has gained in the top Soviet decision-making process in recent years, it is a liability to the Soviet Union's efforts to enhance its vitality and viability through nonmilitary technology. These considerations apply, in part, to the introduction of technological innovations into existing industries and, more significantly, to the adoption of new technologies that may create entirely new sectors of an industry or entirely new industries. In the great majority of new technologies where the Soviet Union has shown considerable progress—such as MHD power, weather modification, or fusion energy—technological advance is in an early stage requiring relatively small resources. These areas of technology have yet to meet the test of the time when advanced development and exploitation will require huge expenditures, and thus come into direct competition with major military programs.

Potential Models of Soviet Society

As one looks into the future, and assuming that there will be no major instability or disruption in Soviet society,[67] it is possible to envision at least five principal models of the Soviet Union: conservative, restricted economic liberalization, pluralistic, ideologically constrained technological orientation, and technocratic.[68]

The Conservative Model

The conservative model is essentially the one that prevailed in the Soviet Union before the full-fledged launching of détente in 1972. It is characterized by a high degree of centralism and control of society by the Party. Ideology is stressed and dogmatically asserted, even though the nation is paying a heavy price in economic terms for its adherence to ideologically motivated policies (e.g., clinging to the collective system of agriculture although it is highly inefficient and expensive). The economic system is highly bureaucratized; truly significant innovations are not normally introduced, but when attempted because of economic or other imperatives, they are usually watered down or emasculated. No extensive technological or economic cooperation with the West is undertaken.

The Model of Restricted Economic Liberalization

Restricted economic liberalization would create a model somewhat similar to, but not going so far as, present-day Hungary.[69] In this model the Soviet leadership would insist on central control by the Communist Party but would introduce limited market mechanisms and incentives, such as those envisioned in the 1965 reforms. The manager of the firm would be given greater freedom of action. New and more realistic incentives would be provided for introduction of technological innovations and increases in productivity in industry. Further reforms would be introduced in agriculture to provide incentives for the peasants and to stimulate production, but the present agricultural system combining state and collective farms would be preserved. Some further concessions would be made to the consumer, but they would fall short of the relative abundance of consumer goods at present available in Hungary.

Whereas trade with the West would be expanded and mutual participation in industrial projects with selected Western nations would be increased, the Soviet leadership would not allow free movement of goods and regional specialization on an international scale in accordance with the principle of the relative economic advantage (another deviation from Hungary). The scope and activity of government-inspired MNCs would be enlarged. The Soviet Union would participate with advanced non-Communist nations (including the United States) in cooperative programs in science and technology; this would help alleviate the pressure of resource scarcity in the USSR. However, the Soviet leadership would be careful about the form and potential consequences of cooperation; it would contain any tendencies to weaken or erode society's commitment to Communism. Soviet society would become somewhat more pluralistic, but, as in the case of Hungary, manifestations of pluralism would be carefully watched by the Party and restricted. Ideological considerations would still be fairly strong in Soviet decision making but not quite to the same degree as in the conservative model; more concessions to pragmatism would be made.

The Pluralistic Model

The pluralistic model assumes that, under the impact of science and technology, Soviet society becomes more complex and differentiated, with new increases in the influence of professionalism and tendencies toward dispersal of authority.[70] Pluralization along professional and economic divisions is further augmented by that along nationality lines, somewhat similar to that in Yugoslavia today. The Communist Party, in an effort to adjust itself to reality, would increasingly incorporate these various groups into its midst and thus transform itself into a forum for the interplay of the various constituencies of the new Soviet society. This development is facilitated by the shortages of resources needed to take advantage of advanced technology and develop the natural resources of Siberia. The Soviet Union continues and further expands its cooperative projects in science and technology with advanced free nations. Japan and other advanced non-Communist nations extensively participate in mining, oil, and electric power schemes in the hinterland of the USSR.

Introduction of market mechanisms and various incentives for technological innovation largely eliminate economic and technological inefficiency in the Soviet economy. The Soviet agricultural system is modified to the extent that the old *sovkhozes* and *kolkhozes* are virtually abolished, although the regime still maintains a degree of control over agriculture. The latter is fairly efficient and satisfies nearly all of Soviet needs in foodstuffs; the remainder is imported.

Whereas the Soviet Union thus succeeds in generating large resources internally, these resources are not primarily geared toward external aggrandizement but are used for consumer goods and for strengthening the various powerful constituencies that exercise a considerable degree of influence over resources. The Soviet regime is still authoritarian but lacks strong central guidance at the top and is characterized by various groups' competing among themselves in the top Party and governmental organs. The Soviet Union is thus characterized by a considerable degree of internal dynamism in the competition of various interests, but these interests tend to balance and neutralize each other so as to emasculate the country of a thrust in policy.

The Ideologically Constrained, Technology-Oriented Model

The ideologically constrained, technology-oriented model would be similar to the "Economic System of Socialism" introduced in East Germany in 1967 and emphasized after the Soviet invasion of Czechoslovakia in August 1968.[71] This model would, however, be more restricted by ideological considerations characteristic of the top Soviet leadership than its German counterpart.

The technology-oriented model of the USSR would be run primarily by Party officials, but they would have a sophisticated appreciation of the potential of technology and would be closely assisted by scientists and technologists. A certain amount of decentralization and incentives would be introduced on the lower level, but the emphasis would not be on these but on the rational structuring and the scientific planning of the economy with the aid of mathematical models and cybernetics. The regime would still, however, lean on the military to a considerable degree, and this would restrict its ability to take full advantage of nonmilitary technologies. The Soviet leadership would also continue its ideological feud—augmented by power motives—with China and thus give the military an additional lever of influence in the allocation of national resources.

The Soviet leaders would take advantage of cooperative programs in advanced technology with non-Communist nations, but they would be cautious and restrictive in this activity. Although reasonably effective in applying advanced technology (including computers) to help solve its agricultural problems, the regime would still insist on self-sufficiency in agriculture and would be ideologically constrained in adopting certain pragmatic solutions when they conflicted with ideological tenets.

However, under this model the Soviet Union would sustain a reasonable amount of vitality at home and abroad. It is entirely possible that, in this model, a strong leader would emerge who would be keenly aware of the principal interests operating in society and would, as appropriate, foster them. But he would have enough power of his own to break the entrenched resistance of the interests that favor the status quo or to break a deadlock among various interests and to move society in a distinct direction.[72] Within the framework of this model the Soviet Union would still be committed to proving the superiority of its system over that of "capitalism" and would lean on both military and nonmilitary technology to advance its interests.

The Developed Regions

The Technocratic Model

The technocratic model would be somewhat similar to that of East Germany under the New Economic System (NES) prevalent between 1963 and 1968. The Soviet technocratic model would, however, be less constrained by ideological considerations than the NES was. Decentralization and incentives (profit motive, rewards to encourage both production and technological innovation, and a price system designed to strengthen the application of incentives) would be introduced on the lower levels of the economy. These measures would be more liberal than in the model of restricted economic liberalization, but, as in the latter model, the command economy (as distinguished from a market economy) would be preserved. A technocratic Soviet Union would, however, lean extensively on mathematical and cybernetic models to run the economy and would require a level of economic development more advanced than the present. In both of these respects the technocratic model would bear resemblance to the ideologically constrained, technology-oriented model.

In the technocratic model scientists, engineers, and managers would rise to the top as the leading elite group in the Soviet Union. They would be sufficiently socialized by Communist ideology to use its language and absorb its competitive spirit with regard to the "capitalist" world, but they would be pragmatic men who would use the clichés and slogans of ideology mainly as instruments to help them to control society. If reasonably important considerations of power or effective use of technology in support of the national interest of the USSR would require a deviation from ideology, the Soviet leadership would so deviate.

The technocratic leaders of the USSR would be effective in developing and using technology on both the macro and micro scale. They would be highly competitive with regard to the rest of the world but would use nonmilitary technology, rather than military power, as the principal instrument of competition. Whenever it was expedient, however, the Soviet technocrats would not hesitate to enter into a cooperative arrangement with advanced free nations, always keeping the competitive end in mind. As with the preceding model it is possible that the Soviet technocratic society could be run by a single strong leader. However, unlike the ideologically constrained, technology-oriented model, the various technocratic and institutional interests would have a relatively greater degree of autonomy, and the leader would have a lesser ability to control them and would have to rely more on the manipulation of the internal balance of power to help steer society.

Internally the technocratic leadership would rely on a number of factors to maintain it in power: a reasonable satisfaction of consumer wants (without going overboard on this—resources would be mainly channeled into state-serving technological programs), dynamism and success of the USSR internationally, and ideological appeal. Externally it would seek a following by capitalizing on the success of the Soviet economic and technological growth and the dynamism and purposiveness of the nation's policies, backed by technological instruments.

The Prospect

Which of these models the Soviet Union is the most likely to employ in the coming decades will depend on both internal developments in the USSR and on external events. At present the Soviet Union is in transition. It does not fully fit any of these models; rather, it combines some elements of each.

Characteristics of the conservative model are still quite strong in the USSR. The Soviet leadership continues to adhere to the system of collective and state farms in agriculture and to autarchy in this area, in spite of the huge costs of this policy. The liberal elements of the 1965 reforms have been largely emasculated. Whatever concessions are being made to consumers, heavy industry and, in particular, the defense sector still receive high priority.

At the same time some elements of restricted economic liberalization and even pluralism are clearly discernible. The USSR is grouping enterprises into "production associations" and giving them a degree of autonomy. It is introducing, albeit slowly, the concept of hiring and firing on the lower level through the Shchekino system. The Soviet Union is studying and introducing Western management techniques and entering into extensive cooperative agreements, involving joint operations, with Western corporations. Institutional pluralism is clearly crystallizing in the USSR. According to some students of Soviet society even a greater degree of pluralization is taking place, especially on the lower level. Along with the "official" economy, there appears to be an economy similar to that of the New Economic Policy (NEP) in the 1920s, in which a sizable amount of private and "semiofficial" goods are sold in accordance with what amounts to the free-market principle.[73]

By expanding industrial cooperation with the West, the Soviet leadership has allowed a degree of interdependence with advanced non-Communist nations. But it is not clear to what extent, if any, this implies the acceptance of the principle of relative economic advantage on an international scale. By offering long-range agreements to export energy and certain other products in exchange for advanced technology, the USSR becomes, at least for a while, dependent on technical assistance and spare parts from the West. But as the Soviet Union upgrades its own technological capability, that dependence is likely to diminish or disappear. Communist China, at a much lower level of technological advance than the USSR and much more dependent on outside technical assistance, demonstrated in the late 1950s and the early 1960s that dependence on foreign technical assistance need not be a decisive factor in constraining the nation's political will.

Thus it appears that the Soviets accept the present degree of dependence on "capitalist" nations as a price to be paid for the solution of their pressing economic problems but not necessarily as a firm commitment to international interdependence. This is not necessarily to say that, in the evolution of its system, the Soviet Union will not accept the principle of relative economic advantage but merely to suggest that, at present, there is no clear-cut evidence that the Soviets have embraced this principle.

The present strategy of the Soviet leadership appears to be to transform their society from what was basically a conservative model into an ideologically constrained, technology-oriented model. This is to be achieved with the help of two principal instruments: Western technology and limited reforms (production associations, the Shchekino system). This strategy suggests two questions: Is the model of an ideologically constrained, technology-oriented society a workable one—or is it an unrealistic fantasy? If it is a realistic model, what is the outlook that the USSR will be able to achieve it?

It is entirely possible that this model could be workable for Soviet society. There seems to be considerable justification for the Soviet expectation that computers, when effectively introduced into the economy, could be of great assistance and might revitalize central planning and thus enable the Soviet leadership again to be in firm control of the "commanding heights" of the economy. Extensive introduction of advanced technology into the economy and its effective application could more than compensate for the decline in the influx of manpower anticipated for the 1980s. However, even with the help of computers, the central planners cannot effectively control and steer all the minutiae of modern complex economies. Thus the USSR will probably have to introduce more incentives and more room for individual initiative on the lower level than it has done heretofore to make this model work.

The second question is more problematical. As we have seen, the USSR has some important weaknesses, not the least of which is its huge, rigid, and inefficient bureaucracy. The Soviet leadership would therefore have to solve the problem of bureaucracy first; otherwise it may defeat the leadership's efforts to steer society. Another obstacle is the sheer magnitude of the undertaking, involving a country covering one-sixth of the earth's land surface and a population of about 250 million. This population is not so well educated and disciplined as that of East Germany, where this model of society had its early test. On the other hand this model is not expected to be implemented in the USSR until some time in the 1990s, and this gives Soviet planners a considerable lead time to solve some of their problems.

Whereas the long lead time might provide an opportunity to solve obstacles on the road toward achieving Soviet objectives, it might also result in diverting society's development to a different direction. Societal trends frequently run at crosscurrents, and it is not easy to steer society in a certain predetermined direction over a period of some 20 years. Thus, when the Soviet leadership in recent years backed mathematical economists along with computers in preference to the more liberal, market-oriented economists like Liberman, it was doubtlessly influenced by ideological considerations. But now mathematical economists advocate more Western techniques and more flexibility in planning, and Soviet leadership appears to be backing them against the more traditional planners relying on concrete quantitative goals.[74] One wonders where the requirements of effectiveness will take the Soviet planners, especially since the journey spans more than 20 years.

Moreover, the USSR is facing some serious questions in the near to midterm future (five to eight years). Will the Soviet efforts to improve efficiency of the economy succeed? As we have seen, the Soviets are making progress in improving labor productivity. The expected beneficial impact of imports of technology from the West is yet to be felt; it may help sustain growth of labor productivity and may increase productivity of capital. It is thus conceivable that the USSR may go rather far in invigorating technologically backward sectors of the economy and stimulating productivity in coming years. The extent of Soviet economic reforms is modest, but, over a period of five years or so, they may bear fruit, especially if strongly backed by top leadership and if aided by the expansion and effective functioning of Soviet MNCs. Certainly the potential benefits from increasing productivity of the Soviet economy are vast, and so are the potential savings in materials from improvements in machining technology and other techniques related to use of materials.

However, there is no guarantee that the USSR will succeed in solving the problem of productivity and generating the necessary resources to maintain its viability and vitality. The Soviets are facing significant problems in this area. If they do not solve them, then the situation in the eighties—as the influx of manpower drops—is likely to become critical. In view of this prospect the USSR may have some very important decisions to make about 1980. The choices will include:

(1) Accelerated and meaningful economic reforms; (2) agricultural reforms or permanent dependence on agricultural imports; (3) cutting allocation of resources to the consumer sector or, alternatively, if resources are obtained elsewhere, giving this sector a truly high priority and making it efficient; (4) limiting military expenditures.

Resort to economic and agricultural reforms is likely to lead to greater institutional and group pluralism in the USSR and may eventually result in a pluralistic model of Soviet society. If the Soviet leadership cuts allocation of resources to the consumer sector, it must be prepared for a decline in productivity and a possible consumer revolt. The alternatives here are repression, a return to the conservative model and economic stagnation, or liberalization and a possible evolution of the pluralistic model. If the Soviet leadership gives high priority to the consumer sector and makes it efficient, at least two divergent potential developments might take place. The USSR may succeed in increasing the productivity of the economy, and society may continue to proceed—within the framework of restricted economic liberalization—toward the ideologically constrained, technology-oriented model. Alternatively consumer appetites and needs may continue growing. The leadership may be yielding to consumer pressure, and society will be moving toward a pluralistic model.

The implications of diversion of resources from the military to the civilian sector would vary, depending on the extent of this diversion. Some limitations of military expenditures and their allocation to the civilian sector would be a prudent course of action for the USSR, even if there were no critical need for resources to improve productivity. After equality with the United States is achieved, the burden of military expenditures would significantly increase if the USSR chooses to go beyond equality to seek military superiority. The Soviet Union would then have to find its own paths to what comprises leadership in military power and not just follow in the footsteps of the United States as it had been doing before. This is a complex task where the total effort, and its potential mistakes, can be very costly.[75]

To meet the requirements of these considerations, the diversion of resources from the military to the civilian sector need not be major. What would be involved is a curtailment of future growth of the military sector, perhaps formalized in an arms-limitation agreement. This development would not have important systemic implications, except in the long run: As the civilian sector continues to grow (while the growth of the military sector is retarded), it will eventually gain in the institutional balance of power. This is especially likely when new advanced technologies, requiring large scale and huge resources, reach the procurement stage.

If, however, the diversion of resources from the military sector were large, then its implications would be significantly different. For such a major reallocation of resources, one of the following two preconditions—or both—would probably be required: (1) a true sense of crisis with regard to the need for resources in the civilian sector and (2) a strong leader, capable of changing the internal institutional balance of power with impunity to himself. This move would have proximate systemic implications, although their course is not entirely clear. On the one hand it might result in giving more power to economic planners, scientists, and industrial managers oriented toward futuristic, nonmilitary technologies. If this group is backed by the principal Party leaders, the USSR may be set on the course toward the technocratic model of society. On the other hand, if major diversion of resources from the military sector is accompanied by assigning high priority to the consumer sector, this might lead to other potential developments, including a pluralistic USSR. But it is also possible that the reallocation of resources from the military sector may simply mean a continuation of the course set by the Soviet leadership toward an ideologically constrained, technology-oriented USSR, except with a significantly diminished military component.

We have seen that a technocratic model of the USSR may evolve in response to a major shift of resources from the military sector. It is also conceivable that the Soviet drive to achieve an ideologically constrained, technology-oriented society may itself eventually bring about a technocratic model of society. It is possible that, given ideological constraints and the magnitude of the task, the Soviet Union may run into serious difficulties in implementing its gigantic computerized All-Union Automated System for Planning and Management, and the mathematical economists may not exactly provide a panacea for Soviet problems. The Soviet leadership may then turn to a less structured, more pragmatic approach, attempting to capitalize on whatever opportunities science and technology provide for enhancing the country's viability and vitality. A technocratic model may thus evolve. It is also possible that a technocratic model may appear after a reasonably successful achievement of an ideologically constrained, technology-oriented model. Society—including its top leadership—may simply get tired of the ideological shackles, water them down, and co-opt technocrats into its ruling circle.

External developments will have an impact on the evolution of Soviet society. If the developed Western nations run into problems of their own—like social instability as a result of the technological impact, difficulties in generating resources to solve multiple societal problems, and poor economic growth—the Soviet economic difficulties may not have major repercussions for the system. If, however, an economically stagnant Soviet Union were to face a viable and vital "capitalist" world, internal pressure for change is likely to be strong, especially as a new generation of leaders rises to the top in the early 1980s.

In summary, the early 1980s will be an important period for the Soviet Union, a period during which a number of critical decisions bearing on the evolution of the Soviet system will have to be made. The number of possibilities is wide. At one extreme is the possibility that the Soviet leadership may fail to make firm choices, may spread its resources thin into various directions, and—the cooperation with the West notwithstanding—may fail to raise economic productivity. It may thus seriously overextend the nation's resources, causing trouble at home and forcing a sharp curtailment of military commitments. But it is entirely possible that the Soviet society will go through a more orderly evolutionary process. Of the various models of Soviet society that might emerge, the most desirable one—from the point of view of the West—would be a pluralistic model. Soviet reversal to an orthodox conservative model might be distasteful to the West on ideological and humanitarian grounds, but it is not likely to provide a serious challenge to Western nations, since it is likely to be characterized by economic stagnation and military withdrawal or semiwithdrawal. The models of the USSR that might produce a new technological challenge to the West are an ideologically constrained, technology-oriented model; a technocratic model; and, possibly, a combination of a technocratic and a pluralistic model. This challenge can be very real, especially if the USSR has a strong leader.

Western Europe

6

MORE THAN THOSE of any other advanced region, the viability and vitality of Western Europe are endangered by the impact of future technology. I shall first discuss some background developments and then address the subject directly.

Post-World War II Resurgence of Europe's Vitality

In comparison with the interwar period, post-World War II Europe has established an impressive record of progress in bolstering its growth and vitality by using the available technological potential. Western Europe's achievements and assets since the war have largely been a product of the organized will and energy of its people, although certain favorable conditions and external factors were also contributors.

One such external factor was the Marshall Plan, which from the point of view of Western Europe was a "given"; whatever credit Western Europe can claim for its success lies in the effective organization of institutions and manpower for the Plan's implementation.[1] The destruction of plant and equipment and the low level of capital formation during the war years facilitated the introduction of technological advances into Western European economies.[2] The long-postponed demand for consumer goods provided a stimulus for investment in new productive facilities.

Other developments also tended to fuel Europe's economic–technological vitality. The rearmament program, especially after Korea, was important in this respect. Historically strong in chemistry and aided by worldwide demand for chemical products, Western Europe was able to restore the primacy of its chemical industry. The full-scale emergence of new technologies—in particular, television—and the sudden blooming of some that were not so new (the automobile and the commercial airplane)[3] contributed directly to Western Europe's growth and also acted as catalysts for other sectors of the economy. The postwar discoveries of petroleum in the Middle East and North Africa also facilitated Europe's economic growth. The later discovery of gas and oil in the North Sea strengthened Western Europe's energy position and also stimulated its growth, although this discovery was not consequential enough to free the Continent from heavy dependence on imports of energy.

Favorable as these factors were, they would not have been effective without deliberate policies and planning by the Western European governments. Management of the economy has become not merely an acceptable but actually an essential feature of governmental activity in postwar Europe; in this respect Western Europe has gone beyond the United States, and the development as a whole represents an important deviation from the past.[4] Economic integration of the European Community (EC) countries, although stimulated by the outside influence of the Marshall Plan, properly belongs to the new rational volitional effort; the integration was important in facilitating the technological modernization and economic vitality of Western Europe. Perhaps the single most radical innovation was the introduction of science and technology policies by Western European governments, the most remarkable example of which is France. From a country that was conspicuous in neglecting the potential of science and technology in the pre-World War II decades, France succeeded in developing the most energetic and purposeful (if controversial in the context of the European political environment) policy in science and technology in Western Europe.[5]

The "Technology Gap"—Illusion or Reality?

By the early sixties, Western Europe had completely recovered from the effects of World War II and attained a gross national product (GNP) that was about two-thirds that of the United States and was growing at a rate at least as rapid as that of the United States. About that time, however, the specter of the "technology gap" began to haunt Europe. Concern about this gap between the United States and Western Europe reached its peak in the late 1960s.[6]

Then, in the late 1960s and the early 1970s, federal appropriations to U.S. science and technology declined. The United States found itself with a surplus of Ph.Ds. in practically all fields of science. The "brain drain" has all but disappeared. At the same time the United States was seriously challenged by Western Europe and Japan in the sale of "high-technology" products—especially consumer electronics, plastics, and automobiles—in the world market and within the United States itself. The term *technology gap* was soon no longer in fashion.

What happened? Was this gap a mere illusion, an ephemeral phenomenon that received more attention than it deserved? Or is it still strongly, yet less visibly, entrenched in the trans-Atlantic relations, perhaps likely to reappear in a much more pronounced form? There is no single or simple answer. The magnitude of the technology gap was exaggerated. The United States was not in quite so strong a position technologically, nor was Western Europe quite so weak, as the technology gap debate tended to suggest. Moreover, in the allocation of resources to research and development (R & D), the United States placed emphasis on outer space, defense, and atomic energy, and very little on economic objectives.[7]

As the debate was developing, European governments undertook to place more emphasis on supporting those commercially relevant sectors of technology that could produce near-term payoffs.[8] Particular attention was paid to increasing productivity in Western European countries, and a measure of success was achieved. These countries continued to gain in productivity—as compared with the United States—into the early seventies.[9] Quite apart from the growth of productivity, transfer of technology to Western Europe through American multinational corporations (MNCs) contributed to the strengthening of Europe's economic position.

Although Western European nations have in recent years been quite effective in closing the technology gap in many products of present-day commercial significance, this is not so in advanced technology proper, that is, the technology whose economic impact is in a relatively early stage or has not materialized at all.[10] These technologies encompass outer space (including the aspects of it that are directly applicable to economic activity on earth, such as satellite communications), weather modification, lasers, magnetic fusion, magnetohydrodynamic (MHD) power, solar energy, and others. The real test of the technology gap will come in the future, when these technologies become economically important. In most cases they will require large resources, the ability to exploit them on a large scale, and long lead times. Unilateral technology enhancement programs undertaken internally by individual European governments have been effective in strengthening Europe's commercial viability in the sixties and will retain a considerable measure of effectiveness through the seventies, but such compartmentalized and stopgap approaches are not adequate in facing advanced technologies of the future.

The foregoing suggests that, unless Western European nations significantly improve their capability with regard to advanced technologies, the technology gap—in its economic dimension—is likely to reemerge, this time

The Developed Regions

in a much larger magnitude and with much more critical consequences. Yet, as we look into Western Europe's technological future, the term *technology gap* as a measure of Western Europe's success or failure as a viable region is inadequate. At best it would indicate the relative ability of Western Europe, compared with that of the United States, to generate or use technology to sustain or enhance its economic vitality. But the potential impact of technology on Europe's future is considerably broader than economic. It must be examined in a global context and from many aspects—among them, the adverse impact of technology that can be exploited economically only remotely if at all. Advances in medical technology that extend the human life span and might lead to intergenerational conflict and social instability are one example. Another is the impact of technological advance on the psychological health and values of European population, which may impair its vitality.

Western Europe's Weaknesses

It may be helpful to appraise the situation in Western Europe in the context of one important technological trend discussed previously. Western Europe is potentially a major beneficiary of disappearing determinism of the location of raw materials and energy sources and of the rise in the importance of scientific and technological capability as the critical factor in the viability and vitality of countries and regions. Western European countries are seriously deficient in a number of resources, but they do possess an impressive technological potential. However, both the form and the extent of volitional effort that these countries have been displaying in recent years have been inadequate to make that potential count. The lack of coordination and unified purpose among the policies in science and technology of the various Western European countries has been a very important handicap.

Governmental Policies

Under the leadership of General de Gaulle, France's science and technology policy was dynamic, but it failed to generate enthusiasm and unequivocal support from her Continental neighbors, since it was aimed primarily at bolstering France's position and prestige and at using European regional scientific and technological programs to that end. Moreover, in her search for power through the development of advanced technology, France badly overcommitted her resources.[11] The resultant difficulties led to a searching review of policy by the Pompidou government and to the shift of priority to economic growth and the strengthening of the competitive capacity of the French industry in the world market.

Although considerations of power and prestige—including support of advanced military technology—have not exactly been neglected, the trend of gearing science and technology programs to promote economic objectives became even more accentuated under the regime of Giscard d'Estaing. France has been successful in her drive for economic expansion. Before the oil crisis she was growing at a rate of 6 percent annually. The oil crisis has reduced the rate of French growth—as it has that of the rest of Western Europe—but she is likely to keep her position as the leading major European nation in economic terms.[12] The government of Giscard d'Estaing appears to be giving greater consideration than its predecessors to the needs of Western Europe, but, on the whole, France's policy is primarily self-centered.[13]

The West German effort in science and technology is not so centralized as that of France, but it has been flourishing. In earlier decades the largest proportion of financial support for science and technology came from the state governments rather than the federal government. Since about the mid-1960s the picture has begun to change: the federal government has been playing an increasing role in science and technology and concentrating on the development of a national science policy. By 1975, the budget of the Federal Ministry of Research and Technology (which excludes defense R & D and the science component of the Ministry of Education and Science) had exceeded $1.5 billion. The rapid and continuous growth of the federal budget for science and technology was not interrupted until 1976, when the government's financial difficulties induced what appears to be a temporary setback.

Originally timid in any activities that might have implied political aggrandizement, beginning with the late 1960s the Germans started to look to fields increasingly important for international trade, politics, and prestige, such as outer space, nuclear power, computers, and oceanography. Economic considerations remain foremost in German science and technology policy, but in recent years it has been partially redirected to areas that at the same time are of major importance for the improvement of the quality of life of the German population. The Germans have been the most ardent supporters of Western European regional cooperation in science and technology, but lack of substantial progress in that sector provided an impetus for Germany to engage in more limited forms of cooperation, including bilateral projects with the United States.[14]

In contrast to her Continental neighbors, Italy has been slow in developing a comprehensive national policy for science and technology, and her present institutions in this field are relatively weak,[15] although she has been doing remarkably well in certain areas of technology, especially in consumer products.

Having emerged from World War II as the only Western European power that did not suffer defeat, Britain tried to maintain the versatile military capability of a minisuperpower for some 20 years. This gave her the leading position in military technology in Western Europe but was a major factor in straining her resources. Over the years Britain has evolved a fairly effective governmental machinery for science and technology; while more decentralized than that of France and Germany, it is not beset by the excessive pluralism of the United States.

The Department of Industry is currently the principal agency for enhancing technological viability and strength of a wide range of industries. The Department of Energy, established in early 1974, is concerned with technological matters related to its field. Basic research, on the other hand, is supported through research councils dominated by nongovernmental scientists. In general, government support of science and technology in Britain is less concerned with national prestige than with commercial exploitation, but the latter's success has not been especially spectacular. As compared with other Western European nations, Britain probably still has the strongest general capability in science and technology; she is especially strong in basic science.[16]

Nationalistic Policies versus the Challenges of Technological Impact

One can distinguish at least four reasons why Western Europe is either seriously handicapped now or is likely to be handicapped in the future by the lack of a common purpose and coordination in governmental policies with regard to technology and its impact.

One reason is the tendency of technology to create a degree of dependence of major magnitude and thus generate serious problems whose solution can best be achieved only through a concerted policy. A case in point is that of energy dependence. In 1950 the six countries that were the original members of the European Community (EC) [17] were satisfying internally 93 percent of their demand for energy, predominantly coal. Between 1950 and 1965 the change in the technological base of Western European nations and the availability of inexpensive oil from the Middle East and North Africa resulted in a major change in the consumption of energy: coal dropped from 74 to 38 percent, while petroleum rose from 10 to 45 percent.[18] By 1970 petroleum made up 64 percent of the total energy consumption of the EC nations (by then increased to nine). Ninety-five percent of petroleum consumed by the EC countries in the early 1970s was imported. Most (65 percent) of the oil in Western Europe is used for domestic and industrial heating. The demand is thus inelastic, so that any interruption of supply or any increase in the price of oil adversely affects Western Europe's competitiveness in the world market and the balance of payments.[19]

In December 1968 the EC Commission submitted the "First Guidelines for a Community Energy Policy." Progress in the implementation of the recommended policy was slow, and by 1973 the situation had become considerably worse than in previous years. The oil crisis of the fall and winter of 1973–1974 thus caught Western European nations unprepared and shook the very foundations of their economies. As a result of the crisis, in December 1974 the EC Council of Ministers approved a target of reducing the dependence of the EC on foreign energy sources from 63 percent of total consumption in 1974 to between 40 and 50 percent in 1985. A great deal of emphasis was placed on nuclear power.

In conservation, Western European nations have made substantially better progress than the United States since the energy crisis. However, a year after the EC Council's decision on energy, EC members were still badly divided on establishing a common energy policy, while all over Europe antinuclear campaigns were flaring up.[20] There is thus no certainty that the EC will produce a policy to achieve its energy target. It is possible—but far from certain—that the International Energy Agency (IEA) established in Paris (but not including France) in November 1974 under the aegis of the Organization for Economic Cooperation and Development (OECD) may eventually develop a viable long-range program for decreasing the dependence of energy-consuming nations on imported oil. If so, the IEA may provide the focus for European energy policy, with cooperation of the EC as an organization or independently from it. It is also possible that no concerted and effective energy policy aimed at decreasing their dependence on oil will be developed by European nations.[21]

Whereas the case of energy illustrates problems generated by dependence on the outside world, that of the environment creates interdependence within Europe itself, and in turn requires concerted policies by European governments. European geography itself makes nearly every environmental problem international. For example, the Rhine has been frequently referred to as the sewer of Europe, but no "ideal" environmental policy on the part of, say, West Germany, can be effective unless also supported by France, Switzerland, Luxembourg, and the Netherlands. Unilateral policies that curtail pollution without cooperation of the other states bordering on the river are likely to penalize domestic industry and decrease the nation's competitive position through higher costs of its products. Similar considerations apply to the Baltic, the Adriatic, and the Mediterranean, all of which are badly polluted.

In July 1973, the EC Council of Ministers finally adopted the Community's first comprehensive environmental policy.[22] Some progress has been made since, but the policy has a long way to go before it is translated, if at all, into an effective system of monitoring and enforcement by individual member states. What exists at present is still largely a patchwork of individual efforts by national governments, intergovernmental organizations, and international nongovernmental organizations.[23]

Whereas concerted policies are required to solve environmental problems, a much greater extent of interdependence and concerted action is essential to take advantage of technology capable of changing the environment. Technologies and programs for weather and climate modification are the principal examples in this regard. Hardly any program for weather (let alone climate) modification can be undertaken by a Western European nation within its own frontiers without affecting the weather in neighboring states. Because of the relatively early stage of technological development in this area, weather and climate modification is not an important consideration at this time, but it carries major implications for economic and political viability of Western Europe in the future.

Last but not least, a certain unity of purpose and a concerted policy are required because of the large scale and huge costs of advanced technologies that make it impossible for any single European nation to develop them.[24] Western Europeans have been attempting to solve this problem by establishing institutions for regional cooperation in science and technology, but their success in regional cooperation has at best been mixed.

The earliest European regional organization in the field of nuclear energy was the European Center for Nuclear Research (CERN), founded in 1953. CERN has had its share of budgetary problems, but, mainly because it is a scientific organization, it has probably been the most successful of European regional undertakings in science and technology. A different picture is presented by Euratom, established by the EC nations in 1958 for the development of commercial nuclear power. In its design Euratom reflected nationalistic tendencies of its members and in its actual operations it was handicapped by those tendencies. Euratom's activity was limited to R & D, and its purpose was to coordinate and supplement national programs. An early effect of its establishment was to instigate the creation by the member nations of a multiplicity of atomic research centers that were more competitive than complementary; many of them were little more than token establishments. A substantial amount of money was spent by the EC nations for nuclear research, individually and through Euratom. A good part of this money was competitive and a duplication of effort. For the resources spent the results were meager.[25]

In addition to regional organization Western European nations resorted to bilateral and limited multilateral cooperation to meet the hurdle of the large scale and large costs of advanced technology. However, the success of the majority of these efforts has not been at all spectacular, and their very existence is frequently symptomatic of the nationalistic rivalries among Western European countries. The Concorde, the cooperative project between Britain and France on a supersonic transport (SST) initiated in 1962 and costing about $3 billion in R & D alone, is perhaps the most ambitious and best known of bilateral programs. After many political and technical difficulties along the way this program succeeded in producing a flying prototype in 1972, and in January 1976 the Concorde began commercial service. However, even in the absence of competition from an American SST, its commercial viability is in doubt.[26]

Among limited multilateral arrangements a tripartite agreement on development of the gas centrifuge technology for the enrichment of uranium between the United Kingdom, West Germany, and the Netherlands, concluded in March 1970, is one important development. The outcome of this undertaking is not completely assured, but so far it has been successful. The French, however, in competition with Uranium Enrichment Co. (Urenco, the company created by the three nations), established Compagnie Européenne Diffusion Gazeuse (Eurodif), an international company which uses the conventional diffusion process for the enrichment of uranium. In an effort to minimize waste of resources stemming from the rivalry between the two groups, the EC Commission proposed that a European enrichment capacity be created by the promoters of both technologies. It remains to be seen whether this proposal will be eventually implemented. At present America's virtual monopoly in supplying enriched uranium to Western Europe continues, although the U.S. position in this regard is bound to weaken. What is at stake for the next 10 years, however, is the extent of U.S. dominance, and not its disappearance.[27]

Another area where the requirements of huge resources and large scale enabled the United States to obtain a domineering commercial position in Western Europe is computers. Unlike the previous case, however, the American ascendancy was achieved, not because of governmental support of a particular technology, but solely because of the endeavors of giant corporations, principally IBM, which reportedly spent $5 billion to develop and introduce the IBM-360. When it was introduced in 1965, the battle for the computer market in Western Europe was, in effect, won.[28] In 1973 European companies undertook to pool resources across national borders to challenge European corporations, and the EC Commission issued recommendations of its own. Observers of the European scene noted, however, that these measures—even if they resulted in a merger of all European computer makers—are mere "band-aid solutions." A serious competition with the outside challenge would require combining computers with telecommunication systems into a unified European data communications system and policy—a step European nations are at present not ready to take.[29]

In the realm of policy another important development, a growing pluralization of interests within European societies, has begun to manifest itself in recent years. The interests have increased in influence, and they have tended to impede European integration. A product of technological advance, these pluralistic demands—by workers for full employment, by farmers for preferential prices for agricultural goods, by industries for various forms of support, and so forth—are not usually satisfied transnationally, but by national governments. The resultant politicization of resource allocation impedes the ability of national governments to take bold initiatives and to make compromises in the interests of the EC as a whole. On the contrary the tendency is toward a more narrow, nationalistic stance designed to protect the short-range claims of domestic constituencies.[30]

This development came to the surface especially strongly during the energy crisis of 1973-1974 and its aftermath. It suggests some disturbing questions for the future of Western Europe. Has the pluralizing influence of technological advance progressed to the point of forestalling the achievement of the minimum of unity among Western European nations essential for sustaining the region's viability in coming decades? Has the balance between the unifying and the divisive forces shifted irrevocably in favor of the latter? The answers to these questions are not clear. It is clear, however, that the pluralizing effect of technological advance finds a particularly fertile soil in the European setting, where national and institutional pluralism was strong to start with. It would require a farsighted and determined volitional effort to overcome.

Other Characteristics of Western Europe
Impeding Technological Dynamism

Aside from the lack of harmony in governmental policies Western Europe has a number of other handicaps that, in their sum total, are probably equally important. They can be broadly categorized as an aggregate of conditions and characteristics of the European scene that impede technological dynamism and growth.

In a number of Western European countries there are distinct gaps between the development of science and its application. The enterprising spirit so characteristic of American business and the availability of venture capital are generally lacking. European corporate life shows strong signs of rigidity, the determinism of social origin playing a very important role in the promotion of top executives. There are no suitable institutions to provide the large-scale financing necessary for a modern corporation's scope of operations. In a number of European countries the educational and scientific institutions lack the flexibility needed to supply the skilled manpower and R & D essential to keep abreast of technological advances.

Moreover, the existing technological superstructure of Western Europe would require a change to meet requirements of the future. Britain, France, and Italy have national grids for distribution of electric power that can absorb, in varied degrees, the moderately large capacity of present-day nuclear reactors. Germany, on the other hand, has only regional grids. A degree of interconnection between European national and regional grids exists at present, but the huge reactors of the future will require a much more extensive interconnection and integration of Western European power distribution systems. Another case in point is the unevenly developed telephone network in Western Europe, quite advanced in some countries, but relatively backward in others. It will require considerable investment of capital, modernization, and standardization of systems if it is to cope with the increased volume and speed of telecommunications that will become necessary as computerized data transmission is added to the presently rapidly growing international telecommunications traffic.

As a result of all these factors Western European countries are facing serious problems on two levels. The first is *timely development* of advanced technology, which requires a large-scale effort and the pooling of resources. The second is the *utilization* of technology on a sufficiently large scale. Even if future technology is developed elsewhere and freely offered to Western Europe, this would not ensure that Europeans would be ready to use it. Because the transfer of technology is feasible, the problem of timely development of technology is ultimately less critical than that of the ability to utilize technology on a sufficiently large scale, especially since the importance of the latter will continue to increase. As things stand, however, Western European nations suffer on both counts and are thus facing the prospect of falling behind in the competition for world markets, potential military power, and political influence.

The Trends of Change

There are some trends and developments that allow for a more optimistic outlook. Most of them are quite recent, rising to importance early in this decade; still others can be traced to earlier years, but they assume a new significance in the context of these more recent developments.

The entry of Britain into the EC in January 1973 is a major development. Since that date, relations of Britain with the EC have had their ups and downs, but in the longer run Britain's membership in the Community is potentially very important. It provides a substantial addition of advanced technology to the know-how of Continental Europe and should, eventually, remove a number of obstacles to regional cooperation in science and technology. The three principal technological powers of Western Europe—Britain, France, and Germany—are basically complementary in their strong and weak points pertaining to scientific and technological development.[31] If the three powers actively cooperate in science and technology, each could exert a beneficial effect on the others. In 5 to 10 years, this could produce a "takeoff" in scientific and technological development, eventually resulting in a truly impressive capability.

A promising development in Western Eruope has been establishment of the European Space Agency (ESA) in 1975. The ESA succeeds two earlier organizations, both established in 1962: the European Launcher Development Organization (ELDO) and the European Space Research Organization (ESRO). The former in particular was plagued by political and technical difficulties. Although the idea of merging ELDO and ESRO into a single European space agency that "could begin to look NASA in the face"[32] goes back to the 1960s, progress had been slow. Ironically it was NASA's inviting Western Europe to join the United States in a post-Apollo program that provided an impetus to more resolute action, resulting in establishment of ESA.

The ESA's mandate is broader than those of ELDO and ESRO combined. Included among its missions is the coordination of national space activities and their integration into the common European space program. The most important project of ESA at this point is construction of the Spacelab, a component of the post-Apollo program with NASA that will cost Europe about $500 million and will enable Europeans to participate in manned space probes. Probably more important in the long run is construction of an independent launcher (the Ariane) and a permanent European Communications Satellite (ECS) system. The latter will create a satellite-based telephone system providing for voice and data transmission and compete with Europe's terrestrial long-distance telephone network. It will also expand Western European TV traffic and its reach.[33]

Another important development was the approval, in January 1974, of a common EC science and technology policy by the EC Council of Ministers. Actually, it is but a modest beginning toward a common policy in this field, but significant nonetheless in that it provides a foundation to build upon. Two new entities established under this policy are particularly noteworthy:

1. A Committee for Scientific and Technological Research, known by its French acronym CREST. Its function is to coordinate national R & D efforts not subject to military or industrial secrecy. This is an ambitious mission that will not be achieved very soon. In its early stages CREST has been assembling information and stimulating voluntary coordination of action among member states.

2. A European Science Foundation (ESF), somewhat analogous to the U.S. National Science Foundation (NSF), but with an initial budget of only $500,000.[34]

The activities of American companies also introduce change into Europe. Possessing large-scale capital, superior technology, and dynamic management, American firms have been actively establishing subsidiaries in Western Europe and, in some cases, have been known to secure greater benefits than European companies have from the advantages of disappearing tariff barriers among Common Market nations. Their impact on Western Europe has been controversial but fundamentally beneficial. American multinational firms have facilitated the transfer of technology and managerial skills to Western Europeans, and they have stirred up considerable resentment among Western European companies, who have begun to use the practices introduced by their American competitors.

Western Europe is making progress toward a common patent system. In October 1973, sixteen countries signed a convention for a European system for patent grants. A separate EC convention covering the nine EC countries was signed in December 1975. These systems are likely to be operational by 1979.[35] At least two separate legislative proposals dealing with company laws are in an advanced stage of consideration by the European Community. When passed (perhaps in the late 1970s), they would create a European corporation and facilitate cross-national mergers.[36]

Even short of passage of this legislation, since 1970 European companies have demonstrated greater propensity to merge or to enter into various cooperative arrangements. European managers show an increased confidence and greater dynamism in competition with American concerns. Aided by devaluation of the dollar, European companies have become more vigorous in establishing subsidiaries in the United States and in competing with American companies in their own territory.[37]

Companies providing venture capital are beginning to develop. Observers expect that within the next few years financial institutions (probably backed by governments) will be established for the purpose of providing large-scale capital for European corporations. Steady progress is being made toward industrial standardization throughout Western Europe. There are bright signs in the energy field for at least two European nations. Recent discoveries of oil in the North Sea indicate that the United Kingdom will be self-sufficient in oil by 1980.[38] Oil and gas discoveries in the Norwegian sector of the North Sea and in the Norwegian continental shelf made Norway a net exporter of these products in 1975 and promise to make her the most important producer of oil in Western Europe in the near future. Britain has, however, the potential of competing with Norway for this honor.[39]

To the extent that Western Europeans are trying to meet the challenge of the large scale and costs of modern technology, the thrust of their effort is mostly Europe oriented and anti-American. However, some authorities have already raised the view that a "European scale" may not be enough, that, to master certain advanced technologies, Western Europe must cooperate with the United States and Japan.[40] It has also been suggested that considerations of survival may lead the European aircraft industry to extensive trans-Atlantic collaboration in coming years.[41] Jean-Jacques Servan-Schreiber, whose book *The American Challenge* produced a considerable impact on Western Europe in the late sixties because of its call to unite against American technological and commercial superiority, has reversed his position in favor of cooperation with the United States.[42]

It is too early to say how influential these voices will be. But they do suggest that Western Europe may find and expand its paths toward a viable future.

Europe at a Crossroad

As we have seen, there is a conflict between two forces in Western Europe: inertia, traditionalism, and nationalistic attitudes of governments, interest groups, and individuals versus willingness to adapt to the requirements of modern technology and to use the benefits it offers. What, then, is the outlook for Western Europe in the next decades?

One may postulate three models determined by differences in the way Europeans meet the challenges of the impact of technology. One model assumes that resistance to change and parochial interests within European societies will prevail and that European governments will pursue basically nationalistic policies with respect to science and technology. Another model assumes that both the potential benefits and the multiple pressures of scientific and technological advance will induce Western European governments to undertake more cooperative policies and will result in changes in European attitudes and institutions without, however, leading to establishment of a unified Western Europe. A third model assumes that, motivated by technological as well as other factors, Europeans will decide to establish common political institutions and to solve their technological problems within the framework of a common government.

These models are not equally probable. None is likely to materialize in the exact form described here. But an approximation to one of them will probably develop, incorporating elements of the others.

Model One: A Technological-Trends-Resistant Western Europe

If Western European governments pursue nationalistic policies in matters pertaining to science and technology without seeking essential changes in this field with sufficient vigor, Europe's vitality and power position will decline. The course of events might run along the following lines:

After the initial onslaught of the energy crisis in 1973–1974 Western Europe's economic position in the world appears to improve. More mergers take place between European companies; most of these are between companies of the same nationality, but some cut across national frontiers. Western European companies establish subsidiaries in Latin America and Southeast

Asia that capitalize on cheap local labor and thus strengthen their competitive position in the world market. Western Europe's confidence in its ability to compete with the United States and Japan grows, especially since the latter develops her own economic problems. But in the midst of recovered confidence, signs of potential trouble ahead begin to appear.

The EC nations cannot reach agreement on a common energy policy and a mechanism for its implementation. This, in part, is caused by the existence of the IEA: The British, content with their growing output of North Sea oil, argue that a strong EC energy action is unnecessary and that Western European nations should act primarily within the framework of the IEA. But the latter does not prove to be an effective agency. Some short-range emergency measures are agreed upon, but the United States fails to exert strong leadership in the more fundamental long-range matters such as conservation of energy, development of alternative energy sources, and energy R & D. The European nations are too divided to display initiative in these fundamental areas. Moreover, major oil companies are obstructing an effective functioning of the IEA by insisting that governments should leave negotiations with OPEC countries to them and by opposing programs for cutting down on energy consumption.

By 1980 European export industries strongly feel the pinch of rising oil prices, and their position on the international market weakens. Then, in 1981, the situation deteriorates dramatically. Having accumulated foreign currencies and being generally stronger economically, the OPEC nations raise the price of crude oil and decide to stop exports rather than accept a compromise price offered by importing nations. Norway's oil production is much too small to be important to Western Europe as a whole; Britain takes a narrow nationalistic stance and uses her favorable oil situation to remedy her own multiple economic troubles. The extensive construction of nuclear generating capacity, contemplated by Western European nations in 1974–1975, never materialized. Finally, their stockpiles of oil nearly exhausted, Western European nations agree, in 1982, to accept OPEC terms. However, later in that year they receive another rude shock: Having completed the construction of a number of refineries, the OPEC nations insist that crude oil be exported only after the available refined petroleum is sold. Although the price of the refined oil—as compared with the increased price of crude—makes this demand economically acceptable, the growing refining capacity of OPEC nations results in the closure of many European refineries. Italy, which has kept and increased its leading position in Western Europe in oil refining since the early seventies, is hit hardest, many workers becoming

Western Europe   105

unemployed. West Germany, which has employed large numbers of foreign workers (including Italians), also curtails production sharply, forcing thousands of workers to return to already depressed homelands. Because of her own shortages America's assistance to Western Europe in petroleum is insignificant.

The acute energy crisis of 1981–1982 is transformed into a continuous, low-key malady. A number of patchwork devices help to maintain Western Europe in a state of precarious balance in energy. The Soviet Union increases its supply of natural gas and oil to selected European nations, but these exports are accompanied by pressure designed to weaken NATO participation and to hinder European integration. The Soviet Union also largely displaces the United States as the supplier of enriched uranium to European nations, especially since the United States is facing shortages of enriched uranium at home.[43] Concerned about European security, the United States proposes to Western European nations selected joint programs and assistance in science and technology and attempts to breathe new life into the IEA. However Western European nations are unable to generate sufficient internal cooperation to take advantage of these belated American initiatives. Moreover, economic interests in the United States are still concerned about Western European competition in the world market; thus U.S. policy with regard to Western Europe continues to have elements of ambiguity and is not strongly focused, with security considerations opposed by concern about the balance of trade. U.S. economic worries appear to be justified by the still-strong position of European multinational companies in Latin America and Southeast Asia, which vigorously compete in chemicals and electronics.

By the time of this second energy crisis the United States has made progress in Project Independence. At the height of the crisis the United States severely curtails petroleum consumption for transportation and accelerates its programs for alternative energy sources. Although domestic hardships are incurred, America's position in international trade is not impaired. With environmental interests weakened by the crisis, the production of gas through nuclear explosions, coal for conversion to electric power and oil, and offshore petroleum grows rapidly. Large-capacity MHD plants based on coal are introduced in the mid-1980s. In the 1990s, with relatively inexpensive energy becoming available and aided by highly automated, worldwide communications and data processing systems, U.S. corporations significantly strengthen their competitive position in the world market.

The Soviet Union is a major beneficiary from the energy crisis. Having ample domestic energy sources, the Soviets increase their exports of petroleum, natural gas, and enriched uranium in exchange for imports of nonmilitary technology and technical know-how. Energy-deficient Japan becomes a major supplier of nonmilitary technology to the USSR; Western Europe and the United States also help. Soviet productivity is improved considerably through introduction of computers into its economy, application of lasers to industrial processes, and better work incentives. By the late 1980s the USSR emerges as a major exporter of sophisticated products, including aircraft, petrochemicals, optical equipment, and advanced ships (hydrofoils and surface-effect).

Western European companies find it difficult to compete with those of the United States and with the USSR. Western Europe is seriously handicapped by shortages and high costs of energy and cannot solve the problem. Some European countries are more successful than others in using advanced technology to generate energy. Having national distribution grids, Britain, France, and Italy assimilate medium and even large conventional nuclear reactors, but highly economical MHD power plants of 10,000–20,000 megawatts are too large to be used on a national scale. These huge plants cannot be used on an all-European scale either, since European nations have failed to integrate their power distribution grids to allow transfer of large blocks of electric power over large distances. Other European nations are having difficulties in using power plants of 4000–6000 megawatts. The general efficiency of the Western European economy is also handicapped by failure to develop a truly advanced telecommunications network combining satellites, telephones, and computers into an integrated, high-speed, high-volume system. The once-powerful merchant marine of Western European nations is obsolescent, compared with the growing fleet of modern, high-speed surface-effect ships of the United States and the USSR.

The moderately promising environmental efforts of Western European nations of the mid-1970s have failed to protect the environment, and the existing system begins to fall apart in the face of economic realities. Italy, whose oil-refining industry was all but wiped out by foreign competition (especially by Middle Eastern and North African refineries), is the first to free its industry from many antipollution restrictions; Austria soon follows. The two nations thus gain a temporary advantage in economic competition, but it becomes of doubtful value as pollution spreads across national frontiers. Moved by the desire to lower costs for their industries and facing the futility of stopping pollution by unilateral measures, other Western European nations rescind or relax enforcement of their environmental laws. Environmental deterioration is damaging the European tourist industry, which negates whatever benefits European nations obtain in terms of the balance of trade as a result of environmental neglect.

Unemployment and insecurity spread in Western Europe. Unrest in a number of countries results in street riots and frequent governmental crises. The large number of aged people, who promote legislation favoring their interests, is a contributing factor. Having lost faith in the ability of national governments to meet their needs, Western Europe's population increasingly shifts its loyalties to ethnic allegiances—Basque, Flemish, Bavarian, and so forth—and this further heightens conflict and instability.

The Soviets gain the allegiance of large social groups in some European countries, and about 1990 a few coalition governments with the Communists are formed. Other Western European states cooperate with the Soviet Union or the United States in certain sectors of science and technology in an effort to improve their weak positions, but this activity is compartmentalized, inadequately funded, and largely ineffective. Germany is still divided, but the demarcation between East and West in Europe is blurred.

Model Two: A Functionally Oriented Western Europe

If cooperation prevails over traditionalism and nationalism, the question becomes how far can Western Europe proceed in augmenting its technological capability short of creating a unified political structure? It is reasonable to envision that, with the active participation of Britain, Western Europe would create all-European institutions providing venture capital and banking facilities making available large-scale, fixed-return capital that stimulate the growth of all-European companies. The ESA develops an efficient satellite-based telephone system that produces a catalytic effect on the existing terrestrial system. The latter is extensively modernized, automated, and can transmit a large volume of data. Both systems become effectively integrated into one all-European telecommunications network linked with computer utilities and a global communications system. Western European nations eventually implement a common program for their aircraft industry and thus save it from collapse under the pressure of American competition.[44]

Western European action on a common energy program is tardy and halting, but the rumbling of another oil crisis in 1980 combined with a series of timely initiatives by the United States acting through the IEA, prods Western European nations into pooling their resources and activities. Euratom and CERN are merged into a European Nuclear Authority (ENA); Urenco and Eurodif are later brought under its aegis too. During the second oil crisis of 1981–1982, Western European nations display a substantial degree of unity; indeed the crisis tends to cement European nations beyond the energy field itself.

The three technologically strong powers—Britain, Germany, and France—exert a catalytic influence on each other, creating a dynamism in science and technology that post-World War II Europe has never experienced. The Big Three increasingly provide support to a European science and technology policy. As a part of this trend they, joined by others, eventually evolve a body that determines priorities in the development of science and technology for Western Europe as a whole. Western Europeans develop their own system of weather and climate modification and join the United States and possibly the Soviet Union in large-scale intercontinental programs for improving the environment. The centrifuge technology program proves to be a success, and Western Europe, possessing adequate supplies of uranium, effectively competes with the United States and the Soviet Union in selling enriched nuclear fuel in the world market.[45] Western Europe feels the pinch of the shortages of energy through the eighties, but a number of measures help in alleviating the problem. An effective conservation and rationing program is a major step. Western Europe increases imports of coal from the United States and Poland and rejuvenates its own coal industry. Combined with introduction of MHD technology, conventional coal plants diminish the need for oil imports. The ENA cooperates with the United States and Japan through the IEA in development of fusion energy. A united policy with regard to OPEC countries strengthens Western Europe's position at the bargaining table.

Greater intercourse and homogeneity evolve in Western Europe. To accommodate nuclear and MHD energy, electric grids cutting across national frontiers develop. Western Europe is crisscrossed by a network of gas and oil pipelines. The Channel tunnel, which opened in 1985, materially contributes to the economic and psychological linking of Britain with continental Europe. By the late 1980s, magnetic levitation trains and/or tracked air-cushion vehicles (TACV) extensively replace the conventional railroad system of Western Europe. Moving at a speed of more than 300 mph, these vehicles assist in further integration of Western Europe and stimulate the economy. The value system of European business becomes more like that of American business, although the influence of American firms in Western Europe declines with the development of large European companies that assimilate American techniques and management practices. Self-confidence and a feeling of security grow.

As this scenario develops, however, a point is reached at which Western Europeans begin to recognize the anomaly of power without political purpose. Western Europe can transmit television programs to the less-developed world, but what should be their message? Economic and political influences frequently go hand in hand, and unless a government follows through with a political strategy, economic influence can decline. Furthermore, mobilization and development of large-scale science and technology inevitably reach a point at which further growth requires decisions that are increasingly political. If Western Europe evolves an institutional mechanism that determines priorities in allocation of resources for science and technology, this in itself is a quasi-political body. Eventually questions of major magnitude arise. Should Western Europe compete or cooperate with the Soviet Union in weather modification technology or in climate modification programs? Such problems require decisions by the political authorities of Western European nations and, to be timely and effective, these decisions necessitate changes in Western European political institutions.

To develop and use future technology in all its ramifications, Western European states must eventually settle on joint political purposes. Short of this, Europeans will be unable to use the full potential of science and technology. In particular they will not be able to develop technology as an instrument of power comparable to that of the United States and the Soviet Union. Ultimately a central political will is essential for a truly effective use of technology as an instrument of policy.

Model Three: A Unified Europe

Viewing technological factors as an independent variable and taking them as the point of departure, one can envisage two sets of circumstances under which the establishment of supranational European institutions will be possible. One set takes off from Model One. Western Europe is basically incapable of taking advantage of advanced technology. The weakening of the European position and the unrest and political instability that continuously plague Europe produce a reaction. By the mid-1980s Western Europeans feel they have had enough turmoil, and in the late 1980s establish a supranational government to remedy the situation. The next 5 to 10 years are devoted to improving the relative backwardness—as compared with the superpowers—of Western Europe in advanced science and technology. Cooperative programs with the United States help to accelerate improvement. It is possible to envision that, with Western Europe's failure to take advantage of advanced technology, the unification of Western Europe will occur under the aegis of a dictatorial regime.

A second scenario takes Model Two as a point of departure. Western Europe uses regional scientific and technological institutions as the principal lightning rod for withstanding the shocks of modern technology and, through these institutions, develops a substantial scientific and technological capability of its own. But by the late 1980s Western Europe, not having central political institutions, increasingly feels restricted in large-scale development and global projection of technology. Besides, the availability of the technology creates the temptation to capitalize on it for political purposes, which in turn requires a further strengthening of the technological potential—a situation somewhat similar to the one in which Japan and West Germany find themselves at the present time.[46] Europe-minded groups gain in influence in Western Europe. Their activities are facilitated by habits of mutual exchange and contact established by European officials and industrialists, who, for the past 15 years or so, have worked side by side in regional institutions. As a result of these factors Western Europe establishes a supranational structure in about 1990.

If these developments were to take place, there is a distinct possibility that, by the year 2000, Western Europe could catch up with and perhaps exceed the United States as the foremost technological power. The success of Western Europe in this respect would depend largely on its ability to capitalize on nonmilitary technology and to forego the option of diverting resources into strategic nuclear capability (at least for the next 10 to 15 years).

Technological-Trends-Resistant or a Modified Functionally Oriented Western Europe?

Models One and Two run neck-and-neck as the most likely outcome of technological impact on Western Europe in the next 10 to 25 years. It is not entirely clear which will become a reality, but it is reasonably clear that options for Western Europe are becoming increasingly polarized. Unlike the situation during the past 20 years, during which Western Europe could afford to procrastinate and proceed by half-measures in the development or use of advanced technology, the alternatives for the future are becoming so divergent that the old policy simply will not do. Western Europe must make some hard choices in the near future. Since Model One is the grim alternative to Model Two, a belief that people act rationally when alternatives are clear-cut and divergent suggests that Europeans will choose Model Two as reasonable and necessary. However, even a deliberate and conscious choice of this model would result in an approximation to Model One if it is accompanied by measures that are too little and too late. Thus a variant of the technological-trends-resistant model could well turn out to be Western Europe's real, albeit unattractive, future.

A Functionally Oriented Western Europe: Directions of Evolution

Suppose we give Western Europe the benefit of the doubt and assume that its future proceeds toward a functionally oriented model. How might Western Europeans achieve it? To the extent that trends exist favoring such a model, they lean toward a unilateral effort on the part of Western Europe rather than strong trans-Atlantic cooperation. Western Europe's collective aspirations are expressed in such goals as "to reaffirm the real independence of Europe" and "to give Europe back the control of its destiny." [47] Viewed in the context of the 1970s, this approach might be a viable one and even has certain advantages.

Western Europe needs a purpose and appropriate symbols to enhance its unity and mobilize its potential. The impact of technology may well be the single most important peril for Western Europe in coming years, but at present it alone is not likely to energize European nations to produce the requisite degree of unity to ensure the region's vitality and viability. In this light the aspirations of Western Europeans to play a greater role in the world would serve a useful purpose. Since the United States still possesses substantial superiority and even domination in nonmilitary technology, the aforementioned aspirations could be conveniently combined with a concerted action aimed at "the American challenge." Aside from being a rallying symbol for action, "the American challenge" would direct the attention of Western Europeans to nonmilitary technology, an area requiring top priority if Western Europe's viability is to be ensured.

Because the catching up, restructuring of institutions (including educational systems), modification of attitudes and values, and development of the necessary manpower require lead times, Western Europe may need the late 1970s and the early 1980s to accomplish these objectives. It may be able to make significant strides along these lines with its own resources, especially since savings will be achieved through the pooling of available assets and reduction of the existing duplication of efforts. However, if one looks beyond the 1970s, a policy of hostility toward American superiority in advanced science and technology is definitely shortsighted and might prove self-defeating. No matter how good its intentions and how well they are carried out, Western Europe does not have adequate resources to take advantage of the technology of the future with its immense scale and costs. For that matter, neither does the United States.[48]

A number of questions arise in this connection. If Western Europe chooses an approximation of the functionally oriented outcome as a reaction to America's technological superiority, could Western European countries later (e.g., about 1985) successfully readjust their policies to large-scale cooperation with the United States in nonmilitary technology to avoid the cul-de-sac of limited resource? This would depend on the specific nature of European action. If, for example, Western Europe effectively unifies its exploration and exploitation of marine resources and triggers an American–European rivalry for the resources of the mid-Atlantic ridge,[49] it might prove difficult to change the course of events to eventual cooperation in marine and other technologies. The rivalry might also spill over into the political sphere with repercussions for Western Europe's security. Accordingly Western Europe's technological progress would seem to depend on the ability to keep options open for cooperation with the United States while initially pursuing an anti-American policy. This is a thin line that might prove difficult to tread.

A different question can be asked: Why should not Western Europe immediately begin to pursue a course of action involving much more extensive cooperation with the United States in science and technology? This would certainly be a rational approach advantageous to both Europe and the United States, since the latter is already feeling the pinch of limited resources. Moreover the energy crisis and the establishment of new avenues for cooperation—such as the IEA—underscore the utility of this approach. This being so, however, a rational approach may not necessarily be the most realistic and hence the best. Suspicions of American technological superiority will probably persist; commercially and politically motivated obstacles are likely to keep arising on both sides of the Atlantic to hinder effective cooperation. In the face of these obstacles, Western European countries might fall back on a policy of inadequate intra-European programs, procrastination, and palliatives such as compartmentalized, bilateral cooperation by individual European countries with the United States. In an age of polarizing alternatives this policy might result in the relapse of Europe into an approximation of Model One, with its concomitant political perils.

The choice of an early and more extensive cooperation between the United States and Western Europe in science and technology does not, however, have to result in the relapse of Europe into this model. It is important to keep the potential perils and difficulties in mind, but the outcome, in the last analysis, would depend on policies and perceptions of self-interest on both sides of the Atlantic. If such perceptions develop and find their way into concrete policies, this would provide the shortest road to a viable and secure Western Europe.

If Western Europe succeeds in adjusting its legal and fiscal structure, scientific and technological institutions, and values to the requirements of modern technology, and if it enhances its technological capability and weathers or avoids the potentially disruptive effects of technological impacts likely to affect Europe in the 1980s and in later years, political changes in the European picture are likely to follow. Although later than suggested in the third route to a politically united Europe, an aggregate of factors—the temptation to translate technological and economic capability into political power, the impact of the quasi-political authority of regional scientific and technological institutions, greater cohesiveness of Western Europe, habits of mutual exchange and contact among its officials and industrial leaders, and ideas of European unification—are likely to exert their influence for a united Europe. Its timing? Perhaps the early 1990s. In any event, at about that time Western Europe may have to make a choice between unifying politically or relegating itself permanently to second-rate status in the face of the overshadowing strength of the superpowers as they develop and utilize technology on an immense scale.

Japan

7

THE RISE OF Japan in the last 25 years is considered by many to be no less than a miracle. From its decisive defeat and wholesale destruction in the Second World War, Japan became the third-ranked industrial power (after the United States and the Soviet Union) in the world by 1968. By 1975 her GNP had risen to $488 billion, about a third that of the United States. What Japan had failed to achieve by military means she succeeded in achieving by nonmilitary ones—of which technology was the single most important.

The energy crisis, however, triggered a number of latent obstacles to Japan's rapid economic rise. In 1974 Japan's GNP declined by 1.2 percent (minus growth) and her projection of future growth dropped from the sustained level of 10 percent or more that had prevailed in previous years, to about 6–7 percent.[1] Given the large economic base, this rate of growth, combined with an appropriate foreign policy, would still give Japan formidable influence in world affairs. I shall explore which of the two types of factors—the volitional or the favorable deterministic—primarily accounted for Japan's rise to power. Given the kind of factors likely to condition the nation's future, we shall raise the question: What are Japan's options in coming decades?

Favorable Conditions and Factors

As in Western Europe the wartime destruction of Japanese industry was not an unmitigated evil. Since there was no obsolescent technological superstructure to contend with, postwar Japan was free to introduce modern technology on a scale suitable to its effective use. American occupation and the reforms it brought contributed to Japan's recovery and growth in a number of ways. The old *Zaibatsu* were dissolved, and the top executives of major industrial concerns were discharged from their posts. This purge made power accessible to the dynamic younger generation of corporate executives and thus injected an element of freshness and dynamism into Japanese industry.[2] Equally important the postwar reforms laid a foundation for competition in the Japanese economy. Competition helped to accelerate technological change, reduce costs, and facilitate efficient distribution of resources. The socioeconomic leveling of the Japanese people eventually stimulated economic growth, since it created a market of some 100 million people capable of consuming the great majority of manufactures mass-produced by Japanese industry.[3]

By the end of 1947 the concern about the Soviet threat induced the United States to abandon its preoccupation with Japanese social and political reforms in favor of a policy aimed at economic recovery and stability. The outbreak of the Korean war in 1950 gave an additional stimulus to the Japanese economy. Korea and its aftermath were also important, because orders for military and related equipment brought with them advanced American technology, production techniques, and quality controls (in which the Japanese had been notoriously lax in the pre-World War II years).[4]

Many American businessmen who later felt the pressure of Japanese competition overlook the fact that foreign (mainly American) commercial initiative provided an early stimulus to the rise of Japan. In the 1950s and after, foreign businessmen came to Japan, bringing designs for manufactures and arranging markets. In this way foreign buyers solved for Japanese producers such complicated problems as design, style, and market channels. In a number of cases machinery, technical assistance, and capital investment were furnished.[5] This assistance helped Japan to capitalize on one of her principal assets—cheap labor, readily available in earlier years.

Japan's economic growth has also been helped by the military protection provided by the United States, which relieves Japan of the necessity to divert large resources to defense. Japan relies on the American nuclear umbrella for strategic nuclear protection, and even in conventional forces she largely depends on the United States. In recent years the share of Japan's GNP devoted to defense has been 0.8 percent, while that of European NATO countries has been, on the average, about 3.9 percent, and that of the United States, 7.7 percent.[6]

Japan has a small land base and is relatively poor in resources, but the technological trend weakening the deterministic effect of the location of raw materials and sources of energy has worked in her favor. A nation of islands, Japan heavily depends on marine transportation, not only for imports of raw materials, but also for domestic (coastal and interisland) traffic. Marine transportation is the cheapest form of transport and it has been declining in cost, which can only help Japan. Over the past several decades the importance of coal has declined while that of oil and hydroelectric power has risen. Oil is a more mobile resource than coal, and the new supertankers have made transportation of oil even more economical for Japan. They have also opened up opportunities for shipbuilding. There is little room for further expansion of hydroelectric power in Japan, which has been important in Japan's industrialization and recovery because water power is one of the few resources with which Japan is well endowed. As late as 1973 hydroelectric power supplied 23 percent of Japan's generation of electricity.

Many other technological developments relaxed or removed constraints and opened up new horizons for Japan. In 1973 the Japanese chemical industry was second only to that of the United States. The surge of consumer electronics in the post-World War II decades—where Japan also became second in the world—is another example of broadening opportunities for a resource-deficient Japan: The important element here is skills (which Japan either had or was capable of developing), and not raw materials.

Organization and Will

Though important, the factors discussed above fall short of explaining Japan's success, nor, curiously enough, can the success be attributed to an unusually effective science and technology policy proper. As we shall see later, this policy compares favorably with that of other advanced nations but leaves room for improvement. Rather, it is a complex combination of historically conditioned cultural characteristics, institutional and social organization, and governmental policies—reinforcing one another—that enables Japan to capitalize on technology in promoting growth.

In Japanese society a close relationship exists between government and business. Unlike the United States, where the remnants of *laissez-faire* are still strong and business circles are suspicious of government interference, in Japan government has historically played a paternalistic role with respect to business, and businessmen look to government for assistance. While the government thus plays an important role in guiding the economy, the Japanese system has avoided the pitfalls of excessive bureaucratization and hierarchical rigidity plaguing the Soviet Union; it is not tightly centralized and is pragmatically flexible.

The Economic Planning Agency presents long-term plans that are merely advisory, intended to indicate the most efficient directions for the economy to move in and the areas where government support should be focused. Actual policies are worked out in the ministries concerned, of which the Ministry of International Trade and Industry (MITI) and the Ministry of Finance play a particularly important role. Each ministry or bureau within a ministry is responsible for control over an industry or a segment thereof. The governmental bureaucracy is competent and efficient; it is trusted by businessmen, who deal with the government through numerous trade associations as well as by frequent personal contacts.[7]

Since the landing of Commodore Perry in 1853 Japan has striven to assimilate Western technology, but in the post-World War II decades the goal was not just to catch up with but to surpass the West. Contrary to the USSR, where this goal is primarily a governmental policy, in Japan it became almost a cultural trait motivating millions of Japanese as well as the government. It was reinforced by other characteristics of Japanese society, such as the strong "love of work" ethic. But unlike American businessmen imbued by the individualistic Protestant ethic that also emphasizes the virtues of work, the Japanese are highly collectivistic, especially with respect to the outside world. Thus, while strongly competitive at home, Japanese firms cooperate abroad and are fiercely competitive against foreign interests. In this competition, profit motives are secondary; expansion, growth, or benefit to Japan dominate. If, in the process of zealous pursuit of these goals, firms flounder, the government steps in with cartelization, credit, or some other remedy. Assistance is usually rendered to the firms capable of strength in the future; true weaklings are usually allowed to succumb. Basically orchestrated by the government, the Japanese economy thus behaves as one huge corporation—Japan, Inc., a flexible conglomerate wherein the stronger components support the less mature ones and one another in an effort to achieve, with respect to the rest of the world, the synergistic effect of a totality greater than the sum of its parts.[8]

In Japan the firm is not merely an organization for production or service but a basic social unit that claims—and receives—strong loyalties from its employees. In terms of cohesiveness and loyalty the company is an extension of the family. In turn the company extends its loyalty to its employees and their families. Employees—be they factory workers, clerks, or managers—are engaged by companies for life. The company arranges group trips and vacations for them, takes interest in their private lives, provides gifts on the occasion of marriage and births, and so on.

Employee loyalty provides an incentive to performance and a deterrent to labor–management conflicts. It also reduces turnover and keeps skilled workers within the firm. However, seniority rather than quality of performance is principally rewarded, since length of service is viewed as a measure of loyalty.[9] Japan's labor force is extensively unionized and its leadership is often left-wing, but strikes are mostly symbolic and short-lived, and production is seldom disrupted by work stoppages. Featherbedding is also rare.

The rate of saving in Japan is among the highest in the world (more than 20 percent of personal disposable income, as compared with 7 percent in the United States), owing to the customary wage pattern of large cash bonuses; the limited coverage of Japan's social security system, which encourages reliance on one's own resources; the habits of thrift; and the low level of U.S.-style consumerism. On top of this savings pattern, governmental policy with regard to business credit allows the nation to stretch its capital resources for growth well beyond what is acceptable in other advanced nations. Since the government stands behind the debt position of major companies, nearly 85 percent of the capital structure of Japanese corporations is financed by borrowing (in the United States the percentage is about 25–30 percent).[10]

The government also promotes education. In the 1960s Japan boasted the highest literacy rate in the world—99.8 percent. In 1965 more than 30 percent of Japanese college graduates were science and engineering majors. High educational level, emphasis on technology, and personal industry have been major factors in achieving high productivity levels.[11]

An unusual, perhaps unique, Japanese institution that has significantly contributed to Japanese economic success is the trading company. There are about 6000 of them, including such giants as Mitsubishi Corporation, Mitsui and Company, and Marubeni Corporation. The most important function of trading companies still lies in finding raw materials for Japanese industry. However, the trading company has evolved into a modern, pioneering organization that conducts what may be called "systems business"—it acts as an intermediary between other companies and organizes every possible type of business or transaction rapidly and efficiently. Through their contacts and know-how, trading companies engage in such activities as helping manufacturers to relocate outside Japan or to go into other fields as needed in the light of technological or economic realities. Their operations are on a huge scale; in 1975 the top 10 companies alone moved $75 billion worth of goods.[12]

In general Japan is very much attuned to the importance of science and technology and continuously builds on its historical experience in borrowing technology from the West and adapting it to Japan's needs. Thus Japan has its "antennae system"—of which the trading companies and the quasi-governmental, nonprofit Japan External Trade Organization (JETRO) with its worldwide network of offices are important components—whereby Japanese abroad closely watch developments in the areas of their competence and relay the information to their home offices. The needed technology is then acquired by obtaining licenses from foreign concerns to manufacture the desired products in Japan. While relying heavily on foreign technology, the Japanese are not mere imitators; in many instances they were able to improve the technologies they borrowed and provide stiff competition to companies of the nation where a particular technology originated.[13]

Although Japan has been successful in using technology as the principal instrument for its economic progress, its governmental organization for science and technology policy and the policy itself have been less than ideal. Responsibility for policy is dispersed among several bodies. Priorities are mainly determined by rivalry among the various groups and agencies within the bureaucracy and outside groups (e.g., the scientists). In this respect the situation is similar to that in the United States, except for two notable differences: (1) Science and technology policy in Japan is somewhat more centralized under the Council for Science and Technology and the Prime Minister than is the case in the United States. (2) Perhaps more important the Japanese government, as well as outside groups influential in science policy, are strongly imbued by the overall motivation to catch up with the West. Although this is insufficient to establish a consensus on priorities, it does give Japanese science and technology policy a sense of purposiveness and narrows the dimensions of differences.[14]

The most important body in Japan's science and technology policy is the Council for Science and Technology (STC) chaired by the Prime Minister and including four cabinet ministers (Education, Finance, Economic Planning Agency, and Science and Technology Agency). Whereas STC engages in frequent deliberations, daily coordination of science policy with regard to other departments is provided by the Science and Technology Agency (STA). The Director-General of STA, who also carries the title of the Minister for Science and Technology, chairs the Executive Committee of STC, and STA provides administrative support to STC. In addition to bureaucratic rivalries among the various departments and agencies involved in technological activities, the problem of providing an overall leadership for Japan's science policy by STC and STA is complicated by two factors.

First, more than half of the total research and development (R & D) budget of the Japanese government is allocated by the Ministry of Education to universities and research institutes. These funds are not subject to coordination by the STA. University scientists, who are represented by a separate body to the Prime Minister (the Science Council of Japan), conduct research along discipline lines, scorn the problem approach to research of

STA, and distrust government. Second, the government finances only about 30 percent of total R & D, whereas industry supports 70 percent (the ratio is usually reversed in advanced Western countries, including the United States). Although the Japanese government thus has little control over most research funds, the situation is somewhat mitigated by the close relationship between government and industry.[15]

In general the thrust of Japanese science and technology policy in the post-World War II years has been in support of Japan's economic vitality and growth. As we shall see later, the areas of "big science" involving large expenditures—such as advanced defense research, outer space, nuclear energy, and oceanography—have not been pursued at a level of funding comparable to that of other advanced nations.

Japan's Technological Capability

What has been the product of the interplay between favorable conditions and factors on the one hand and Japan's effort on the other? What kind of technological capability have the Japanese developed? In some respects the picture is impressive indeed; in others Japan displays distinct weaknesses.

Japan is the largest and most technologically advanced shipbuilder in the world. In 1974 she launched 17.6 million gross tons of shipping, 50.9 percent of the world's total for that year. Japanese shipyards routinely build tankers of 400,000 deadweight tons, and some are capable of building tankers of 1 million deadweight tons, although none of this size has been built.[16] The Japanese automobile industry is the second in the world (after the United States). As noted in chapter 3, by 1964 Japan had succeeded in building the second largest steel industry in the non-Communist world, which has provided a backbone to her heavy industry and which offers competition that is a serious source of concern to major manufacturers abroad. The Japanese chemical industry mushroomed into a $10 million giant in the 1960s, began to approach a saturation point with respect to the domestic market, and, in the early 1970s, was preparing to export chemicals and construct plants abroad on a large scale.[17]

In the postwar decades Japan introduced stricter quality controls and upgraded quality in photographic equipment. By 1964 the value of Japanese camera exports exceeded that of West Germany. Japan is no longer an imitator in photographic equipment—she has produced a number of technological "firsts" in this field. The trend in recent years toward cooperation between the foremost German manufacturers of photographic equipment and Japanese companies (e.g., Zeiss–Yashika, Leitz–Minolta) has strengthened Japan in the production of equipment of the highest quality.

Japan placed heavy emphasis on the development of a domestic computer industry, and, with the help of governmental subsidies ($100 million for fiscal years [FY] 1972–1974) and benign regulation, progress has been achieved. After the United States, Japan is the most computerized nation in the world. She has one of the most advanced telecommunications systems, which facilitates the integration of computers into the economy. IBM Japan, Ltd., an American company, is still by far the most powerful in the Far East, but the M Series of computers produced by the Fujitsu–Hitachi Group (1974) and the ACOS System introduced by Nippon Electric and Toshiba in 1977 are rivaling the largest IBM machines. By the end of 1975 Japanese manufacturers had succeeded in controlling 54 percent of the domestic market, and in the following year they launched an offensive to expand their small (3.4 percent in 1975) share of the world market.[18]

The other side of the coin in Japan's science and technology policy is its relative weakness in "big science." The largest of all "big science" activities in Japan is the nuclear program, whose budget in FY 1976 (ending on March 31, 1977) was $330 million. However, in 1974 and 1975, Japanese nuclear facilities (including the nuclear-powered ship *Mutsu*) had multiple troubles, resulting in a number of suspended operations. Japan's pride in the energy field other than nuclear is the broad-based Sunshine Program, to be completed about the year 2000. But the allocation of resources for the entire program is about $3.5 billion, while the U.S. allocation to energy R & D (including nuclear) is moving toward $3 billion *annually*.

Allocation of resources is even smaller in other "big science" fields. The budget for outer space for FY 1976 amounted to $293 million; for ocean development, $60 million; and for military R & D, $50 million—only a fraction of U.S. multibillion-dollar budgets in most of these areas. For FY 1975 the Japanese government allocated 1.7 percent of the nation's GNP to science and technology. This indicated an improvement as compared with 1.3 percent in FY 1965, but Japan was still behind other advanced nations, which have customarily allocated from 2 to 3.5 percent of GNP to science and technology.[19]

In spite of her weakness in "big science," Japan claims some notable achievements. In February 1970, the Tokyo University outer space program (using home-made technology and somewhat to the chagrin of its governmental counterpart, the National Space Development Agency [NASDA]), succeeded in placing into orbit its first scientific satellite, OSUMI, and thus made Japan the fourth nation (after the USSR, the United States, and France) to reach into outer space. NASDA pursues a broader mission—to develop satellite capability for practical applications. Leaning on U.S. technology supplied by American industry (not NASA) under a 1969 U.S.-Japanese agreement, NASDA launched two relatively small (130–135 kg) engineering test satellites in September 1975 and February 1976. In 1973 NASDA had deviated somewhat from its policy of developing its own launching capability by requesting NASA to launch three satellites about 350 kg each in 1977 and 1978. This shortened NASDA's timetable by about three years.[20]

The Japanese university system is highly hierarchical and rigid, and this hampers academic research. Industrial research (which, unlike university research, is mostly applied rather than basic) is more dynamic and has been growing in recent years, but it is still relatively small by American standards. Successful in adapting new techniques and products, Japan is not known for the originality of her own research. Imports of technology from abroad had reached $715 million in FY 1973 and their continuous increase was not broken until 1975.[21]

Another weakness of Japan's research is its compartmentalization among the three principal sectors: the universities, industry, and the government. Unlike the United States, Japan does not have a history of "think tanks" that could facilitate application of basic research findings to national needs and help problem solving in Japan. Around 1970, think tanks began to appear in Japan, and in the fall of 1973 the government helped by establishing a large think tank, the Comprehensive Research and Development Corporation.[22] However, the economic recession put a damper on the fledgling evolution of think tanks, whose utility has never been generally accepted in Japan.[23]

The Benefits and Challenges of the Future

Many prospective developments in technology of the future are very promising for Japan. Poor in energy resources the Japanese could be major beneficiaries from progress in fast breeder reactors, magnetohydrodynamic (MHD) power, fusion energy and superconductivity. Although Japan has been relatively slow in developing technologies for exploitation of marine resources, she is a major potential beneficiary of these technologies. In fact apparently major deposits of offshore oil have already been discovered in the vicinity of Japan, in the area of the Senkaku Islands, situated between Okinawa and Taiwan.[24]

Japan is likely to benefit strongly from the exploitation of manganese nodules. They are especially rich in quantity and metal content in the Pacific Ocean and could supply Japan with extensive amounts of manganese, iron, copper, nickel, and cobalt.[25] Since in the future Japan will have to rely on long-distance transportation even more than now—to bring in raw materials, to export finished goods, and to maintain connections among her worldwide enterprises—such developments as very large surface-effect ships, with both overland and overseas mobility and high speed, would be particularly advantageous to Japan. They could, for example, move finished products from Japanese cities into the hinterland of Australia without stopping and unloading at the shore and could bring back raw materials necessary for Japanese factories. The ships' high speed and large size would not only reduce the cost of Japanese economic operations but also increase their tempo as well. Introduction of robots—which Japan emphasizes—and other labor-saving devices can be very important in maintaining the nation's high productivity and neutralizing high labor costs.[26]

Communications satellites are important for expanding Japan's internal communications network to accommodate vast quantities of computerized data, for facilitating the functioning of her global transportation system, and for expanding Japanese enterprises and marketing on a global scale. Land satellites (Landsats) could provide Japan with an instrument to locate in the developing countries resources needed for locally situated subsidiaries, and could also become a means of strengthening Japan's relations with the less-developed world.

Although future technology is very promising for Japan, her ability to capitalize on it hinges on Nippon's success in coping with four interdependent problems. One problem is to strengthen Japan's science and technology policy and to eliminate present technological weaknesses. Second, Japan must generate or otherwise obtain the necessary resources to take advantage of advanced technology of the future. Third, she, as other nations, will have to cope with the impact of science and technology, which may affect her stability and value system. Lastly, Japan must develop a successful national strategy of which a farsighted science and technology policy would be a part.

Science and Technology Policy

Improvements in Japan's science and technology policy would require creation of a central body to determine priorities in science and technology and to allocate resources accordingly. This will be especially important in the near future since the development of Japanese science and technology has reached a point where Japan cannot merely imitate the West and borrow technology; she will have to demonstrate her own foresight about which future technologies will be most advantageous to her and develop them largely with her own resources, placing appropriate emphasis on those deserving high priority.

A corollary of this development—indeed, a necessity—would be a much greater allocation of resources to R & D by the Japanese government. As Japan faces the alternatives of future technology—alternatives that would be extremely costly—she will have to forego some of the pride of "going it alone," the kind of pride she displayed (with, at times, rather painful results) in her outer space program. Rational choices will have to be made regarding which sectors of science and technology Japan will develop unilaterally, which she should develop jointly with other advanced nations, which she should let be developed abroad and then import, and which she should abandon altogether. Such decisions cannot be wisely reached by tug and pull among the various bureaucratic institutions and groups.

A closely related problem is the need to upgrade the quality of research and to overcome the compartmentalization in R & D among the government, industry, and the universities. This may not be easy to accomplish, especially in view of the distrust of the government on the part of most university scientists. The growth of nonprofit research corporations could alleviate this problem. However, to ensure that this growth be of sufficient scope to meet Japan's needs, the government would have to go beyond the establishment of a joint "think tank" with the private sector and actively promote the spread of research corporations. If Japan is successful—and there are some cultural obstacles, since the Japanese are not accustomed to "give and take" discussions—the establishment of nonprofit research corporations might also help to clarify national priorities.

Resources

The problem of the adequacy of resources for future technology has already been alluded to. Large resources will be needed, not only for R & D, but also for initial outlays in the use of advanced technology. In comparison with other advanced non-Communist nations of corresponding GNP, Japan is capable of generating considerably larger capital resources, and this is her major advantage. But it may not last if serious economic difficulties force Japanese companies to borrow abroad, where tolerances of large debt-to-equity ratios are much less generous. Moreover, Japan will also need greater expenditures of resources for areas other than science and technology. With the economic growth rate shrinking to some 6–7 percent, resources will be more difficult to generate.

Japan faces major requirements for resources for social needs. In her pursuit of economic growth Japan neglected her social infrastructure, and this neglect cannot be tolerated much longer. The nation's urbanization has been proceeding rapidly. Housing, sanitation, transportation, schools, recreational facilities and the like have not kept up with the rapid growth of Japan's cities in the past 20 years. Much of Tokyo lacks sewers; Japan has less than a half of the ratio of paved roads per automobile than the United States and the United Kingdom, which creates severe congestions; [27] the Tokyo subway system carries four times as many passengers than it was designed for; and because of the unusually high concentration of industry in the Tokyo–Yokohama complex, air and water pollution is serious. In 1972 Premier Kakuei Tanaka undertook an ambitious plan, to cost $1 trillion and to be completed by 1985, to relocate and restructure Japan's industry, transportation, and communications throughout the islands, but the plan was shelved for economic reasons.[28]

Social requirements also have a claim on Japan's resources. In spite of improvements in recent years Japan remains behind other advanced nations in social security: In 1975 Japan allocated about 8 percent of GNP to social insurance payments, less than half of West Germany's.[29] As we shall see later, the lifelong system of employment by large companies—perhaps the single most important pillar of security in Japanese society—is being seriously undermined. Thus the government has to step in to compensate for the declining role of the large companies in providing security to urban employees. Moreover, the potentially destabilizing impact of technology accentuates the need for a greater government role in this area. But greater security may adversely affect personal savings and capital accumulation for investment.

The future may see a greater allocation of national resources to the military sector, although at present (FY 1976) they are still kept at a low level of 0.9 percent of GNP. As we shall see later, perhaps more important for Japan's future is a greater allocation of resources to foreign aid. Between 1963 and 1974 Japan's official development assistance (ODA) rose from $140 million to $1.126 million or 0.25 percent of GNP. However, with Nippon's economic problems, foreign aid appears to have become stagnant and even more oriented toward Japan's economic interest than before.[30]

The Impact of Science and Technology

A remarkable characteristic of Japan's modernization was her success in absorbing technology within her own culture without significantly disrupting that culture and its social organization. Indeed the Japanese have been very effective in welding their cultural characteristics with modern technology and economic organization to support growth.[31] However, there is no certainty that this will continue. Sufficiently comprehensive data on Japanese society are not available, but there are distinct signs that technological impact is changing Japan. This may result in major discontinuities in the nation's sociopolitical trends.

The Institute of Mathematical Statistics, an autonomous but government-supported organization located in Tokyo, conducted a series of national surveys of various age groups for the years 1953, 1958, 1963, and 1968. There is a definite trend away from the feeling of collectivity and toward individualism and self-interest, especially among the young. There is a decline in the sense of duty and a leaning toward self-fulfillment and/or hedonism. There is also a decline in the value assigned to money, although it still remains strong. The younger generation is less inclined now to leave decisions to political leaders, even if the leaders are good.[32]

Family cohesion is weakening in Japan, and so is loyalty to the company. Ambitious Japanese white-collar workers are beginning to resent the Japanese system of lifetime employment that rewards seniority rather than ability. Moreover, in 1975 unemployment in Japan reached 1 million, in spite of the companies' effort to accommodate their employees. Bankruptcies exceeded 12,000, a postwar record.[33] Projections of surplus labor for coming years will also likely affect the relationship between companies and employees. The trend toward greater investment abroad accentuates the weakening of company loyalty. Its traditional symbol—commencement of each workday with singing of the company anthem—has been abandoned in the case of Japanese companies abroad. The collapse of the system of lifetime employment is probably inevitable; this may generate insecurity and friction in society, with imponderable results.

Although technology favors equalitarianism, Japanese women have been unusually docile and traditional in their role in society. Most are content to be housewives and do not seek employment after marriage. One wonders how long this condition can last. There is now no women's liberation movement in Japan, the the emergence of one is likely. What this would mean for social and political developments in Japan is not clear, but, in light of the magnitude of the pent-up change, the impact on Japanese society could be far-reaching.

The likely increase in human life expectancy by 20 years in the next 25 years is likely to put to severe test the Japanese respect for age and the influence that older persons enjoy in Japanese society; it may also undermine the traditional stress on consensus in decision making. As the impact of advanced technology on the Japanese people and the nation's institutions grows, Japan will increasingly become subject to the hazards afflicting other highly advanced societies: alienation of the individual, feelings of helplessness and insecurity, and general social discontent.

One technologically induced development in particular has already become a source of turmoil in Japan: the pollution problem. Although the government established an environmental agency in 1971, it has been slow to cope with pollution. Public reaction has been strong indeed. What is unusual in this development is the willingness of the man in the street to form public interest groups and to oppose the establishment head on—via boycotts, picketing, and litigation. In a collectivistic Japan where consensus and mutually satisfactory compromise after discussion characterized the nation's political life and where it was almost unthinkable to seek legal redress from the state, antipollution activity represents a fundamental change.[34] A similar development, accompanied by consumer pressure and demonstrations, is taking place in the field of credit. Although individuals provide 40 percent of bank deposits, they receive only 7 percent of bank credit because banks give preferential treatment to large manufacturing and trading companies.[35]

In summary, under the impact of technology the Japanese appear to be losing their most important asset: cohesiveness of their society and their collective spirit. In this regard the Japanese are out of phase with inchoate trends in Western societies and with requirements of the future. As we shall see in chapter 14, emergence of the postindustrial society—which Japan has technically entered in 1975, when her labor force in the service sector exceeded 50 percent of the total—calls for communal cohesiveness. Advanced Western societies, where individualism has been traditionally strong and where the impact of technology has further pluralized society, are beginning to realize the value of greater cohesiveness and—in scholarly analyses and the sentiments of the younger generation—beginning to grope for a communal spirit. Japan, on the other hand, is moving in the opposite direction.

The individualism and pluralization of the Japanese people carry serious potential dangers in this stage of the society's development. The cohesiveness and collective spirit sustained and nourished by institutions such as the family, the company, and the nation provided an important psychological and material shield against insecurity, despair, and alienation for the Japanese in post-World War II decades. But this shield is breaking up precisely when the tempo of a highly advanced society, its potential rigidity, and impersonality are significantly increasing. Not having had a history of individualism and its hardening experience, the Japanese may be less immune to this form of technological impact than Western societies.

Individualization and pluralization of Japanese society are likely to affect the decision-making process and the ability of the nation to move in a clearly defined direction. The potentially growing deadlock between and among the various interests and groups may not only have a paralyzing effect on the national will but may also result in a significant drain of Japan's energies and resources.[36] In short, the impact of technology—the instrument that made Japan a great power—may become the most important factor in causing her stagnation or decline.

National Strategy

It is essential for Japan to have a farsighted national strategy, with science and technology policy integrated into it. This is so because, more than any other great power, Japan depends on the outside world for her viability and vitality. The question of national strategy is especially important now, since Japan is in a state of transition.

In post-World War II decades, Japan's national strategy had a short span of vision and could largely be described in two words: economic self-aggrandizement. Its purpose was twofold: to increase the economic well-being of the Japanese people and to satisfy a combination of the nationalistic and psychological craving for prestige and power. Their success notwithstanding, the Japanese have a strong inferiority complex vis-à-vis Westerners, and to be the greatest economic power in the world was an important means of quenching this complex, a desire that motivated factory workers and high government officials alike.

As we have seen, Japan developed a highly successful machinery to pursue her goals, but both the machinery and the strategy were not without flaws. Those already mentioned need not be repeated, but an additional aspect should be pointed out. In non-Western countries—especially in Southeast Asia—the aggressive tactics of Japanese businessmen and, at times, their arrogance, have gained for Japan a reputation of being an "economic animal," a reputation that strongly persists.[37] It is precisely in this context that a broad-based development assistance to less developed countries (LDCs), free of obvious short-term economic benefits and including a strong technological component, could be an important part of Japan's national strategy.

President Nixon's trip to China, his new economic policy, and the energy crisis of 1973–1974 produced a succession of severe shocks to Japan leading to an agonizing reappraisal of her national strategy. Japan has begun to act more independently on the diplomatic front. She has sought closer relations with China, and stepped up economic (Siberian resources) and territorial (the Kurils) negotiations with the Soviet Union.[38] But her increased activity in world politics is hardly commensurate with her economic power.

Japan has been more agile and substantive in making economic readjustments. The new goal is "stable growth" at about 6–7 percent annually. More importantly the structure of the Japanese economy and its geographic distribution are undergoing change. The so-called higher value-added industries—like computers and electronics, which generate little pollution—will be developed primarily in Japan. Highly polluting industries like steel and chemicals are expected to be built abroad. The stress is on a much greater rate of investment outside Japan. Accordingly the Japan Economic Research Center projected that the total direct investment abroad will increase from $15 billion in 1975 to $71 billion in 1985. In both investment and trade Japan strives toward multilateralism so as to spread risks.[39] Japan's official circles stress the need for curtailing pollution and allocating larger resources to science and technology. Whereas progress in both areas is taking place, the rhetoric appears to exceed progress by a considerable margin.[40]

In short, Japan's well-oiled momentum and mentality for economic growth appear to be continuing, which raises a number of questions. Will Japan be farsighted enough in coming years to neutralize the potentially countervailing effect of the image of "economic animal"? Will the projected growth be achieved? It is possible, although the potential impact of technology brings imponderable elements into Japan's future. But if achieved, will it satisfy economic aspirations of the Japanese people? It is also possible, but it still leaves a void unfilled: the aspirations for superiority, which can no longer be satisfied by making Japan the greatest economic power of the future.

Japan is a highly vulnerable nation whose vitality and viability heavily depend on a favorable world order. Can she afford to assume that it will remain favorable without a strong commitment of her resources (as distinguished from her rhetoric) to sustain it? This brings us back to science and technology. Japan has capitalized on nonmilitary technology, especially in the consumer sector. It may be that, in a changing world and given stalemate in the military sector, nonmilitary technology is the most promising vehicle for national vitality, influence, and power. But nonmilitary technology can be effective only if the stalemate in the military sector, on the nuclear and subnuclear levels, persists. Especially in view of the growing military power of China, Japan will have to determine the particular mix of technologies, military and nonmilitary, that will best serve her interests in the next decades. This task applies not only to reevaluation of the balance between the military and nonmilitary sectors but also to a reexamination of the components of the nonmilitary sector. Again the question may be, not which technologies are most effective in promoting the economic interests of Japan, but rather which would best serve her national strategy, of which an important component would be the need to build earnestly for a world order favorable to Japan's future.

The Prospect

As the above discussion implied, there are many imponderables in Japan's future. The Japanese people also possess certain cultural characteristics that further complicate the picture.

Japan's Cultural Characteristics and Foreign Affairs

The Japanese are inward-oriented people whose values and judgments are formulated in a collectivistic fashion, through interaction in society. They are thus a product, not of rational norms or logic, but of relativism—what society thinks or accepts at a given time.[41] The dominance of relativity in value judgments makes the Japanese adaptable and pragmatic, which qualities have served them well since World War II. On the other hand, what may be called "collectivistic relativism" makes the Japanese respond and adapt well to circumstances, but does not make them very adept at anticipating developments and guiding their own destiny rationally and skillfully. When the Japanese learned the lessons of World War II the hard way, they made an about-face, and turned to pacifism, rejection of nuclear weapons, economic pursuits, and the tutelage of the U.S. in political matters. It took another shock—President Nixon's trip to China—to dispel Japanese complacency and cause a reappraisal of policy. But short of such telling experiences, Japanese decision making is painfully slow. For years the Japanese have recognized that the image of "economic animal" is hurting them, but little has been done within the government–business alliance to modify the policy. Similar examples exist in other areas.[42]

Japanese inward orientation, self-centered attitude, sensitivity, and insecurity make it difficult for the people to relate to other nations, to see clearly the nature of change in the world and how they fit into it.[43] This is not to say that keen perception and appreciation of world developments cannot be found among certain Japanese, but individualistic, Western-type leadership does not yet exist. Decisions must pass through the consensual process, and here cultural characteristics impose their constraints. This is a serious handicap for a nation as dependent on the outside world as Japan is.

Cultural characteristics of the Japanese people, combined with the setbacks on the economic front and the intricacies of détente, have tended to produce two opposing views in Japan: "passivism" and "activism." The former, of which Professor Chie Nakane is a leading representative, stresses limitations of the Japanese people in world politics and favors low-profile, *ad hoc* diplomacy combined with an orientation toward internal problems like pollution and inflation. According to her, this policy would minimize the potentially dangerous tendency of the Japanese to pursue single-mindedly a certain course of action once they are set on it. The activist argument, represented by Professor Kei Wakaizumi, recognizes Japan's limitations in world politics but emphasizes the need for a constructive approach based on longer range principles and objectives, including support of a desirable world order.[44]

Alternative Strategies in Japan's Future

One alternative strategy proposed is a triangular détente among the United States, the Soviet Union, and Japan.[45] It has potential attractiveness on economic grounds (development of Siberian resources), but it is not very realistic politically. The Japanese look down on and distrust the Soviets, an attitude that has strong historical roots. The necessary prerequisites for a smooth functioning of a tripartite cooperation among the three "superpowers"—one of whose superpower status is strictly economic—do not exist. The picture is further complicated by the rise of China and, in particular, the availability of her large oil resources. A second option, a national strategy aimed at establishing a community of interest in Southeast Asia under the leadership of Japan, would not be very realistic either, since it is not likely to be acceptable to Southeast Asian nations in the light of World War II memories.

A third option—a close Sino-Japanese cooperation—has a number of attractive features. The Japanese people feel a cultural and racial kinship to the Chinese. The two countries are economically complementary and will be more so in the 1980s, when China is expected to move full speed ahead into her industrialization drive. Such a cooperation might reopen the possibility for Japan—this time in cooperation with China—of rising to preeminence in the world and thus satisfying nationalistic and psychological aspirations of the Japanese people. The desirability of such cooperation for Japan would depend on the availability of other options for her national strategy and on the nature of the cooperation itself (see chapter 8).

A fourth option is an *ad hoc*, "muddle through" national strategy, similar to the passivistic approach of Chie Nakane, except with stronger economic overtones. This is similar to the national strategy Japan is pursuing at present. Some of the key questions it presents are: For how long can Japan afford to enjoy a "free ride" on a favorable world order? If things go wrong, to what extremes could Japan find herself subjected?

A fifth option is a national strategy that is not narrowly geographic in orientation but is geared toward supporting a desirable world order. It is a strategy similar to that proposed by Kei Wakaizumi. United States and Japanese interests in a future world order are parallel, and such a strategy would be advantageous to both nations. It is a national strategy that enjoys only minority support in Japan and that, given Japanese cultural characteristics and her bureaucratic system in foreign policy,[46] would be difficult to implement.

Technological Change and the World Arena

part 3

Technology and Distribution of Power

8

HISTORICALLY THE IMPACT of technological change on societies has seldom, if ever, been the same. Some states tended to benefit from a particular technological innovation or a particular technological trend more than others; still others were affected adversely or not at all. This differential impact of technology has been especially true in terms of power (see chapter 1).

Just as the impact of technology on the distribution of power was differential in the past, so also is it likely to be in the future. I shall now examine the technological trends and projections I have already discussed with respect to the following questions: What are the factors that will play a role in changing the world distribution of power? What might be the extent of such a change, on what levels, and under what conditions?

Change in the World Distribution of Power through Military Technology

Barring the unlikely outbreak of a strategic nuclear war or a major miscalculation in the development of strategic weapons systems by one of the superpowers, the four most important potential changes in the military sector are (1) the acquisition of nuclear weapons by third states, (2) a growing stalemate between the United States and the Soviet Union at the subnuclear-umbrella level, including its spreading to various parts of the globe, (3) potential erosion of U.S. alliances through global projection of Soviet "conventional" military power, and (4) a keen appreciation and timely exploitation of those military technologies which cannot be readily stalemated but which can provide an edge of superiority.

If strategic nuclear weapons are acquired by such great powers as Japan or West Germany, the change in power would not be radical, since the potential effectiveness of the weapons would be neutralized in the overall nuclear umbrella. A similar consideration would apply to China, insofar as the Chinese conduct a rational military policy and abstain from risky adventures. But because China claims leadership of the less developed world, an effective nuclear capability provides the Chinese with a somewhat special dimension of power in the political sense. Perhaps the largest differential impact of nuclear technology is likely to be produced by the acquisition of nuclear weapons by such smaller states as Israel, Egypt, and Argentina primarily for regional purposes. Even here the initially radical change in the regional balance of power will be eventually neutralized (unless actually triggered) by the establishment of regional nuclear umbrellas with links to the global nuclear umbrella or, more likely, with autonomy under it.

Since modern technology facilitates global projection of "conventional" military power, the Soviet Union enjoys a differential advantage in countering the earlier projection of America's military power into various parts of the globe, provided that the Soviets are willing to allocate resources for this purpose. It does appear that the Soviets are willing, and are likely to produce a major change in the world distribution of power should they persist in this policy. If the United States resorts to energetic measures to counter this change, she can slow it down, but she can not forestall it altogether.

In some respects the growing projection of Soviet military power on a global scale goes beyond merely stalemating the United States in various regions; it has the potential for eroding American alliances. This is especially true in the case of the expanding Soviet naval power. The growing number of Soviet naval ships in foreign ports—accompanied by the declining deployment of U.S. ships—might make some U.S. allies wonder if a more neutralist position would not be wiser. The Soviet long-range logistic and amphibious capability is quite limited at present; however, as the Soviet Union develops its capability in these areas, its eroding influence on U.S. allies will increase. The future extent of this influence is difficult to determine. Among other things it would depend on the extent of allocation of resources by the Soviet Union for this purpose, on the ability of the United States to provide a convincing margin of security to its allies, and on the existing international climate. For that matter, changes in the international climate may significantly modify the meaning and nature of alliances and hence the impact of Soviet military power.

Certain advanced military technologies provide an opportunity to keep some technological areas or geographic regions from becoming stalemated and may provide the necessary margin of security to prevent erosion of U.S. alliances. It would be possible for either the United States or the Soviet Union to capitalize on the potential of these technologies, although the United States possesses certain differential advantages over the Soviet Union in this area. The United States has a larger economic base, a richer technological mix, and greater flexibility for the development of new technologies, which, given willingness and effective leadership, can provide the United States with the necessary lead. America's geographic characteristics and strong base in ocean-related technologies give it a differential advantage in capitalizing on the power potential of the oceans—a highly promising area where military and civilian technologies overlap and reinforce each other. Closely related technologies are those pertaining to intermodelity (development of a vehicle that, ideally, can operate in the air, on land, on water and under it), high-speed air-cushion vehicles, hydrofoils, a global, real-time detection and surveillance system, and electro-optical weapons (in the early stages of growth). At any rate, a superpower or a great power able to discern and capitalize on the areas of future military technology that cannot be readily stalemated and that might provide differential advantage will be in a position to change the distribution of power in its favor.

Nonmilitary Technology and Power

Although there is still room for effecting important changes in the world distribution of power through military technology, a far more promising potential is presented by the nonmilitary technological sector. The two major areas where physical expansion is possible and will take place are the seabed and outer space. Expansion and national influence, if any, can be effected in these areas only through science and technology and farsighted policy—not through wars. More importantly, in modern societies military power has always been primarily an outgrowth of the industrial foundation built by civilian technology. To a large extent because of the stalemate in the military sector (both nuclear and conventional), the scope for major changes in power through nonmilitary technology has considerably broadened. If and when these changes take place, over a period of years they would restructure both the present military balance of power and power potential and probably significantly alter the meaning and impact of the nuclear umbrella.

A number of trends and technological developments that have potential for contributing to the differential impact of nonmilitary technology were discussed at length in chapter 3. Others will be discussed here.

Since the deterministic effect of the location of natural resources and sources of energy on economic power centers is declining, opportunities are now emerging to change the distribution of power by organized human will, farsighted planning, and the transfer or indigenous development of technology (see chapter 3). Nations able to capitalize on volitional aspects of the differential impact of technology—which involves the ability to modify or discard obsolete values and technological superstructures in a timely fashion and to restructure institutions in order to take advantage of new technological opportunities—will be both instrumentalities in changing the future global pattern of power potential and its major beneficiaries.

Also important is the unprecedented rapidity of present technological change. History is being accelerated by modern technology, which underscores the importance of foresight and timely effort in staying abreast of power-relevant areas of technology. There is a price to be paid if one fails to do so. The Europeans have discovered this in such fields as computers and telecommunications. Depending on American policy in the immediate future, the United States may—or may not—find itself in a similar, or worse, predicament in such areas as weather modification, magnetohydrodynamic (MHD) power, fusion energy, superconductivity, the air-cushion principle of propulsion, and development of marine resources.

Large scale and high costs of modern technology are factors in technology's differential impact on the world distribution of power. Some states may be more successful than others in taking advantage of the economies of large scale. Western Europe may never succeed in this regard and thus decline in relative power. Owing to their size and advanced technological development the United States and the Soviet Union are in a better position than other states to take advantage of the large scale of advanced technology, but for varied reasons, the potential benefits might be differential.

Because of their large size American companies have been more successful than their foreign counterparts in exploiting modern technology. However, the growing scale and costs exceed the capability of private enterprise and increasingly require governmental initiative in the development of technologies and in the organization of their use. Moreover, an international effort may be needed. Certain characteristics of the United States government and American business attitudes—which generate apprehension of governmental interference—are not conducive to the development of the necessary initiatives and to close government–business cooperation essential for effective exploitation of technology.[1] Thanks to their central planning and greater purposiveness the Soviets are likely to be more effective in marshaling resources for large technological programs. On the other hand the Soviet Union has the potential handicap of getting bogged down in the bureaucratic maze— especially if the program in question lacks a high priority. Moreover, if broader international cooperation is required, the government-owned enterprise of the Soviet Union is not likely to be so flexible and adaptable as American companies.

Since the superpowers are committed to allocating substantial resources to the largely stalemated military sector, new opportunities exist for other advanced nations to capitalize on nonmilitary technology to improve their relative power position, provided they minimize their own military expenditures. As discussed previously, Western Europe—provided it gets organized for this purpose—can significantly rise in the global balance of power in the next decades, possibly overtaking the United States. West Germany and Japan, enjoying the benefits of American military protection, supporting relatively small military budgets, and energetically capitalizing on nonmilitary technology, have done very well in increasing their relative power potential in the last 25 years. In 1968 Japan moved into second place (after the United States) in the non-Communist world in GNP, while West Germany fell back into third. Between 1968 and 1975 the economic gap between these two countries and the United States has narrowed further from 16.5 to 32.2 percent and from 15.6 to 27.9 percent, respectively, of U.S. GNP.[2]

Although these nations are successful, the real question is: How far can they go individually in relative power before they become seriously handicapped by the technological requirement of large scale and high costs of advanced research and development (R & D) and of its implementation? It may be that, once Japan and Germany begin to feel these handicaps, they will increasingly enter into cooperative scientific and technological programs with one of the superpowers, between themselves, or with other states and thus continue to rise for a while. Japan's economic problems of recent years notwithstanding, she still has a significant potential for sustaining and augmenting her vitality and power through advanced technologies. Even if Western Europe as a whole fails to establish its own vitality in science and technology, individual European states could strengthen themselves temporarily by undertaking piecemeal programs with the United States or with each other.[3]

Less Developed Countries and International Power

A question worth exploring is the role of the less developed countries (LDCs) in affecting the world distribution of power in coming decades. Is there any indication in existing technological trends that the LDCs might gain in relative power as compared with the advanced nations? Or is it likely to be the other way around?

The fact that the economic gap between the less developed and the advanced countries is widening rather than narrowing has been broadly publicized in recent years and has been largely responsible for a rather pessimistic view of the future of the less developed world. However, economic factors or trends cannot be narrowly equated with power factors; present trends may not continue; and even if attention is focused on economic factors alone, the picture is not particularly discouraging. We shall look at economic aspects first.

In absolute terms the margin between the total GNP of advanced countries and that of the LDC's is increasing rather than decreasing. One important reason is that the total GNP of the advanced countries is so vastly larger than that of the LDCs (82 percent of the world's GNP in 1973) that even a relatively small percentage gain in the advanced world results in large additions in absolute terms. However, as we look at the growth record of the developing nations in the past decade, a number of bright signs become discernible.

In the 1960s the average annual growth of GNP in the LDCs reached 6 percent, and this rate continued until 1975, when the impact of the energy crisis, the economic recession, and a difficult food situation adversely affected most non-oil-producing LDCs.[4] Brazil has been growing at a rate of about 10 percent annually since 1967, until the world-wide "stagflation" caused a drop to a 4 percent growth in 1975. This high growth has been achieved by capitalizing on a favorable resource base, an effective and growth-oriented government, and a strategy for growth that, among other elements, includes purchasing-power escalator clauses to compensate for a high rate of inflation. Brazil is paying a price for economic growth in terms of individual freedom and concentration of wealth, but her case does suggest what a determined developing nation can achieve.[5]

Even if we disregard Brazil and focus on the average growth of 6 percent of GNP in the developing countries, it is a considerably more rapid growth than the 2–4 percent per annum sustained by the industrialized nations in the early stages of their development. The 6 percent growth of the LDCs also compares favorably with the present rate of growth of the advanced countries, which is, on the average, 3.5–4 percent. Because of rapid population growth, GNP per capita has been increasing only at 2.5 percent annually. Although in this regard the LDCs lag behind the advanced ones, by historical standards it is nevertheless a remarkable growth. The effort so far has gone into basic facilities (the infrastructure—roads, railroads, power plants, and communications), which take a long time to complete and whose contribution is not yet fully reflected in growth figures. The construction of these facilities, however, made it possible to accelerate economic modernization and has increased the capacity for future growth.

Of profound importance has been the so-called Green Revolution, a major breakthrough in food production that has significantly postponed, although it has not removed, the prospect of starvation for the less-developed world. It was made possible through the introduction, in the first half of the 1960s, of new high-yield varieties of wheat, rice, and corn. These "dwarf" varieties allow profitable application of up to three or four times as much fertilizer as traditional varieties; when combined with pesticides, irrigation, and other improvements in farming techniques, they make possible a doubling or tripling of yields. The impact on the LDCs has been dramatic: In the span of a few years the output of food has increased anywhere from 10 to more than 500 percent.[6]

The Green Revolution shows that science and technology can render substantial assistance to humanity's search for food, and there is no reason to believe that the potential of science and technology in this regard has been exhausted with it. Moreover, population growth in the LDCs has declined—significantly in some nations—between 1970 and 1975. If the trend continues, it would result in less need for food and higher per capita growth in GNP in coming years.[7] There are other promising developments. If the U.N. Law of the Sea Conference finally agrees on a new regime for the oceans, LDCs would benefit from a number of its provisions, inasmuch as they distinctly favor developing nations (see chapter 10). The U.N. World Science Conference, planned for 1979 or 1980 and intended to consider application of science and technology to economic development, may not have a strong immediate impact, but eventually it is likely to be of benefit to LDCs.[8]

I do not suggest by this discussion that the economic future of the LDCs as a group is necessarily bright or that, as a result, they will significantly rise in relative power position in coming decades. A certain amount of economic growth creates a prerequisite for an increase in power, but the question whether the LDCs will rise or decline in power must be viewed in a context broader than economics alone. In fact the outlook for the LDCs runs at crosscurrents. Some factors suggest that the LDCs (or at least some of them) may decline in power while other factors indicate that, in terms of power (if not necessarily economics), the gap between some LDCs and the advanced countries will significantly narrow. Which factors will prevail will largely depend on how successful LDCs are in coping with the impact of technology and in employing technological instruments for enhancing their power potential.

Some of the factors that may jeopardize the rise of the LDCs—in particular, their failure to cope with certain destabilizing influences within their societies—are discussed in chapter 9. The present discussion focuses on factors that might narrow the gap in power between the LDCs and the advanced countries.

One such factor is that the law of diminishing marginal returns applies not only to economics but also to power. In view of their already impressive technological capability further increments of technological advance add relatively little to the power of the highly advanced countries. But much smaller increments may have important impact on the distribution of power among the LDCs. A hundred thermonuclear weapons added to the arsenal of either the United States or the Soviet Union would mean little, but a few rudimentary nuclear bombs in the hands of Egypt or Israel would have a dramatic impact on the balance of power in the Middle East and in the world at large.

Less dramatic but analogous results can be obtained from increments of nonmilitary technology. To be sure, a certain minimum of economic-technological growth must be achieved for the increments to be relevant in terms of power and to produce appropriately larger marginal returns. The LDCs are achieving this minimum of economic-technological advance, making it possible for them to benefit from the law of diminishing marginal returns as it applies to power. Accordingly, in the future the economic gap between the LDCs and the advanced nations may continue widening, while the power gap narrows.

Alternatively we may witness a somewhat different situation. If at least some of the LDCs succeed in growing faster economically after their basic economic foundation is built, the economic gap may stop widening; it may even start to narrow. Furthermore, if the economic gap becomes stationary, the power gap will narrow much more rapidly than it would if the economic gap were still widening.

Also contributing to the potential rise of the LDCs in power is the present—and growing—technological backlash in the advanced societies. The developed nations will have to pay many billions of dollars for the lack of foresight and planning in environmental matters. These costs will tax the advanced societies' capacity for growth, while the LDCs can learn from their experience, avoid such pitfalls, and thus conserve their resources for further growth. Moreover, the LDCs are not encumbered by huge investments in the technological superstructure built in previous decades that inevitably becomes obsolescent and that may be difficult to restructure to meet future requirements. Similarly these nations can skip stages of technological development and thus significantly reduce costs and speed development. This has been illustrated earlier (chapter 3) by the case of India and her satellite-based community television system.

Certain technologies differentially favor the LDCs rather than the highly developed societies. This has been true with development of the new high-yield varieties of cereals that made the Green Revolution possible. Progress in fertilizer production resulted in significant reductions in costs over the past two decades, which also benefited primarily the LDCs. However, since nitrogen fertilizers are based on oil, and the price of oil has increased significantly in recent years, the trend toward lower costs in fertilizers seems to be at least partially reversed. It may be that the introduction of land satellites, to the extent that it leads to new discoveries of natural resources, will favor LDCs: These nations appear to have been explored less for natural resources than the advanced ones. On balance, however, the totality of future technology will probably benefit the advanced nations more than the LDCs; the differential benefit of certain technologies for the LDCs will be important primarily in conjunction with the decreasing marginal utility of technological advance to the power of the highly developed nations.

The energy crisis, on balance, favors the development of the LDCs, although some will suffer. The oil-producing countries realize that the present oil bonanza is a relatively temporary phenomenon incapable of providing a solid foundation for their economies unless translated into industrial development. Most of them are vigorously investing petrodollars in their own industrial growth. However, a sizable portion of the estimated $250–300 billion of surplus petrodollars by 1980 is likely to find its way into the more stable LDCs as investment, loans, or grants, [9] especially since the more secure advanced nations are reluctant to allow Arab countries to purchase their enterprises and real estate on a large scale. Technical assistance, on a reimbursable basis, will come from advanced nations. If this takes place on a large scale, the OPEC-induced high prices on oil would in effect amount to a form of taxation on energy-poor advanced nations to finance international development.

Both the economic and the power gap between the LDCs and some—if not all—highly advanced nations may narrow in response to changing values in the advanced nations. The old faith in science and technology has lost many of its adherents; quite a few are turning against technological progress. If this phenomenon increases, it can seriously hamper the growth of the nations thus afflicted. Perhaps more important, although less tangible, is the decline of the desire to be productively useful among the young people in affluent societies. If it does portend the value system of the future, it will have significant implications for the world distribution of power.

No single factor mentioned above is likely to be all-important in the power relationship between the LDCs and advanced nations. However, to the extent that these factors prove to be important, their aggregate effect may meaningfully contribute to narrowing of the power gap.

New Candidates for International Power

Not all developing nations will benefit equally from differential impacts of technology. Some will struggle with the problem of high population growth for many years to come and go through periods of famine and starvation. Others will do better. What are some of the candidates among the present LDCs that may close the power gap in coming years?

Brazil has already been mentioned as one such candidate. Her role as a potential superpower of the twenty-first century is especially important in the light of the relative power vacuum in Latin America. The lack of coal and petroleum has traditionally been considered to be Brazil's principal weakness. However, one of the largest coal deposits in the world has been discovered in the State of Amazonas, and substantial reserves of offshore oil have also been found, and these brighten Brazil's long-range prospects for growth.[10] Brazil is also likely to be the principal foreign beneficiary of the apparently large reserves of natural gas recently discovered in Bolivia.

The energy crisis gave OPEC nations significant influence in world affairs, but it is not likely that this influence can be translated into great power. In view of geographic, population, and resource characteristics of oil-producing countries, their ability to convert their huge oil revenues into industrial and military might is limited. Perhaps the only exception in this regard is Iran, which has undertaken a major scheme of self-aggrandizement.[11] Even if this scheme is successfully executed, Iran is too overshadowed by the proximity of the Soviet Union to effect a power change of more than regional significance.

Paradoxically Egypt—devoid of petroleum resources—might present a better prospect for a rise in power, thanks to Middle Eastern oil revenues. It seems that Arab oil capital, through American banks and American companies, is moving into that country. Egypt has a more literate population than the rest of the Middle East as well as a large potential market. If Arab oil revenues are invested in Egypt on a large scale, the prospects for its rise in power appear to be good [12] and, in light of that nation's geographic position, would be politically important. Moreover, if, on top of its own development, Egypt eventually succeeds in inspiring and unifying the Arab world, the change in power could be major.

China's natural resources, population, and governmental policies suggest that she is narrowing the power gap with the advanced world and may even effect a major change in the distribution of power. China has been said to have "the natural resources of a superpower." [13] Recent discoveries of petroleum have radically changed China's outlook in the energy field. The exact extent of her oil reserves is not known, but Western estimates place China somewhere between tenth and first place in the world.[14] China's output and export of petroleum have risen sharply and enabled her to use oil as an instrument for discouraging Japan from undertaking joint ventures with the Soviets in exploiting Siberian sources of energy.[15]

An important limitation of China's natural resources is her land. Only 11 percent of her total land area is suitable for cultivation, and some of the currently used land is of marginal value, with little room for further expansion.

China's population is both an asset and a liability. Estimated at about 850 million (1975), it is perhaps the best in the world from the point of view of development and rise to power: It is homogeneous, industrious, adaptable, and frugal. However, because of the limited arable land and a high rate of population growth (about 2 percent in the 1960s), a large population can be a serious liability. The government dealt with the problem by placing priority on agriculture rather than industry in its development strategy. Throughout the 1960s food production was kept ahead of population growth; by the mid-1970s China was on the verge of achieving self-sufficiency in food.[16] In the early 1970s, after years of oscillating policies with regard to population growth, China embarked on a determined and energetic birth control program. Recent estimates indicate dramatic results: a drop of population growth to 1.1 percent by 1975.[17] The combination of agricultural and population policies therefore promises to make China's masses into a major asset.

In spite of two major aberrations that disrupted the economy—the Great Leap Forward (1958–1960) and the Cultural Revolution (1966–1969)—China's economic growth between 1957 and 1972 has been estimated at about 5 percent annually, with industry growing at 8–9 percent.[18] Moreover, unlike the situation in other LDCs, growth was achieved without the benefit of foreign aid and with no internal or external debt. By 1975, China had apparently reached a point of not being able to absorb effectively the large imports of Western technology prevalent in the early 1970s, but the new regime of Hua Kuo-feng announced its intentions to resume large-scale imports of technology in 1978. These imports enhance China's prospects for development in coming years.

As China changes her priorities from agriculture to industry, her growth will benefit from the shift of her huge manpower to industrial production, where labor productivity is high. A lesser emphasis on the military sector in coming years would also improve China's longer range aspirations to power. But a true "Great Leap Forward" in China's drive to power may lie, not so much in her development policy proper, but in her foreign policy geared to her developmental needs. In particular China's horizons for power would broaden immensely if she were to enter into economic partnership with Japan. Although this is not very likely in the near future, the outlook may change considerably by the early or mid-1980s, at which time China will feel more secure militarily, her emphasis on industrial development would make cooperation with Japan especially attractive, and she may be less rigid ideologically. Japan also may go through a series of frustrations and problems with trade and investment in the West, which would make China, with her large natural resources and manpower, look very attractive for Japanese investments.[19] Other stresses may occur, bringing the two nations together. If the two eventually join hands in economic cooperation, this would be one of the greatest discontinuities in the existing trends with regard to the future world distribution of power.

The Sociopolitical Impact of Technology on Advanced Societies

In the previous century, the road to national power was largely limited to a few sectors of technology—steel production, railroads, shipping, and the machine-building industry. Now, the scope of technology is very broad and expanding rapidly. In the future, it will not be possible—not even for a superpower—to develop and use effectively all the potentially available technology, and some nations may therefore select, either by miscalculation or by deliberate choice, technologies that are power relevant, and others may select those that are not.

The selection of technologies will no doubt be a product of a number of factors influencing a given society at the time. One particularly relevant phenomenon, however, should be pointed out. The present scope of technology, its mature stage of development, and the concomitant affluence generate pressures for allocation of resources into technologies, products, and programs that are, at best, indirectly relevant to national power. In some states affluence stimulates demands for eradication of poverty (the United States); the population of other countries where a basic industrial foundation exists asks for conveniences, if not luxuries (the USSR). But these pressures—and the willingness to yield to them—are likely to differ and thus account for change in the distribution of power.

The advent of leisure [20]—a product of technologically advanced societies—is also likely to result in differential power impact. Some nations—especially those with democratic government—are likely to take full advantage of it, while others—for example, autocratic states like the USSR—may take advantage of it only partially or not at all. Cultural characteristics may play a role in whether or not a nation uses leisure to its full extent. (The United Kingdom, for example, is more likely to take full advantage of the development of leisure than Japan.) Over a period of 15 or 20 years different uses of leisure can create important differences in power.

Leisure can be relevant to national power in a somewhat different sense: Depending on the form of its use, it can be disruptive to a given society, neutral, or mildly conducive to national vitality. Large numbers of individuals with plenty of time on their hands and a degree of alienation from society can become a disruptive force. On the other hand, if the time released from work is used for such pursuits as chess, art, or improved physical fitness, the overall result might be a mentally and physically more vigorous society. The need for giving a wholesome content to leisure will become a major problem of democratic societies in the near future.[21]

Independently of the potential problems of leisure, technology can weaken some nations by producing unrest and social instability. The impact of technology in this regard may take varied forms; it may be indirect or direct. The rapid technological change combined with a situation where technology is induced into or projected on a given society from the outside is likely to be unsettling, as it uproots old values and produces a feeling of impotence against external technological power. Unless Western Europe restructures its institutions to develop—or at least effectively adopt—future technologies, it is likely to become subject to this phenomenon. It may be added that many of the LDCs are also candidates for a similar experience.

Technology is also changing values in those highly advanced societies that are prime sources of new technology. In such societies the existing hierarchy of values is restructured by virtue of making new options available through new technology. Previously unattainable ideals are brought within the realm of choice and thus become realizable values. Similarly new technology alters the relative ease with which different values can be implemented by changing the costs associated with them and thus making some values more attractive than others.[22] Lastly, rapidly changing technology alters the nature of a particular society, the world as a whole, and the relationship between the two. In the face of new reality, human values are apt to change, too. In this regard the difference between the highly advanced and the less advanced societies is but one of degree; both are "transitional."[23] The likely increase in human longevity in the next 25 years exacerbates the destabilizing potential of the change in values, since more generations and more potentially conflicting values will have to be accommodated at each given time.

Depending on the configuration of other factors and on governmental policy, changes in values may or may not produce a serious disruptive effect. However, they are likely to have an important impact on the vitality and viability of nations and on the world distribution of power.

The decline of the Protestant ethic or of the desire to be productively useful among American youth is another change in the value system with major implications for international power. As Peter Drucker pointed out, demographic factors—in particular, the unusually large numbers of teenagers in the 1960s—played an important role in magnifying this phenomenon; it has suffered a setback in the 1970s as the students of the previous decade entered the limited job market *en masse*.[24] Nevertheless it is quite possible—indeed, likely—that the decline in the productive pursuits as a value is not a flash in the pan and that it will grow in the increasingly affluent societies of the future. If correct, then a number of questions pertaining to the viability of the highly advanced societies—especially the democratic—arise.

Is it inevitable that, once an advanced society reaches a certain level of development, its value system undergoes a fundamental shift away from productive pursuits? If so, then the most advanced democratic societies are in serious jeopardy—they may experience a decline long before the growth momentum of the less advanced autocratic states slows down, and this may result in a major shift in the world distribution of power.

Even if we assume such a fundamental shift is not inevitable, a partial decline in productive pursuits as a value raises disturbing questions. How can a society that has historically relied on material rewards to induce occupational shifts to sustain productivity and strength regulate itself in the future when neither material rewards nor satisfaction from being productively useful constitute a value? What will be the balance between those who are productively useful to society and those who are less directly useful or not at all? Will this balance sustain a given nation's capability to function effectively in the international arena and to ensure its survival? In a somewhat different category but nonetheless relevant is this question: Will government policies succeed in countering the increasing alienation of the individual from society and other developments that may eventually evolve into something akin to mass neurosis?

None of these questions can be answered with certainty. As emphasized in chapter 3, success in solving the emerging problems would depend on their timely analysis and its translation into concrete policies. It would also depend on recognition of the extent to which the change being introduced by technological impact is inevitable and legitimate and on the adjustment of institutions and the creation of new ones to meet the requirements of change.

One can speculate on the possible differential impact of this development on the United States, Western Europe, Japan, and the Soviet Union. The United States, as the most technologically advanced nation, has been affected by this development earlier, and apparently to greater extent, than the others. However, this does not mean that it will be the most disadvantageously affected in the future; with its reasonably strong federal government, it is certainly in a better position to cope than, say, Western Europe.

The paradox of Western Europe's position is that it faces a double—if not triple—jeopardy. If Western Europe fails to strengthen its scientific and technological institutions and to generate its own advanced large-scale technology, it will become a victim of technology projected from the outside, with all the concomitant unsettling effects. If it strengthens appropriate institutions and successfully generates its own advanced technology, it will still be subject to its unsettling impact, similar to the one emerging in the United States. However, Western Europe is likely to be more severely affected because the resultant instability will probably not be handled effectively (or, at least, not with equal effectiveness) by the fragmented European political structure. However, the extent of the instability is likely to be less than in the previous variant. Lastly—and not very likely—if Western Europe succeeds in achieving political unification and developing strong technological institutions, it would still not escape potential turmoil. In this case, however, it will roughly be of the same magnitude as in the United States, with equal odds for its resolution.

Japan presents a stronger case than Western Europe and approaches the United States in the degree of her potential resistance to the adverse impact of technology. However, her strong points—in particular, the historically successful assimilation of modern technology into her culture, strong national spirit, and a unitary form of government—are countered by some weak ones by which the United States is much less encumbered, if at all: the limited territorial base and the relative congestion it creates, an underdeveloped or badly run-down social infrastructure, the changing values of Japanese society, the potentially destabilizing problem of the decline in lifelong employment by Japanese companies, and the unique issue of rearmament and nuclear weapons.

Of the four major advanced areas, the Soviet Union is likely to be least affected by the destabilizing and weakening impact of technology. This does not mean that she will not be significantly affected or that major surprises in the country's future might not occur. Through its internal control apparatus the Soviet leadership is in a strong position to maintain the cohesiveness of its system and to suppress any manifestation of discontent and alienation. In their controlled economy the Soviets are not likely to suffer seriously from occupational imbalances that might undermine their productive capacity. On the other hand if the country moves, as it may, in the direction of greater decentralization, it will undoubtedly experience some of the discontents making themselves felt in the West, discontents whose degree and outcome would be difficult to anticipate.

Moreover, as compared with the United States and Western Europe, Soviet society has so far been relatively little affected by the impact of advanced technology. It may yet show some now-unforeseen weaknesses. To go to one extreme—and to paraphrase Lord Acton's famous dictum—if power tends to collapse under the impact of modern technology, absolute power may collapse absolutely.

The Changing Importance and Nature of Power

Our discussion of the impact of technology on the distribution of power would not be complete—and, indeed, in some respects, would be misleading—if we did not take up the matter of the changing importance and nature of power. This is a profound development of recent years whose exact shape and meaning have not completely crystallized. The following discussion is therefore merely an attempt to provide a preliminary and hypothetical assessment of the change and its causes.

More specifically it appears that three developments have taken place: (1) There has been a change in the relative importance of the various factors of power in the international arena, (2) the importance of power in general has declined, and (3) a shift is taking place in the nature of power.

Military power has declined in relative importance, while political, economic, technological, and psychological factors of power have risen. This phenomenon apparently finds its root causes in several developments. The destructiveness of nuclear power, the tendency of military power toward stalemate and mutual neutralization, and a deliberate use of these attributes of military power by the superpowers as a matter of policy to achieve international stability all played a role.

A considerable degree of international stability has been achieved, which has further tended to decrease the importance of military power. After all, once crime subsides, one begins to wonder if a large police force is really necessary. Moreover, with the introduction of a measure of stability, other developments have begun to take place that have tended to diminish the role of military power even further. Political and economic relations and problems have risen in significance and have begun increasingly to occupy the attention of nations. Along with them political and economic factors have grown in importance as instrumentalities of influence and power. The law of diminishing marginal returns has played a role in reducing the importance of the predominant power centers. The decline in the importance of military power has become reflected in a change in the international system. Whereas, in terms of military power, the world is still largely bipolar, it is no longer so in terms of the totality of relevant power factors.

Under the impact of technology both economic and technological interdependence have grown. The scope for employment of military power, and even for its threat, has narrowed. Military power is unique in that it is highly disruptive to a multiplicity of other relations and exchanges.

Thus, with increased interdependence, the cost of resorting to military force has increased correspondingly. This does not mean that military force can no longer be resorted to; it can and likely will be used in coming years—but only where the stakes are high enough to justify the increasingly multiple (and not only military) costs of its use.

The reasons for the decline in the importance of power in general are also complex. In part power has declined because of the decline in the importance of military power, which is perhaps the only "true" form of power in the sense that it is geared and organized toward a single end—potential coercion. Although the military is governed by rules and regulations in the application of force, every member of the military is ultimately trained to abide by a single value—effectiveness in the exercise of coercion.

But the application of political, economic, technological, and psychological factors of power is usually an "imperfect" multivalue process in which values frequently run at crosscurrents. The political process is basically a bargaining process in which considerations of long-range accommodation may be more important than an immediate power gain. Economic factors are primarily concerned with production of goods and satisfaction of human wants; when they are employed for power purposes, their role, in a sense, is artificially distorted. The same is true of technological and psychological factors of power. They are not fully mobilizable for power purposes unless they are used for war. Individuals involved in their application are multivalue individuals, and this fact is accepted and tolerated. Moreover, as instruments of power, these factors can be reasonably effective under one set of circumstances but only at a considerable discount—or not at all—under a different set of circumstances.

Also contributing to the decline of the importance of power is pluralization and diffusion of power inside societies and internationally. The nation-state or another international actor frequently finds it difficult to mobilize an adequate amount of power—of whatever nature and for whatever purpose—to be truly effective in the contemporary multifocal environment of world politics. Still another factor is the partial change in the international political process from the zero-sum game to the positive-sum game. If the trend in the political process among nation-states is away from competition and toward cooperation then, to a degree, "power" in its conventional meaning is bound to decline in relative importance. A closely related development has been a change in values, especially in highly advanced democratic societies. As social concerns shifted, considerations of power seemed to be at best less important, at worst as evil, anachronistic remnants of bygone days.

Along with the decline in the importance of power in general and military power in particular a certain shift in the nature of power seems to be taking place. "Power over" (more characteristic of the zero-sum political process) is declining. "Power toward"—the ability to lead people and mobilize resources for the solution of contemporary problems and toward new frontiers of human achievement—is growing. Clearly, these two aspects of power are not completely separate; they overlap and, in some respects, are complementary. However, whereas "power over" emphasizes control and domination, "power toward" emphasizes rational decision making, problem solving, goals, and ideals. In its modern setting "power toward" has not been adequately explored, understood, or used. Although it is potentially ascending, we are not simply witnessing a shift from one aspect of power to another. Insofar as pluralization of power is a definite trend, it impedes effective employment of any power, "over" or "toward."

The implications of this discussion to our analysis of the distribution of power can be summarized as follows:

1. Power is becoming less hierarchical and more issue relevant and issue oriented.

2. Transition in the nature of power carries potential dangers of inadequate assessment of what true power is in a given period of time. A nation still inordinately impressed by "power over" may overcommit its resources to the military sector and ultimately decline. Conversely, a nation might overlook the extent to which "power over" is still significant, may move too far in the opposite direction, and jeopardize its viability. A third alternative is possible: A heretofore vital nation, unable to understand the importance of "power toward" and incapable of taking advantage of it, may start decaying.

3. The balance of power has never been a perfect instrument of statecraft, and it is even less perfect in a period of change in the nature of power. If the trend in the direction of "power toward" is sustained, the balance of power will increasingly become irrelevant.

4. The trend in the direction of "power toward" is not inevitable and cannot be taken for granted. International instability, especially in the military sector, can reverse the trend and heighten the importance of military power. This reversal may be temporary—lasting only a decade or so—but it may result in substantial damage to individual nations and the international community in general.

Technology and International Stability

9

TECHNOLOGY CAN AFFECT international stability in at least three ways.

1. It can contribute to the development of an international system that is either basically stable or unstable. For example, to the extent that technology contributes to the growth of an international system characterized by a high degree of interdependence among nation-states and by the emergence of influential global functional institutions, it strengthens world stability. In a somewhat different category, nuclear weapons have been a major contributor to the bipolarity of the post-World War II years. Some students of international relations maintain that a bipolar system is inherently less stable than a multipolar one, but others disagree.[1] A relationship between international stability and polarity appears to exist, however. After review of a number of subsystems over the last three centuries, Michael Haas has suggested that "multipolarity entails more violence, more countries at war, and more casualties; bipolarity brings fewer but longer wars."[2]

2. Technological impact can produce serious internal instability within nation-states, which, in turn, can affect international stability. For example, the uprooting of the traditional value system in a less-developed country (LDC) and the resultant domestic instability may lead to involvement of the superpowers or of neighborhood states[3] and thus transform internal into international instability.

3. In a given international system technological developments can be either stabilizing or destabilizing. For example a slow rate of change in nuclear weapons technology in a bipolar situation characterized by the balance of terror can be relatively stable. However, if the rate of technological change is rapid, with a distinct possibility that one side or the other will obtain a relatively invulnerable first-strike capability, the situation could result in serious instability.

With regard to world stability the impact of technology can range from a dependent to an independent variable with gradations in between. If the advanced countries decide to pool their technologies and resources to assist the developing nations and thus forestall the outbreak of serious instabilities, technology will be a distinctly dependent variable. If the differential impact of technology in the next decades favors the rise of the Soviet Union as a highly advanced industrial power while the other power centers experience a relative decline, the Soviet leaders may be tempted toward greater political (and perhaps military) aggressiveness.[4] In this case the impact of technology is, at best, that of a partially independent variable: It is indirect (through the augmentation of power) and provides merely a temptation (not a compulsion) to act; an *intent* on the part of the Soviet leaders would be necessary to use the available power aggressively. A different case arises in the example suggested previously, whereby a rapid change in nuclear weapons technology produces a highly unstable situation. Here technology produces an impact as an independent variable; the prospect of the enemy's achieving a decisive superiority combined with invulnerability may provide the threatened power, not with just a temptation, but with what is thought to be a *compulsion* to act.

The impact of technology in the next decades will be in some respects stabilizing and in others destabilizing. The line between the two is not always clear-cut. The same technological impact may have both stabilizing and destabilizing elements at the same time, or it may be stabilizing in one phase of its development but destabilizing in another.

Strategic Nuclear Weapons, the Superpowers, and World Stability

Two aspects of strategic nuclear weapons may affect world stability: technological change within the existing nuclear umbrella and proliferation of nuclear weapons to third states.

In the 1950s the low level of the nuclear arsenals combined with very rapid technological change (thermonuclear weapons, long-range strategic bombers, IRBMs, ICBMs, Polaris) to create considerable instability in strategic nuclear weapons. By the mid-1960s, the strategic situation achieved a measure of stability. The single most important reason was that the United States and the Soviet Union had reached a point where neither could eliminate the other's nuclear weapons by a first strike and effectively protect itself against a retaliatory strike.

Then, around 1970, new elements appeared in the strategic picture—the multiple independently targeted reentry vehicles (MIRV)[5] and the antiballistic missile defense (ABM). MIRV and ABM could be combined to give a superpower a first-strike capability, whereby missiles equipped with MIRV could annihilate a large percentage of the enemy missiles before launch, while the remainder of the enemy missiles would be intercepted by a limited ABM system. These two innovations introduced potential instability into the nuclear umbrella that could result in a major arms race.[6] This was, however, prevented—at least for the time being—by Strategic Arms Limitations Talks (SALT) and the resultant two agreements of May 1972: The Treaty on Anti-Ballistic Missile Systems and the Interim Agreement on Limitations of Strategic Offensive Arms.

Viewed in the context of technological developments preceding the two treaties, the Treaty on Anti-Ballistic Missile Systems was a significant stabilizing factor. By agreeing to limit their respective ABM systems to two sites (reduced to one site in 1974) with 100 interceptor missiles each, the superpowers in effect agreed not to use their ABM systems to achieve a first-strike capability. Yet, when both treaties are viewed in the long run, they fall considerably short of guaranteeing stability in the nuclear strategic arms sector. The treaties provide for escape clauses: Six months' notice. The Interim Agreement on Limitations of Strategic Offensive Arms introduced a measure of stability by freezing the U.S. and Soviet nuclear missiles in quantitative terms at a level considered to be that of approximate parity, but it contained a number of loopholes. It was limited to five years and covered land-based and submarine-based missiles only (and thus excluded long-range bombers, cruise missiles, and launchers placed on mobile sea platforms other than submarines). Moreover, the Agreement did not limit qualitative improvements, including the right to MIRV the existing missiles.[7] Therefore, unless the first SALT agreements were followed by others, there was plenty of room for a potentially destabilizing arms race.

In Vladivostok, in November 1974, President Ford and Secretary General Brezhnev reached a 10-year agreement "in principle" to limit all strategic delivery vehicles (including bombers) to 2400 on each side, 1320 of which may be MIRVed. At its face value this agreement still leaves substantial room for an arms race in terms of quality and missile size. The outcome of the current SALT negotiations intended to bring the Vladivostok accord into a concrete and binding form will show its true value. In the meantime the arms race is going on, the Soviets having a significant margin in the size of the missiles (the "throw-weight") and the United States leading in the number of warheads (because of advance MIRVing) and in the precision of the weapons.[8]

The rumble of an arms race within the strategic nuclear sector is not likely to be the last one. As technological advance continues to introduce changes into the strategic sector, there will be changes in strategic doctrines, and arms races of varied magnitudes are likely to flare up.[9] One such change is already discernible. With the introduction and perfection of electro-optical guidance, the delivery of strategic nuclear warheads can be made precise enough to destroy individual military or militarily relevant targets while sparing nearby population centers. The thrust of strategic doctrines and weapons development may thus shift toward a counterforce strategy. Whether or not such a development will be accompanied by an arms race, a strategic arms limitation agreement, or both, remains to be seen.

But the question whether or not developments in the nuclear umbrella will seriously undermine world stability depends not only—and not primarily—on whether an arms race will or will not take place but on whether, even if it does take place, it will "spill over" into political relations outside of the nuclear umbrella itself or lead to outbreak of nuclear war. There is no definite evidence so far that the present escalation in nuclear weapons between the United States and the Soviet Union affects political stability. It is, however, conceivable that a major armaments race in strategic nuclear weapons could affect stability outside the military sphere.

Although the nature of the new weapons may provide an occasional temptation for strategic planners to think in terms of a preventive first strike, it is very unlikely that it would be acceptable to the political leadership of either superpower.[10] In summary, although we may witness occasional arms races on the strategic nuclear level, they will not necessarily change the essentially stalemated conditions in this sector. Their impact is likely to be both economic (the cost burden that they will impose) and political; if the equation of nuclear weapons significantly changes, certain hitherto nonexisting political constraints will likely be imposed on the side unfavorably affected.

Nuclear Weapons and Third States

In some respects the outlook for international instability because of proliferation of nuclear weapons is greater than that generated by strategic nuclear weapons of the superpowers. First, proliferation is favored by a combination of technological, political, and economic factors, and the prospects for its control are not particularly good. Second, the possibility of a "small" nuclear war involving Third–World states is greater than a large-scale nuclear holocaust triggered by the superpowers.

Rightly or wrongly, many nations feel, in varied degrees, that possession of nuclear weapons is important, if not essential, for their national security. Still others—like France—are more concerned about the prestige value of the weapons than their military value. The potential for using peaceful nuclear explosives for economic purposes aroused interest in acquiring nuclear technology. The energy crisis also stimulated commercial interest in selling nuclear technology.[11]

The principal politico-legal instrument for controlling the spread of nuclear weapons is the Non-Proliferation Treaty (NPT), which came into force on March 5, 1970. It was reviewed by the Review Conference of the Parties to the NPT in May 1975, but left without change.[12] The Treaty binds nuclear weapons parties not to transfer nuclear weapons and other "nuclear explosive devices" to any other country. It also commits nonnuclear weapons parties not to manufacture or otherwise acquire nuclear explosives.

The NPT is not a particularly effective instrument. France, the People's Republic of China, and India—which have military nuclear capability (incipient in the case of India)—have not signed the Treaty and are free, if they choose to do so, to assist other nations in building up nuclear forces. Several nations that have varied degrees of nonmilitary nuclear capability are also nonsignatories. These include Argentina, Brazil, Israel, Pakistan, South Africa, and Spain. Moreover, the NPT contains a positive obligation (a concession to nonnuclear states) to promote nuclear technology for peaceful uses, while conversion from peaceful uses to weapons is difficult to control. India, which obtained nuclear technology from Canada before the NPT but bound by her agreement with Canada, chose to define the term "peaceful" to include explosives allegedly intended for mining and earth moving and built a nuclear device exploded in May 1974. Last but not least, any party to the Treaty can withdraw on three months' notice if it decides that "extraordinary events" jeopardize its "supreme interests." [13]

Given the various shortcomings in the Treaty's effectiveness and the spread of nuclear technology, a newly acquired nonmilitary nuclear capability could eventually be converted into nuclear weapons. It is thus likely that, among developing nations, Israel, South Africa and perhaps Brazil and Argentina, might develop nuclear weapons in the near future, while Pakistan, Egypt, Spain, Iran, South Korea, and Turkey are possible candidates for the more distant future.[14] To counter the trend, seven advanced nations, including the United States, the USSR, Japan, and Western European powers (including France), reached an understanding in January 1976 designed to discourage the spread of nuclear weapons.[15] It is not very likely that this move will stop the process, but it may slow it down.

The potentially highly unstable phase in the spread of nuclear weapons is that in which one protagonist in the developing world acquires them before his enemy does. It is worth recalling that nuclear weapons have been used before; it was not an easy decision for the United States to bomb Hiroshima and Nagasaki in August 1945, but it was made much easier because Japan had no retaliatory nuclear capability. In view of the international instability of the developing countries and the tensions that already exist or may arise among them, is is conceivable that nuclear weapons could be used against an enemy who does not have them.

Once the other side also acquires nuclear weapons, a considerably greater degree of stability could result. There is an important difference between conventional and nuclear weapons in terms of their effect on international stability in an adversary climate. A moderately small superiority in the quantity and quality of conventional weapons could spell the difference between victory and defeat and thus provide a temptation for preventive wars or fueling an arms race. This is not usually the case with nuclear weapons, especially of the rudimentary variety such as is likely to be available in the Third World; margins of superiority are not important, since the use of nuclear weapons would be suicidal for both sides. Nuclear weapons could thus introduce a balance that might dampen the arms race in a given region of the Third World and eventually introduce a measure of stability.[16]

In summary, an international system characterized by a global nuclear umbrella sustained by major powers and a number of smaller regional umbrellas created by smaller states could prove to be reasonably stable.

The "Conventional" Military Sector and World Stability

There are two possible sources of instability in the area of "conventional" military weapons: technological change, which might lead to an arms race, and the possibility of further Soviet efforts to enhance their capability of stalemating the American forces globally, of eroding U.S. alliances, or otherwise weakening the U.S. position in certain geographic regions.

We have seen in chapter 2 that technological trends in the "conventional" military sector favor a stalemate or mutual neutralization of power. However, in the early phases of the development of electro-optical weapons, the offense is favored over the defense. Because of the extraordinary power of these weapons neither side may feel it can afford to stay behind in their development; accordingly an arms race is a possibility. Just as in the case of strategic nuclear weapons, however, an arms race in electro-optical weaponry may not spill over into international instability. Eventually, as the defense aspects of electro-optical weapons gain, a mutual stalemate is likely to be brought about and thus contribute to greater stability.

It is paradoxical that, with the development of electro-optical weapons, a truly effective stability on the strategic nuclear level may, in overall terms, prove to be destabilizing. A resort to tactical nuclear weapons provides about the only means of breaking the deadlock brought about by electro-optical weapons. If the nuclear umbrella is highly stable and unlikely to be triggered by resort to tactical nuclear weapons, then it is conceivable that these weapons may be resorted to for breaking a stalemate brought about by electro-optical warfare. Thus the interest of international stability in coming decades may require that the strategic nuclear level be kept but moderately stable, well short of a foolproof stability. Of course this conclusion is narrowly technologic–military. Given a substantial degree of political stability in coming decades, the potential technologic–military instability may not be important.

Whereas the trends in subnuclear-umbrella military technologies, by and large, favor stability, the nature of Soviet intentions presents an important imponderable. If the Soviet Union continues to invest large resources into the subnuclear-umbrella military sector, is not satisfied by merely stalemating U.S. influence but instead strives to achieve superiority over non-Communist forces by projecting its expanded naval and amphibious capability, and progresses in eroding U.S. alliances, instability would arise.

Transnational Violence

In recent years a new phenomenon has emerged: increasing violence by terrorists on an international scale. The monopoly of the state on the means of violence appears to be eroding.[17] This phenomenon in part reflects general pluralization of power in the world and the fact that the nation-state no longer occupies the overriding position of power it once had. Transnational violence is also a product of technology. Modern means of transportation and communications allow terrorists to plan and execute an act of violence hundreds or thousands of miles away from their base, as was the case with Arab terrorists striking at Israeli athletes in Munich during the 1972 Olympic games. Moreover, modern technology provides such groups with unusually effective weapons. Thus the vulnerability of an airplane at 30,000 feet makes it possible for even one individual to control the lives of many people with a knife pointed at the pilot. The possibility that nuclear weapons might be available to groups or individuals further raises the specter of transnational violence.

Not subject to any rules or conventions as regular warfare usually is, transnational violence, in some respects, may be more objectionable than interstate conflict. In some cases it might be instrumental in triggering interstate conflict. So far nations have had limited success in coping with it, and it remains to be seen how effective future international and regional agreements, if any, will be in controlling this phenomenon.

Transnational violence frequently represents intrasocietal instability amplified through technological means and projected on the international scene. Thus, whereas it is important that national and international mechanisms be available to control this phenomenon when it erupts, it is even more important to prevent its eruption through appropriate policies designed to maintain social stability. Insofar as nonmilitary technology is often a cause of social instability, technology thus provides both the initial impetus to, and instrumentality for, transnational violence.

Nonmilitary Technology and World Stability

In examining the stabilizing and destabilizing effects of nonmilitary technology, it is convenient to differentiate between its impact as a dependent and as an independent variable.

Technology, *as a dependent variable,* can be a truly major instrument for achieving international stability. It could, if appropriately applied, solve a number of world problems that either produce, or are about to produce, serious destabilizing effects.

Technology now available and that expected in the near future (medical and nonmedical, such as audiovisual means of instruction for demonstrating birth control techniques) is, or soon will be, adequate to help solve the problem of population explosion. Still other technologies will be able to expand agricultural, mineral, and other material resources and increase human productivity, all of which are necessary for raising the standard of living of the billions of people who live in poverty or on the verge of starvation. Poverty is a source of major instability in both the advanced and developing countries now; the instability produced by poverty and concomitant problems is likely to become especially critical in the developing world, where many millions are flocking to the already overcrowded big cities characterized by high rates of unemployment and substandard housing facilities.[18]

None of these problems is insoluble through technology backed by effective planning, organization, political support, and adequate allocation of resources. These conditions, however, do not exist at present. There is no effective coordination of development assistance policies among the three principal donors of the Free World, the United States, Western Europe, and Japan.[19] Concessional aid to LDCs amounts to only about .30 percent of the GNP of developed nations, considerably short of the U.N. goal of .70 percent. Technology, on the whole, is not effectively tailored to the development needs of LDCs because of the failure of both advanced nations and LDCs to address themselves to this matter in a concerted and constructive way. As a result, the opportunity to accelerate economic growth and to skip stages of technological development—a major potential asset of LDCs in the development process—is not adequately capitalized upon.

The process of resolving the polarization between the so-called "Group of 77" (LDCs demanding a new, more favorable international economic order, for the achievement of which commodities are used as a leverage of power) and developed nations principally represented by OECD may, as its byproduct, bring about a more effective application of technology to development. But, at best, it will be a protracted process in which only certain leading LDCs (e.g., Brazil, Mexico) are likely to continue to apply technology effectively to their economic growth, while the outcome for the poor developing nations remains much less certain.[20]

Technology, *as an independent variable,* will, to an extent, produce a stabilizing effect on world society in the next decades. The value systems of governmental elites of the advanced countries are, in some respects, becoming more similar. This appears to be true even in the case of such erstwhile antagonists as the Soviet Union and the United States: The Soviet elite is more managerially inclined and less revolutionary, both countries are concerned with international stability (in particular with regard to nuclear weapons), and both are facing similar problems of science and technology. Some of these problems (like global ecology) require mutual cooperation. Eventually, as China moves farther down the road of development, it might become subject to the same phenomenon. International rivalries will not disappear or perhaps even greatly diminish, but they are likely to be directed into less explosive channels and externalized in less aggressive forms and thus contribute to international stability.[21]

There is a related development that, in particular, affects autocratic or totalitarian societies embarked on forced industrialization. After a certain economic and educational level of development has been reached, the population of these societies increasingly demands to share the benefits of growth; it demands consumer goods and decent living conditions.[22] This phenomenon is manifesting itself in the USSR at present. Although its exact political ramifications are controversial,[23] it is clear that the Soviet regime, albeit reluctantly, is giving way to this demand. Further stimulated by societal pluralization caused by technological advance, this phenomenon tends to "mellow" external aggressiveness and hence curtails potential instability, inasmuch as it imposes constraints on resources otherwise available for external pursuits and makes society more inward oriented. This is further reinforced by the technological backlash, since the educated strata increasingly demand, not only consumer goods and decent housing, but also a decent environment.

The rise of the "superculture," which penetrates elites of the advanced and the developing world alike, tends to create an element of commonality in outlook among the upper strata of the world's population; to that extent it is likely to exert a stabilizing effect. If the scenario envisioning the growth of functional international organizations materializes, these organizations, by crisscrossing national boundaries and by solving, through international cooperation, the multiplicity of problems facing mankind, will also produce a stabilizing impact.

Whereas, in some respects, nonmilitary technology has a potential for stabilizing world society, in other respects its impact can be highly destabilizing. In this regard technology's impact is primarily as an independent variable.

We have seen earlier (chapters 6 and 7) that Western Europe and, to a lesser degree, Japan are vulnerable to destabilizing impact of technology. If our projections of human longevity materialize, most advanced nations may begin to feel their destabilizing effect at the end of this century. Technological change is also likely to introduce serious instabilities in the developing world. The introduction of new technology on a large scale will uproot traditional values; induce a high degree of physical and, potentially, social mobility into previously highly static societies; and thus generate a considerable amount of personal insecurity in the population and threaten the security of the ruling elites. Given these conditions, the governments may attempt to strengthen their position by diverting the attention of the population to external factors and thus extend societal instability into international instability.[24] Moreover, internal instability may generate international instability by virtue of various factions' seeking external help to support their aspirations to power or desire for survival.

A similar type of internal instability, but with different international consequences, might arise from new audiovisual techniques of mass education combined with large-scale exposure of the poor masses to the mode of life and the standard of living of the advanced countries; this exposure is likely to become available through international television in the next decade. It is one thing to be aware in the abstract that America is rich but another to be able to see personally on the screen of a community TV receiver the standard of living of the average American and to compare it with local poverty. The full implications of this development cannot be foreseen, but one outcome might be pressure by the population of the developing nations on their national leaders to demand greater assistance from the advanced countries. In turn this may result in a greater rift between the North and the South.

Although the superculture produces a stabilizing effect by creating a measure of similarity in outlook among national elites, it can also create internal instability by broadening the division inside nations between those who share in the values and attitudes influenced by the superculture and those who do not. Some of this phenomenon can be discerned in the differences between the internationally oriented, better educated, and highly mobile Americans and the "silent majority" of the more conservative and nationalistic strata of middle-class America. The exact impact the superculture will produce on the internal cohesiveness of the developing nations is not clear, inasmuch as these nations will be subjected to multiple crosscurrents of technological effects, but it is likely to exacerbate internal instability.

Not all internal instability in the developing nations will translate itself into international instability. It is possible that some serious problems in the LDCs—such as major famines, resulting in the death of hundreds of thousands of people—may be virtually completely internalized, with hardly any destabilizing effect on the international forum.

The campus unrest of the late 1960s in the United States demonstrated that some commonly available technological instruments—like the duplicating machine, the bullhorn, the walkie-talkie, and the private car—can be very effective in assisting the dissenters in challenging established authority. Their long-range effect can perhaps be even more profound in autocratic states. It has been noted that the availability of duplicating means in the USSR has created, since about the mid-1950s, a widespread practice of *samizdat* (self- or home-publishing) that publishes not only prohibited works of Soviet writers but also translations of foreign publications, political pamphlets, court hearings of political trials, and so forth. In spite of vigorous suppression by the regime *samizdat* managed to survive and has given birth to a form, however rudimentary, of political opposition to the Soviet government.[25] The effectiveness of such activity against central governmental control (and the multiple technological means that favor central control) is problematic at best, but this development does provide a medium for potential erosion of the regime that, in conjunction with other factors, could be a source of major instability.

The development of technologies for exploitation of marine resources may become a source of major international instability inasmuch as it can generate international rivalry for vast resources of the oceans whose ownership is yet uncertain. It should become clear fairly soon whether the various nations concerned will succeed in agreeing on international instruments for stabilizing this area of activity. International instability from exploitation of marine resources may also arise, not from rivalry over them, but from their availability. For example exploitation of manganese nodules—which might be possible about 1980—may flood the world market with large quantities of manganese, iron, copper, nickel, and cobalt, rendering serious damage to the economies of the raw-materials-producing countries, principally LDCs.[26]

Lastly, the rapid pace of technological change vastly complicates the task of international law as a stabilizing element in the international community. New developments in technology have necessitated establishment of hitherto nonexistent branches of the law. Thus aviation has produced air law; telecommunications have effected the growth of laws to cover this sector of technology; outer space law is the product of man's successes in developing the ability to move outside and beyond his planet. In some individual and relatively rare instances the law has succeeded in outpacing technological developments and in establishing regulations for a particular activity before it had been achieved. A prime example is provided by the Outer Space Treaty prohibiting military activity on the moon, which was signed before military capability for the moon existed. In a number of other areas, however, technological developments have produced new requirements for regulation, before the law could catch up with the change.

Over the last hundred years or so conventional (treaty) international law has largely supplanted customary international law as a source of "new" law, largely because customary law was too slow to evolve; but now even conventional international law cannot catch up with many technology-induced developments. It takes years for states to agree on a convention establishing new rules of international law, and by the time they do, the convention may already be obsolescent. The 1958 Convention on the Continental Shelf agreed on the 200-meter isobath limit of the national continental shelf and then introduced the clause whereby states could extend their jurisdiction beyond that limit if technology permitted exploitation. The signatories of the convention expected that the technological ability to exploit resources beyond the 200-meter isobath would not be achieved for many years, perhaps many decades. But when the convention came into effect six years later, exploitation of offshore oil was already racing toward the 200-meter depth. The wording of the convention thus set the stage for marine technology, as an independent variable, to become an important factor in molding a future regime of the oceans. Unless and until the 1958 Convention is superseded by a new treaty, the development of marine technology, probably, will remain the principal factor in extending national economic jurisdiction over the seabed.[27]

In addition to international law's being incapable of catching up with the development of a number of sectors of technology and of regulating their impact in a way conducive to stability, the rapid pace of technological change seriously weakens international law in its broader, yet perhaps more important, role. One of the principal functions of international law is to contribute to international stability by building a consensus among the policymakers on the nature of international community; it thus serves as an instrument for "socialization" of international society.[28] With the accelerating speed of technological advance this is a function that international law will find increasingly difficult to perform—and at a time when, because of the integration of world society, its growing interdependence, and the resultant need for its orderly functioning, the law's effectiveness in this regard will be much more critical.

In summary: Nonmilitary technology being viewed as an independent variable, its destabilizing potential for the future appears to outweigh its stabilizing potential by a substantial margin. This still leaves considerable room for organized human will to curtail the destabilizing impact of technology and to augment its stabilizing potential. Whether or not such a will is developed remains to be seen. If no adequate effort to stabilize the impact of nonmilitary technology is produced, it is possible that, whereas the military sector proper will get increasingly stabilized in the next decades, nonmilitary technology, perhaps accompanied by transnational violence, might generate serious instability in foreign affairs that would be difficult, if not impossible, to control.

Technology and World Society

10

HISTORICALLY TECHNOLOGY has been the most important factor in creating what can be called "world society." Through developments in such areas as transportation, telecommunications, and weapons, technology has provided for interaction of human activity on a global scale and has thus been instrumental in establishing a substantial degree of interdependence among communities and nations. I now turn to the political impact of technology on world society. This impact can be summarized as follows:

1. By contributing to change in the world distribution of power, certain forms of technological impact may modify, or even radically alter, the international political and legal system.

2. Still other forms of technological impact affect world society by promoting regional integration.

3. Technological impact can induce greater cohesiveness of, or interdependence within, world society by strengthening existing international institutions and by stimulating the development of new ones.

4. Technological impact can contribute to changes in the nature of the international political process.

5. Technological impact can contribute to a change in the loci of authority among the four levels—global (universal), subsystem, national, and subnational—and within each level.

World Distribution of Power and the Nature of International Society

The world distribution of power has always been a major (if not the major) factor in determining the nature of the international political system. The "chandelier" system of the balance of power, wherein several Great Powers balanced each other without firm commitments to a particular ally or bloc, was founded mainly on the approximate equality in power among the principal European continental states in the first 50 years or so after the Congress of Vienna (1815). This system was undermined when Prussia's power became disproportionate to that of her major continental neighbors and when she capitalized on this fact in the Franco–Prussian War (1870–1871).[1] The bipolar system of the immediate post-World War II period reflected the fact that the only great and viable centers of power at the time were the United States and the USSR. Whereas ideological differences between the two superpowers were instrumental in contributing certain characteristics to the international system, they would not have produced an important impact on the system if ideology had not been backed by physical power.[2]

The present global system is still significantly influenced by bipolarity. However, regional subsystems (Western European, Asian, Latin American, Middle Eastern) now enjoy a measure of autonomy from the global system and, at the same time, produce an effect on the system in general and on the relations between the superpowers in particular. We have seen that with the achievement of a measure of stability in the military sphere between the superpowers, the military factor has declined in relative significance while political, economic, and technological factors have gained. This development has further reduced the importance of bipolarity, since it was principally in the military sphere that bipolarity was most persistent and most pronounced. The prominence of bipolarity as the predominant characteristic of the international system has been further blunted by the rise of economic and technological competition among advanced non-Communist nations, including the United States. In this larger and more complex setting the superpowers began to discover areas of common interest—especially in the realm of security and world peace, but also in economic and technological matters. The rivalry between the superpowers persists, but it has become more long-range and less militant, and its incidence has shifted to political, economic, and technological spheres of activity.[3]

The world distribution of power has been important in molding the nature of world society, not only by having established a particular pattern of political relationships, but also by playing an instrumental role in determining the dominant values and legal norms of the world community. Thus international law as it originally evolved from early beginnings in the sixteenth century was mainly a product of Western European nations and reflected their interests as the then-dominant group in the evolution of world society.[4] Great Britain, with her maritime superiority, was a major factor in molding the law of the sea in the eighteenth and nineteenth centuries. The emergence of protection of citizens abroad as a branch of international law reflected the growing need for international intercourse in general, but much of it reflected the interests of the advanced—and, hence, the more powerful—nations in conducting economic relations with the less developed areas. These interests required that certain reasonably reliable rules of law protecting citizens and property of the advanced nations be enforced in a geographic area where the standards of conduct were different and the laws either different, uncertain, or both.[5] The rule of international law with regard to nationalization (illegal, unless "just and adequate" compensation was provided) largely developed in response to the interests of those (advanced) nations that prevailed on the international economic scene and had capital to invest abroad.

Whereas international law has thus, in part, evolved as a "superstructure" of the existing power relationship, it has also tended to produce an influence of its own by helping to mold a consensus, albeit imperfect, of the community of states.[6] Accordingly the role of technology in changing the distribution of power in the future must be viewed in the context of the interplay among power, values, and law.

World Distribution of Power and Some Possible Scenarios of International Systems

One could envision a number of scenarios in which the impact of technology modifies the existing world distribution of power and, hence, the international system. A rather radical scenario would result from an outbreak of strategic nuclear war between the superpowers, in which case the now less developed areas would significantly rise in power. As a result their values and preferences for legal norms would be likely to prevail—at least, for a while. An entirely different set of potential scenarios could be projected if, along with peaceful changes in the distribution of power, rivalry between the United States and the Soviet Union ceased in the next decades.

The discussion below focuses on two scenarios—one envisioning the rise in power of the United States and Western Europe and the decline of the Soviet Union, and the other visualizing the reverse. These scenarios assume a degree of rivalry between the superpowers. The selection of scenarios is not necessarily intended to suggest that they are the most likely ones. I present them because they illustrate the interrelationship between the three variables (technology, power, and the international systems), which in turn implies that there may be room for the development or refinement of policy designed to bring about one alternative international system in preference to another.

Relative Decline of Soviet Power

The oil crisis of 1973-1974 and the rumblings of another about 1980 finally impel the United States and Western Europe to rationalize their respective science and technology policies and to enter, in the early 1980s, into extensive cooperation in this field. Both succeed in avoiding the potentially destabilizing effects of the postindustrial society, and they develop, and stay in the forefront of, advanced technology of the time. Although the Soviet Union does not experience severe shortages of energy and although it benefits, to a degree, from technological cooperation with non-Communist nations initiated in the first half of the seventies, it still has difficulties in raising its technical efficiency and is not particularly effective in solving its economic problems. Ideological élan—especially among Soviet youth—flags. This development is combined with periodic flares of political dissent and growing unrest among consumers.

Soviet resources are clearly overextended. The Soviet Union finds it difficult to reconcile a troublesome internal situation with the extensive military buildup and other worldwide commitments developed in the late 1960s and the 1970s. Retrenchment begins to take place.[7] This development is accompanied by an internal reshuffle in the Soviet leadership. The more conservative group that comes to the top curtails Soviet technological and economic cooperation with advanced non-Communist nations. The USSR becomes largely inward oriented.

The withdrawal of the Soviet Union to its own sphere of influence and its orientation inward do not provide a solution to Soviet problems. The growth of China and her renewed territorial claims are of serious concern to the Soviet leadership, and it allocates substantial resources to military power along the Chinese frontier. The efforts of the USSR to integrate the satellites through COMECON fail as pluralization proliferates in Eastern Europe. Soviet ability to control Eastern Europe is further impaired by growing divisions along nationality lines in the USSR itself.[8]

The United States, Western Europe, and Japan lead in development of multinational corporations, but the earlier efforts of the Soviet Union in this area have failed to blossom. Pressed by resource shortages, poor economic performance, and various domestic problems, the Soviet leadership decides, in the second half of the 1980s, to resume active cooperation with advanced non-Communist nations in technological, economic, and environmental projects. It also abandons its reservations about functional international institutions and supports their activities. Eventually, as a result of contacts with the

West through bilateral and regional programs in science and technology, participation in global functional institutions, and the transfer of Western managerial and technical skills, Soviet economic efficiency and productivity improve. Although the Soviet economic and technological capability now becomes stronger, relative to the advanced West, it is still markedly weaker than it was in the mid-1970s. The Soviet Union is no longer intent on pursuing its rivalry with the "capitalist" countries and on proving the "superiority" of its political system through gain in power. The USSR actively competes, however, with other advanced nations in the world market. Some further reforms (including a major "liberalization" of the collective farm system) are undertaken to improve the nation's economic efficiency. The Soviet Union is distinctly "mellowing."

The international law that evolves in the process is essentially "Western," although it incorporates some of the influence of the less developed countries (LDCs) and certain principles advocated by the USSR. The Soviet Union abandons its claim to a separate "socialist" system of international law.[9] International law gives increasing recognition to entities other than nation-states as its subjects; they include multinational corporations (MNCs) and the law increasingly extends to individuals.

Although the United States and the USSR still possess great military power, "bipolarity" as a characteristic of the international system has lost much of its old meaning. Antagonism between the two superpowers has practically disappeared. Pluralization of power, in the world at large and within the superpowers, appears to be by far more prominent than its polarization or centralization. The existing international system is basically stable, although pockets of instability exist, and new ones arise and eventually subside. World stability is significantly strengthened by the growing influence of worldwide and regional functional organizations, by MNCs, and by bilateral cooperation of the United States and the USSR in trying to reduce military conflict as well as social instability. China is no longer isolated as she was in the 1960s. She has improved her economic position and largely abandoned her militancy to the West, but she maintains a strong military posture and is not reconciled with the USSR. China is a factor in the Soviet desire to continue to cooperate with the West. Although the functional agencies of the U.N. family have grown considerably, the United Nations as a political body has not gained in influence.

The Decline of the United States and Western Europe and the Rise of Soviet Power

The toll of the energy crisis has not exactly been negligible for the United States: It has tended to drain the nation's resources, contributed to inflation, necessitated a relaxation of environmental standards, and imposed restrictions on economic growth. The United States increasingly feels the problems of the postindustrial society: the growing costs of cities and pollution; popular recoil from the highly complex, computerized environment; increasingly available leisure without meaningful content; new flare-ups of unrest at the universities and in urban centers. America's vitality on the international scene declines, and she becomes inward oriented.

In comparison with the United States, Western Europe is much more severely affected by the impact of technology. In fact the scenario that eventually evolves is that of a technological-trends-resistant Western Europe, described in chapter 6.

While the United States becomes increasingly inward oriented and Western Europe is plagued by instability, Japan, on top of the severe blow dealt by the energy crisis, encounters the difficulties of the postindustrial society, suffers from internal dissension, and loses the momentum of growth of the 1960s and the early 1970s. The Soviet Union fares much better. After some further difficulties with technical efficiency and productivity in the nonmilitary sector the Soviet Union succeeds in solving most of these problems by 1985. This is accomplished principally with the help of additional reforms, gains in the use of computers, and imports of technological know-how from the United States and Japan.

Scientists and engineers make up most of the top Soviet leadership. While keenly sensitive to the potential of technology, they have been assimilated into the power-oriented milieu of the Kremlin and develop a technological strategy of both national and global dimensions. The Soviets capitalize on key sectors of advanced technology, such as fusion energy, magnetohydrodynamic (MHD) power, computers, lasers, marine technology, and weather modification and are successful in their development and application. The requirements of large-scale technology help the Soviet Union in integrating its Eastern European satellites through the framework of COMECON; the latter is also conducive to mobilizing resources for large projects essential for taking advantage of the new technology. COMECON extends its technological and economic programs westward. Soviet-originated gas and oil pipelines and high-power electric transmission lines—the agreements on some of which were concluded as early as the late sixties and early seventies [10]—increasingly entwine Western European nations. Gradually COMECON expands its membership and establishes associate membership for states (including LDCs) that meet certain political and economic qualifications.

To be sure, the Soviet Union faces some problems of the postindustrial society—popular insecurity because of rapid change, manifestations of discontent and restlessness in urban populations, and some student unrest. The Soviet leadership either thwarts or successfully copes with these problems by the expansion of social security, strict legislation to enforce discipline in places of employment and in the universities, development of new instruments of social control with the assistance of social and "hard" sciences, and resort to repression.

In general, Soviet power is rising. While still cautious and subtle in using available leverages of power in some areas, the Soviets are becoming bolder and more aggressive in others.

The USSR generates more aid to the LDCs and strengthens its influence there. It also makes inroads into Western Europe through MNCs, cooperative programs in science and technology, and use of diplomatic, psychological, and political leverages to change internal conditions in selected European countries. The Soviet Union has a number of major cooperative programs in science and technology with Japan.

Initially the cooperation with Japan is beneficial to the Soviet Union because of the transfer of "conventional" technology from Nippon. Later, as cooperation extends to "big science" (outer space, oceanology, etc.), Japan becomes increasingly dependent on the Soviet Union, especially in view of her internal problems and the limitations of her resources.

Soviet military power has grown considerably. The Soviet nuclear arsenal greatly exceeds that of the United States, and the Soviet Union develops a superior navy. Officially the Soviet Union still pays lip service to "peaceful coexistence" with the United States, but now that Soviet internal economic and technological problems have been largely solved, the USSR is less careful about jeopardizing its relations with America. Leaning on a strong military base, Soviet diplomacy grows bolder. Using a combination of economic and military aid, political and subtle military pressure, and propaganda, the USSR succeeds in establishing influence in Latin America. The Brezhnev Doctrine has been revived and now resembles the Monroe Doctrine as it was applied in the early decades of the twentieth century. It has become a semilegal concept, increasingly accepted by a number of states. The doctrine is invoked as a warning to the United States when a number of senators call for a strong U.S. stance in support of "anti-socialist" elements in Argentina. In the face of U.S. internal problems, the lethargy of the American public, and the power of the USSR combined with Soviet avoidance of blatant provocations, the United States acquiesces to these developments.

The new international system that evolves in the wake of the changes in the world distribution of power is not radically different from its predecessor, but there are significant variations. It is still characterized by bipolarity—in the sense that the United States and the USSR are clearly the two foremost military powers on the world scene. However, unlike the case of the 1950s and the 1960s, world influence strongly gravitates toward the Soviet Union rather than the United States, with new axes of relationship established between the USSR and various subsystems, especially those involving the LDCs.

The Soviet Union is successful in using international law as an instrument for gaining influence in the less developed world. Tiers of relationship have been established with many LDCs whereby, in varying degrees, these states are influenced in the direction of joining the "Socialist" system of international law. Operating from a strengthened power base and an international climate generally more favorable to it, the Soviet Union is more effective in employing its version of international law as an instrument against "capitalist" nations. Within the "Socialist" camp, the web of treaties and regional organizations plays an important role in maintaining a system akin to federalism, in which the USSR enjoys unquestionable primacy. Sino–Soviet rivalry continues, but China is having difficulties in developing her economy, and the Soviet Union does not experience serious problems in containing Chinese power.

There is considerable instability in the political, economic, and technological spheres. With its growth in economic and technological power the Soviet Union gains confidence in the success of its system, and, given favorable circumstances, it does not hesitate to translate its strength into specific political gains. The Soviet activity attempts, however, to avoid producing major crises and does not usually trigger a concerted response.

Technology and Regional Integration

Technology also affects world society and the international system through regional integration. Regional integration modifies forms of conflict in a given region, makes conflict less violent, and thus contributes to international stability (provided, of course, that integration of a particular region does not lead to conflict on a broader scale, between various regions); it also changes the world distribution of power.

Even with the vastly growing importance of large-scale technology, technological factors are not sufficiently compelling to induce regional integration in the absence of a favorable political climate. Thus, whereas cooperation in technological projects is entirely conceivable between the USSR and Western Europe, regional integration between them is not likely. The only area where technological factors may play a major role in regional integration, with important consequences for the international system, is Western Europe. Whatever degree of integration is achieved in other areas as a result of technological cooperation might, to a varying extent, contribute to international stability by diminishing the outlook for intraregional conflict.

As discussed in chapter 6, technologically induced integration is likely to increase in Western Europe in the next decades, but it is more likely than not to fall short of political unification. If so, a tripolar world is not a strong possibility in this century. However, the pooling of resources by Western European nations to exploit advanced technology would inevitably involve some political decisions on how the newly acquired technological capability is used; to that degree the integrative process in Western Europe will contribute to the erosion of bipolarity. The interplay between the process of integration and the necessity for making political decisions with respect to the existing technological capability will eventually reach a point (perhaps about 1990) when Western Europe will have to decide between political unification and not taking full advantage of the advanced large-scale technology of the time. If Western Europe chooses unification, the resultant tripolar world may not necessarily be more stable than its predecessor. However, if Western Europe decides not to unite and thus permanently relegates itself to a second-rate status in the use of advanced technology, it may develop serious internal instability having global repercussions.

Technology, Functionalism, and World Society

Whereas regional integration can create a new political unit out of several nation-states, it does not fundamentally alter the present international system based on the existence of, and interaction among, nation-states; the new entity is simply a larger state that supersedes a number of smaller ones. It is therefore legitimate to go a step further and inquire whether or not the impact of technology in the next decades is likely to alter some of the fundamental principles on which the nation-state system is based. Since technology is a functional (as distinguished from political) factor in international affairs, it might be helpful to discuss it with reference to functionalism as a concept and the extent of its effectiveness in the world forum.

David Mitrany, the principal exponent of functionalism,[11] envisaged the rise of "nonpolitical" international institutions in the economic, social, and humanitarian fields that, by fulfilling basic needs of societies, would expand horizontally (as distinguished from the basically pyramidal structure of the state's authority) and would gradually take over many of the functions currently carried out by nation-states. In this way the functional international institutions would narrow the scope of the state's authority and, eventually, undermine and do away with the nation-state.

Actually, at least since the theory was first advanced in the early 1930s, the form of functionalism Mitrany espoused has not proved to be so effective a force as its theory suggested, especially when it conflicted with political realities.[12] One particular aspect of the theory of functionalism is of relevance to our analysis of the future: The theory did not address itself to the question of which functional areas—economic, social, humanitarian, or technological—were most promising in making functionalism work, and this neglect deprived its adherents of concrete guidance and a sense of priorities in terms of instruments for bringing about the world they would have liked to see. Mitrany viewed functionalism as creating a broad, rather undifferentiated web of relationships producing an impact as a whole, whereas, in fact, political implications for the international system of economic and technological activities might be considerably greater than, say, the humanitarian activities of an agency like the International Red Cross. Technology in particular received relatively little attention in the functionalist approach.[13]

Certain characteristics and trends in technology appear to have increased markedly its importance as a factor in functionalism (see chapter 3). This, in particular, is due to phenomena such as the integrative trend of technology, the requirements of large scale, the large costs of technology's utilization, and its rapid development. In conjunction with economic considerations and forces, technological developments might significantly strengthen the validity of certain forms of functionalism in the following decades. However, to understand the potential impact of functionalism, it must be examined in the context of its evolution as a practical factor in international affairs. This brings us to the consideration of functionalism as a dependent and as an independent variable.

Mitrany essentially viewed functionalism as an independent variable, having a dynamism and logic of its own and, in the process of its development, producing a political effect. Functionalist theory made little, if any, allowance for the possibility that the state itself would, as a matter of policy, use functional agencies as a dependent variable to achieve political goals.[14] On the contrary functional agencies and state authorities were expected to cooperate when faced with pressing functional needs, and the political objective of curtailing sovereignty and cementing the world was to be achieved, in the last analysis, *against* and *in spite of* the state. In practice, the United States began, after World War II, to adopt some forms of functionalism as a part of its policy, since the political advantages it offered—for example, greater stability in the world, the possibility of "mellowing" the Soviet bloc and of molding the value system of the less developed world—appeared to outweigh possible disadvantages, including restrictions on America's sovereignty.[15] This policy has not always been adequately articulated; at times it was hesitant, not uniformly consistent, and lacked high priority in terms of resource allocation, but it did signify an effort to use functionalism as a dependent variable in pursuit of a particular international system however hazy its vision was.

The experience of the past decades had also revealed two other characteristics of functionalism important for the assessment of its relevance for the future:

1. Although functionalism has proved ineffective against strong political forces (e.g., the East–West conflict during the Cold War, the new nationalism of the former colonial areas), it has been able to affect political developments once a certain minimum of favorable political climate existed (e.g., East–West relations in a period of détente).[16] Thus, in the last analysis, political considerations and factors have to lead the way for functionalism by establishing an environment wherein the interplay of functional and political factors can take place, with functional forces translating themselves into politically consequential results.

2. Functionalism, more as a dependent than an independent variable, has turned out to be more important on the subsystem than on the global-system level. The minimum of favorable political climate necessary for functional forces to produce political results was much more readily found on the subsystem than on the global-system level, and states used functional instruments to achieve political goals in such favorable settings. The United States resorted to the Marshall Plan to achieve recovery of Europe, to stimulate its integration, and to strengthen U.S.–European relations within the Atlantic Community subsystem. The Soviet Union uses functional instruments within COMECON to increase its ties with the satellites. Func-

tional forces (economic and technological), acting as a dependent, "semidependent," and independent variable, are important in effecting integration within the Western European subsystem. The Organization for Economic Cooperation and Development (OECD) is another example of functionalism at work on a subsystem level. In fact political considerations and rivalries themselves stimulated functionalism on the subsystem level, as, for example, when the East–West conflict energized the United States to use foreign aid as an instrument of policy directly and also indirectly within the framework of the United Nations. It is not usually clear, however, what the relationship is between a specific application of functionalism on the subsystem level and the evolution of the international system, and this consideration is seldom a matter of a purposeful policy.

Keeping these trends in functionalism in mind, we can now address the role of technology within their framework. There are three possible vehicles through which technology may produce significant impact on the future international system: (1) specialized international institutions dealing with certain aspects of science, technology, or their impact; (2) the MNC; and (3) marine resources development. More specifically the following questions can be raised: Will the technological impact necessitate the growth of specialized international institutions that will materially encroach on the authority of the state and thus reduce its significance? Will the further growth of the MNC produce an authority that, along with other international institutions, will become an important factor in changing the present international system? Does the development of marine resources—which opens up some 70 percent of the earth's surface to human exploitation, with a strong possibility for internationalization of vast areas of the seabed—eventually promise a major modification of the nation-state system?

Global Functional Institutions and World Society

Looking into future technology, one can envision at least four areas that will require control or regulation on a global basis: activity endangering global ecology, weather and climate modification, outer space, and marine resources development outside national jurisdictions. There is little doubt that, ideally, problems generated by each of these areas, as well as benefits to be derived through international regulation of them, are truly major and would best be dealt with on a global scale by an appropriate international specialized agency. However, the establishment and functioning of such a specialized agency would encounter serious political obstacles, and it would be an oversimplification to condense these obstacles into a single term: state sovereignty.

A salient feature of the present world society is its evolving a political process of its own that, in a number of respects, is not unlike what exists within nation-states and that is gaining recognition in the international relations literature under the term *transnationalism*.[17] Whereas the nation-state is a highly important unit in this process, other important participants and interests determine the outcome: bloc interests (such as the North versus the South); specialized agencies of the United Nations (which have vested interests and political consciousness of their own); bureaucracy of the U.N. secretariat; and, perhaps less directly, various economic groups (e.g., the MNCs) and government agencies. The establishment of a major international specialized agency to deal with any of these four areas would have to survive the realities of this transnational political process. Whereas nation-states are reluctant to establish specialized agencies, effective opposition also frequently comes from the bureaucracies of existing functional organizations, which are concerned that their own organizations' power might be curtailed by the new entity.[18] In short, international functional activities themselves have become politicized.

Another significant factor is that the evolving international decision-making process is not mature enough to solve many problems as a matter of administration or management. The problems that, within nation-states, would normally be viewed as purely administrative or managerial are still treated largely as political or legal [19] in world society, with resultant delays and slowness in the decision-making process. In turn this creates reluctance to establish international institutions to cope with functional problems that require expeditious decision making and effective management. In the areas where international institutions do exist, the interested parties are hesitant to refer important problems to them for solution.[20]

As a result of all these factors there has been a tendency to relegate highly pressing technology-created problems of global significance to the subsystem level. The case of nuclear weapons has been "solved," however imperfectly, within a U.S.–USSR subsystem (with others, like France and China, eventually joining) through an elaborate deterrence structure. Some secondary aspects of the problem were referred to the global-system level: the Non-Proliferation Treaty, verification through the International Atomic Energy Agency, and so forth. The tendency to solve compelling international problems created by technology on the subsystem, rather than the global-system, level is likely to persist for decades. There will, however, be important differences. Although the incidence of cooperation for control and problem solving in science and technology will fall into the subsystem level (mainly regional, but in some cases bilateral) [21] at the expense of the narrowly national level, cooperation and control on the global level will also grow in importance, although less so than on the subsystem level. It will be increasingly recognized that this cooperation in certain technology-related areas is a necessity, not a luxury.

It is thus likely that specialized agencies with worldwide scope will grow in the future. The case of the environment suggests a possible pattern. In response to the recommendations of the U.N. Conference on the Human Environment of June 1972—more generally known as the Stockholm Conference—the General Assembly of the United Nations established a separate agency to deal with the environment, the U.N. Environment Program (UNEP). Consisting of a Governing Council of 58 nations and an Environment Secretariat headquartered in Nairobi (a concession to the LDCs), this organization has limited funding—a total of $100 million for the first five years, raised through voluntary contributions by governments.

The function of UNEP is double: coordinating and catalytic. More directly, it is expected to coordinate environmental activities of the existing U.N. specialized agencies, but through some of its programs (like Earthwatch) UNEP also provides a framework for coordination and synthesis of activities of relevant governmental and nongovernmental organizations. UNEP's coordinating role is relatively weak. As the international environmental field keeps evolving, it is entirely conceivable that certain important global environmental activities may coexist with UNEP and not be subject to its coordination, while some environmental organizations on the subsystem level will remain largely independent from UNEP.[22] In its catalytic role, UNEP stimulates international organizations and national governments to undertake action of major importance to the environment. In practice, a number of important international environmental activities were initiated independently from UNEP.

A central activity of UNEP is the monitoring of environmental conditions in the oceans, the atmosphere, and key ecological systems. This is to be accomplished through Earthwatch, a program that would include some 100 monitoring stations distributed throughout the globe. Most of these stations—especially those in advanced countries—would be funded nationally, but those LDCs that do not have requisite facilities and cannot support them would be assisted by the Environment Fund. Thus the total cost of the Earthwatch network would run into many millions of dollars, but the contribution of the Environment Fund would make up only a part of the total.[23]

Communications satellites and Landsats gave a boost to a potential United Nations role in outer space.[24] However, a U.N. specialized agency in this field is not likely to be established until sometime in the 1980s, if then. A similar pattern, with delays and reluctance to delegate functions—even coordinative, let alone those involving true authority—to the global level, is likely to prevail in other fields.

In summary, global functional agencies will be emerging and worldwide activities within the framework of existing specialized agencies will be increasing, but their growth will be relatively slow and intertwined with national and subsystem organizations and activities. A substantial part of authority and actual control over resources and facilities will be maintained on the national and subsystem levels. However, in varied degrees, the global activity will impose constraints on the nation-state. The extent, scope, and relative influence of the activity of global institutions would depend, not only on the impact of technology, but also on political factors; for example, a strong pressure of the LDCs for certain specialized agencies relevant to their problems might compel advanced nations to act. The extent and influence of global specialized institutions would also depend on the world distribution of power: If the Soviet Union rises in power, Western Europe significantly declines, and the United States withdraws, the influence of global functional institutions will be much smaller than the reverse scenario. In any event it is not likely that powerful, technology-oriented functional international agencies will arise in the next decades to rival the authority of the state and the political institutions of the United Nations.

The Multinational Corporation

The MNC is basically a product of modern technology, although its origins can be traced to the nineteenth century. Modern communications and transportation, and, lately, computers have made it possible for a single enterprise to operate on a global scale. The combination of superior technology and managerial skills provided the economic advantage necessary to establish subsidiaries in foreign countries with promising markets or raw materials. The growth of the MNC was facilitated (if not necessitated) by the requirements of modern technology for large scale and large costs.

The significance of MNCs to the future international system lies in the scope of their development and the rapidity of their growth. The figures on MNCs are not precise, but available estimates provide a reasonably satisfactory picture of their magnitude. According to U.N. estimates, total world output of MNCs produced abroad amounted to $330 billion in 1971.[25] The U.S. Department of Commerce estimated that sales by majority-owned foreign affiliates of U.S. companies totaled $291.5 billion in 1973. Their average annual growth between 1967 and 1973 was 17 percent.[26]

The growth of MNCs led some students of international relations to envision a major change in the structure of the international system.[27] The importance of MNCs should not be exaggerated, but their rise does affect the international system in at least four ways. They stimulate economic growth of the areas they operate in and thus potentially contribute to such divergent political phenomena as social stability or, depending on local conditions, instability and to change in the world distribution of power. The MNCs, motivated primarily by profit calculations, facilitate the impact of technology as an independent variable. In part MNCs themselves act as an independent variable and thus tend to curtail state sovereignty. Lastly they contribute to global pluralization of power.

The MNC assists the transfer of technology from one geographic area or nation to another—in some cases, against the explicit wishes of the parent government and in circumvention of the laws then in effect.[28] When so transferred, the particular technology becomes an independent variable with regard to the originating nation. Inasmuch as such transfer of technology is not controlled on a worldwide scale in accordance with a broader societal design, it exerts an impact, for good or for bad, as an independent variable on the world at large.

Through their economic activities MNCs produce an impact that can have important political consequences. They commingle human and material resources of many nations and formulate problems and solutions with little regard for national boundaries. While tying economies and policies of advanced countries more closely together and restricting their means of maneuvering against each other, MNCs also increase tensions and areas of conflict. They may bribe foreign nationals to facilitate sales, which may lead to government crises (as in Japan in 1976), or they may energize European governments and companies to counter "the American challenge" of U.S. affiliates abroad.

The impact of MNCs has further restricted the areas that traditionally have been under state sovereignty. However, as a reaction, individual nation-states (especially LDCs) may well resort to nationalization to reassert their sovereignty.[29] Conversely, major powers might effectively use MNCs to promote their national interests.

Efforts to establish international mechanisms for controlling or regulating MNCs have been slow in bearing fruit. The principal reason is that states find it difficult to agree on a common position on this subject. But work in this direction continues on both the global and the subsystem level. The United Nations, OECD, the European Community, and the Organization of American States, among others, have been involved in this activity. Their concern has been focused on formulating a code of conduct for MNCs and on requirements for more extensive public disclosure by them of their activity and financial data.[30] These organizations are not systematically concerned, if at all, with the total impact of MNCs on the evolving world society.

In the context of world society it would be a gross oversimplification to characterize the rise of MNCs primarily in terms of their relationship to the state and its authority. To the extent that the activity of MNCs produces an impact as an independent variable, it imposes constraints on other actors in world politics. The MNC is thus but one new element—important as it is—in the pluralization of international power. By contributing to this pluralization the MNCs—which themselves are pluralized, not having a unified will of their own—contribute to a profound change in the international system.

Marine Resources Development

Marine resources development could be readily covered within our discussion of international institutions and the MNC, except for its one important aspect: the prospect for internationalization of the seabed outside of the national jurisdiction. It suggests that a potentially huge area subject to an international regime and involving exploitation of natural resources may be established.

Under the auspices of the United Nations and in various committees and conferences, the international community has been considering a future regime of the oceans for nearly a decade. Finally, at the Geneva Law of the Sea Conference (March–May, 1975), the various individual items have been pulled together into an Informal Single Negotiating Text (ISNT), resembling a draft treaty and changed in 1976 to a Revised Single Negotiating Text (RSNT).[31] Several provisions remain to be clarified and agreed upon, but at least there is a prospect that agreement on a treaty covering the future regime of the oceans may be reached as early as 1977.

The RSNT provides for territorial waters of 12 miles and a national economic zone, subject to the jurisdiction of the coastal state, extending at least 200 miles beyond the territorial waters. It is possible that, in the final version of the treaty, the economic zone may extend to the outer edge of the continental margin, if, in a given case, this margin goes beyond 200 miles.

An International Seabed Authority (ISA), whose jurisdiction would extend over the international area (i.e., the residual seabed outside of the national economic jurisdiction), would be established. No consensus has emerged on the authority and function of ISA—it may conduct deep sea mining through its proposed operating arm, the Enterprise; it may issue licenses (at a fee) to national companies to do so; or it may collect a percentage of revenue from enterprises engaged in the exploitation of mineral resources of the seabed. The LDCs have a clear majority vote in ISA. The RSNT includes provisions for promoting the transfer of appropriate marine science and technology to the LDCs and for their effective participation in the activities in the international area. Moreover, the needs of the LDCs are to be given particular consideration in the distribution of economic benefits derived from the international area.[32]

The establishment of an international regime of the seabed by itself may not necessarily be significant. What is important is the exploitation of resources within the international area on a scale sufficient to produce a meaningful political impact. It would principally depend on the economic potential of the international area.

The most important mineral resource of the deep ocean areas (the areas expected to be internationalized) are manganese nodules. The nodules contain potentially attractive quantities of manganese (up to 41 percent), copper (up to 3 percent), iron (up to 26 percent), nickel and cobalt (up to 2 percent each). Only a few years ago their exploitation appeared remote, but the rising cost of raw materials and improvements in technology greatly increased their economic potential. According to the U.S. National Oceanic and Atmospheric Administration, manganese nodules are likely to be commercially exploited about 1980.

Since the nodules lie on the bottom of the ocean floor, they can be mined from considerable depths by dredging or by suction. Their total reserves—especially in the Pacific—are vast. It has been estimated that if only 10 percent of the nodule deposits prove economic to mine, the many metals they contain will be sufficient for thousands of years at the present rate of consumption.[33]

Other sources of metals from the deep ocean area are much less known but no less intriguing. An example of potential further discoveries is provided by the sediments on the bottom of the Red Sea. These were formed by currents of hot (up to 56.5°C [134°F]) brines whose origin has not been determined with certainty, but the sediments contain large percentages of metals. They are located at a depth of about 2000 meters (6580 feet) and may be as thick as 300 feet (91 meters). Only the upper 30 feet (9.1 meters) have been studied in detail. It has been estimated that the gold, silver, copper, zinc, and lead in the upper 30 feet of sediments alone should be worth about $2.5 billion.[34] The sediments are soft and, in spite of considerable depth, mining would not present an unsolvable technological problem. But technology aside, exploitation of the Red Sea deposits is handicapped by a dispute about their ownership.[35]

On the assumption of an agreement on a national economic jurisdiction over the seabed of at least 200 miles, the Red Sea deposits will not be under an international regime, since the Red Sea is less than 400 miles wide. However, there are geological reasons to believe that similar deposits may be found in the deep ocean areas.[36] In fact geologically similar deposits, formed by hot brines carrying dissolved metals, were found in 1973 in the median valley of the Mid-Atlantic Ridge. Unlike those in the Red Sea, the Mid-Atlantic Ridge deposits are rich in manganese but very poor in iron, nickel, cobalt, copper, and chromium. However, scientists who discovered the manganese deposits believe that they are only "the icing on the cake" and that massive metallic deposits, including copper, may be found under them.[37]

Oil is not likely to strengthen the prospective international regime of the oceans in the near future. To be sure, it is possible—indeed, likely—that large quantities of oil exist in the deep ocean area,[38] and there is oil in the outer edges of the continental margin outside of the 200-mile zone. However, its exploitation is unlikely for some time, because there are substantial deposits of oil at lesser depths. This would not, however, critically diminish a potential political impact of the prospective internationalization of the seabed. If agreed upon in the late 1970s, an international authority will be collecting revenue from exploitation of manganese nodules perhaps in the early 1980s, and from other sources in later years. The revenue collected may significantly buttress the economic growth and influence of at least some LDCs.

The importance of internationalization of the seabed for this century should not be exaggerated. At most what is likely to evolve will be an international subsystem, embracing a large area but relatively few individuals and activities. It will exist along with other subsystems of world politics, but will not be one of the most powerful ones. What the emergence of this subsystem might mean for the next century is a different question.

The International System: Directions of Evolution

In concluding our discussion of technology and world society, it might be helpful to place it in a perspective of the evolution of the contemporary international system.

Until relatively recently the international system consisting of nation-states was basically "deterministic"—in the sense that it was a "given" and that there was no conscious and consistent will to steer it in a certain direction. If the system had a purpose at all, it was quite narrow; limited to the system's self-preservation, it was expressed in the concept and operation of the balance of power.

The Elements of Purposiveness in the International System

One of the earliest efforts to articulate a *directional* purpose for the present international system was displayed in the early decades of this century by people interested in international organization. Imbued by the interest in keeping international peace, they traced the emergence of the League of Nations and, later, the United Nations, to an evolution of international institutions that began with the Concert of Europe (1815), if not earlier. Characterized by the expanding regulation and structuring of the international system, with the achievement of international peace as its central goal, the evolution, presumably, would culminate in a world government. The concept of an evolving world government also obtained support from world-minded international lawyers who were seeking a structural support to the growing system of international law.

Both the theories and practical efforts aimed at world government received some support from, and in part were inspired by, technologic–economic forces that were integrating the world and transforming it into an interdependent geotechnical system. Universal functional organizations that began to grow as a result of increasing interdependence gave inspiration to a parallel evolutionary concept, that of functionalism. The functionalists were similar to their more politically inclined world government colleagues in that they viewed the international system as quasi-deterministic; though aided by the efforts of enlightened individuals, the system was to evolve naturally. In the case of functionalism, however, the nation-state was to succumb, not to a growing political authority of the international community, but to powerful functional agencies that would deprive the state of its *raison d'être* by taking away its role in the functional sphere.

Neither the world government nor the functionalist groups, separately or jointly, represent the dominant forces in world politics at present. However, they do reflect new elements of purposiveness that have emerged in the international system. This does not mean that the present international system is imbued with a clear-cut directional purpose to which it consistently and consciously strives. A good part of the evolution of the international system proceeds in response to the interplay of various forces that carry it into directions not guided by a central purpose, not clearly perceived, and not always foreseen. But it is also clear that the system's evolution tends to focus on two growing forces, one political and the other functional: the striving for peace (or, more broadly, international stability) and the satisfaction of human wants. In recent years still another force, also functional, began to take a distinct shape: maintenance of the planet's ecological system. These forces tend to endow the international system with a measure of purposiveness and direction, regardless of whether or not their end product will be the demise of the nation-state.

In the old international system, the political process was viewed as a zero-sum game; that is, one's gain was necessarily somebody else's loss and vice versa. The central and domineering position of (primary material) power in traditional world politics tended to accentuate the zero-sum approach: One nation's power has meaning only when compared to that of others. However, as international peace and stability began to emerge as a value in its own right, they marked an inchoate trend in world politics toward a non-zero-sum game. This trend did not become fully convincing until the spread of nuclear weapons. The superpowers realized that gain in power by one state could result in mutual suicide. Optimization of national security required cooperation in certain areas, not competition.

Whereas technology, through nuclear weapons, was instrumental in providing an impetus to the non-zero-sum game in the political sphere, its role became even more pronounced in the functional area. As, under the impact of internal demands of societies, satisfaction of human wants emerged as an important goal value in the international forum, science and technology offered an instrument for achieving a minimum desirable material standard—however defined in different societies—in a way not necessarily in conflict with, or at the expense of, other nations. This has been dramatically illustrated by the Green Revolution, which greatly increased agricultural production in a number of countries. Science and technology thus offer a new and important dimension to the international process, a dimension in which the principal competitor of national systems is nature itself, and not each other.[39]

Under the circumstances each state has the theoretical option to compete on its own—with nature as it were. But characteristics of modern technology being what they are—in particular its large scale and large costs—cooperation with other states emerges not only as a desirable, but also, in the case of most states, as a necessary alternative. Here the international political process is viewed primarily as a positive sum, in which every participant gains.

The backlash of technological advance, with its threat to the environment, also induces a positive-sum approach to world politics. In this regard each state has a common enemy who presents a danger, not only to itself, but also to world society as a whole.

Some scientists tell us that there is a reasonable possibility that, by the year 2020, the accumulation of carbon dioxide in the atmosphere will produce a greenhouse effect that will melt the polar icecaps and inundate the coastal cities. Still other scientists point out that the accumulation of particulate matter in the atmosphere will reduce the amount of sunlight absorbed by the earth and ultimately cause a new ice age.[40] Although one could take solace in the hope that the last two problems will be happily resolved by the two opposite effects' neutralizing each other, a more serious approach suggests that, the uncertainties about the exact timing and precise effects notwithstanding, nations will have no choice in the future but to develop joint policies and to pool resources for the prevention or solution of forthcoming problems.

The impact of science and technology, in both its positive and negative aspects, is likely to strengthen purposiveness in the international system, but this is far from saying that we can expect a millenium of international cooperation in the next decades. The concept of national power is changing; it is being modified by regional and other influences, but national power and struggle for power are still important—indeed, prevalent—aspects of world politics. Thus what we are likely to see in the next decades is a world politics that is a "mix" of a zero-sum and a positive-sum game, with the latter gaining in importance in interstate relations. However, the picture is getting complicated by the growing emergence of new actors on the international scene who are not necessarily pursuing a positive-sum game. Insofar as they contribute to the expansion of the totality of available resources, the MNCs contribute to the positive-sum aspect of the international political process. But they are also competing among themselves for certain limited resources and power; in some respects they compete with nation-states as well. They thus contribute to the international political process as a zero-sum game, as the functional international organizations do when they engage in competition among themselves and with other international actors. We therefore seem to be witnessing a paradox: While the positive-sum game is gaining in the relations among nation-states, the gain may well be at least partly neutralized in the increasingly politicized character of international activities.

The Nation-State and the Changing Pattern of Power and Authority

We are not likely to see a radical change in the nation-state system of world society in this century. One way to appraise the viability of the nation-state is to examine the functions it performs and to raise the question whether these functions could be realistically performed by other entities. These functions are basically as follows: (1) to protect human values and to help satisfy value goals; (2) to provide organization for change, in response to changing values or material, political, or other conditions; and (3) to provide organization for stability. In fulfilling these tasks, the nation-state requires loyalty from its subjects; indeed, loyalty itself has become a part of the value system of its population.

In none of these three functions is the nation-state completely self-sufficient and, under the impact of science and technology, it is becoming increasingly less so. However, the nation-states have shown a remarkable degree of adaptability to changing conditions by providing makeshift solutions (some of which have worked well and others less so) without sacrificing themselves to a supranational authority. Clearly the nation-states have taken calculated and grave risks in facing some of the most pressing problems presented by science and technology. The nuclear powers did so when they resorted to deterrence as a solution to their ultimate inability to protect the survival of their national substance. So far, however, the measures resorted to have proved workable, though far from perfect. Individuals who point to global ecology as necessitating a strong supranational authority might be underestimating the nation-state system's adaptability and resiliency, at least in the more immediate decades. They may also be overlooking the fact that many of the measures necessary to preserve global ecology will be undertaken at the national level for more immediate reasons—such as the reaction of the population against the proximate hazards and discomforts of the technological backlash. Internal measures will not serve as an adequate substitute for international activity and institutions, but they are likely to significantly reduce the necessity for granting global institutions a truly supranational authority.

Thus, of the three levels on which technology and its impact can be controlled—the national, subsystem, and universal—the incidence of control is likely to fall on the national and subsystem levels. Only what cannot be solved on these two levels will be referred—reluctantly—to the global level. This particular way of coping with the impact of science and technology may not be the ideal one, nor even perhaps quite satisfactory; but it is one that can realistically be expected in the remaining decades of this century.

I do not imply that the nation-state system persists primarily because of its clever adaptability combined with skillful efforts on the part of nation-states to keep their authority in spite of good reasons to the contrary. The problem is much more fundamental. Coming back to our three functions of the nation-state, we discern that there is no realistic substitute in sight in this century to perform the first—and the single most important—function: the protection of human values.

Although the world is being integrated by science and technology, integration is far from complete. There are vast divergencies in economic development, standard of living, cultural, and other characteristics among the various nations. Even if world society were uniformly permeated by advancing science and technology, there is no assurance that this would result in sufficient uniformity in values to permit a world government or some other institutional arrangement to supplant the nation-state's function in protecting group and individual values. For that matter even uniformity of values would not eliminate group interests and power aspirations likely to lead to interstate conflict. But especially in view of the divergencies in world society continuing throughout this century and beyond, the nation-state's function as the ultimate protector of values is likely to persist.

Since there is no viable alternative, value formation itself gravitates toward the nation-state. Some departures from this phenomenon should be noted. The "superculture" induced by science and technology does tend to enhance international values at the expense of nationalism, but it lacks a symbol to command loyalties. Certainly the United Nations has not so far succeeded in attracting loyalties of large numbers of people, and this is even much more true of its functional agencies like UNESCO or ILO. Furthermore the superculture embraces only a relatively small segment of population in the advanced countries, and it is less than skin-deep in the developing world. Whereas, under the umbrella of the nation-state, loyalties in the advanced countries are showing signs of change—some shifting to smaller groups and others reaching beyond national frontiers—nationalism is a vital force in the developing countries, which embrace two-thirds of the world's population.

Although the nation-state is still in a relatively strong position in terms of protecting the values of its population (on the assumption that deterrence works and the single most important value, human survival, is thus protected), the nation-states, in varied degrees, have to look to the international community to help them to satisfy value goals of their population. This, in particular, applies to the developing countries. All countries will have to resort increasingly to extranational arrangements to help them provide organization for change and stability. The trend does not, however, seem to be in the direction of centralization of authority and power in an international system; on the contrary it is in the direction of *pluralization* and *diffusion of power*, which is taking place inside the nation-states as well. The scope of state sovereignty is likely to decrease, but this is not inconsistent with the concept of the state as a variable (as distinguished from a constant) whose degree of authority and other characteristics are subject to change in response to changing conditions and forces of the evolving world.

The present trends can be summarized by a diagram (figure 10-1), dividing the totality of power and authority (including loyalty) in the world into four levels: universal, subsystem, national, and subnational. These trends run at crosscurrents, but, on balance, the subsystem level appears to be the single most important net gainer.

Figure 10-1. Changes in the Relative Incidence of Power and Authority

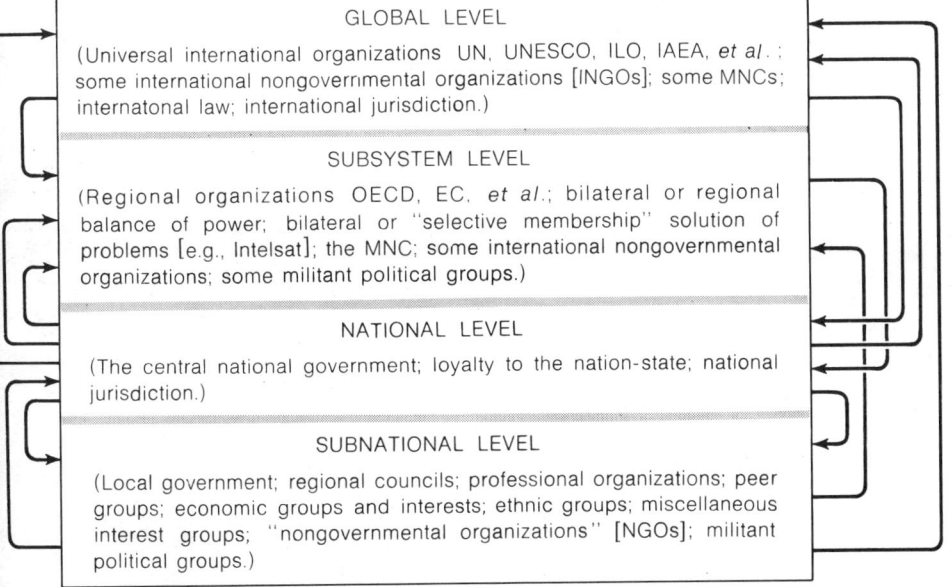

The preceding discussion provides a sufficient explanation of this diagram, except, perhaps, for the subnational level, which requires elaboration. The rise of importance of the subnational level, a relatively recent phenomenon, manifests itself in a number of ways. There is growing recognition that neither the central national nor the existing local government can adequately solve emerging problems of societies; thus regional councils, for example, emerge, which detract from the actual or potential authority of the national government. In a different category are the growing—and, at times, obstreperous—claims of various groups for a voice in decision-making on national resources and on other policies.[41] They tend to influence, curtail, and at times to deadlock, the ability of national authority to act.

American Policy and the Impact of Technological Realities

part 4

Concepts, Institutions, and Values

11

THE FOREGOING DISCUSSION of the potential impact of technology on the world distribution of power, international stability, and the international system has a number of implications for American policy. In particular the scope of the policy, the adequacy of America's institutional setup and decision-making process, and some specific policy problems are considered here.

Control of Technology and National Strategy

To meet the requirements of the following decades, American policy with respect to science and technology must comprise an integral continuum from the kind of technologies that might be developed at home to their possible impact on a remote part of the globe and on the international system at large. In safeguarding the future of America in a world permeated by technological factors and influence, the distinction between "domestic" and "foreign" policy with respect to technology is not "disappearing" or "blurred"; it simply no longer exists. The extent of world interdependence produced by modern technology is far-reaching. However, institutionally and procedurally, the nation is committed to the distinction between "foreign" and "domestic" policies and issues. For example a particular policy or issue is referred to one congressional committee or another (or one executive department or another), depending on whether it is classified "foreign" or "domestic." Whereas this procedural distinction will continue and whereas it may be essential in practice to decentralize policies designed to control the impact of technology, at least in the fundamentals of their conceptual design these core policies would have to be comprehensive and integrated.

For reasons discussed in the introduction no single nation can effectively control technology and its impact; such control can be achieved only by a central international machinery, but even then it cannot be complete. This suggests at least two criteria for American policy. As circumstances warrant the United States should be earnest and persistent in its efforts to subject technology to control (defined as the maximization of its benefits and the curtailment of its potentially harmful effects in the context of social purposes and goals) in cooperation with other nations. At the same time—and because of the existing limitations of the world political system, which, at best, can produce but incomplete results in controlling technology—we as a society must concentrate on improving our own capability in this regard. There can be little hope of achieving global control of technology if we are incapable of providing the requisite instruments of control within our own nation.

What might be the "ideal" model of activities designed to control technology and its impact? Political realities of our society being allowed for, such an ideal model cannot be completely achieved and function perfectly, but it might provide a useful pattern to strive for in the evolution of our institutions and policies for the future. Indeed some components of such a model have already evolved and are operational, while others merit greater emphasis or development. For completeness the entire spectrum of such a model policy is briefly discussed, even though some of its components are obvious or presently applied.

Where this cannot be effectively done by private enterprise, the U.S. government would have to assist in the development—or develop on its own—those technologies that are instrumental in meeting desirable sociopolitical goals, external and internal. If existing legal, administrative, or economic regulations (national, state, or local) inhibit the development or introduction of a particular technology beneficial to the nation, the government would be expected to act so as to remove, modify, or neutralize the restrictive effect of such regulations. Where technology is developed and applied by private enterprise and where this is beneficial to society, governmental policy would be confined to a careful watching of developments and to letting technology take its course as an independent variable.[1]

In still other cases a deliberate action will be necessary to restrict the development, spread, or application of technology. Such action may be necessary because some technologies are harmful to society or, on balance, more harmful than beneficial. Some examples are the spread of nuclear weapons, certain techniques of catching fish that are so effective as to threaten extinction of the species, and use of certain pesticides. A more complex case is presented by those technologies that are normally beneficial but in a particular application are injurious to the public or national interest. In those cases decisions would have to be made on the merits of each case.[2]

Furthermore the government must be prepared to intervene and, as necessary, assist the private sector in phasing out obsolete or obsolescent technology and thus minimize the constraints and rigidities imposed by an earlier technological development. This activity may involve assistance in phasing out an entire industry, if it is obsolete.

Lastly a policy would have to be developed to cope with undesirable effects of what is generally regarded as "normal" or irreversible technological growth. In this category can be mentioned the insecurity generated by rapid technological change, pollution, the problems of leisure, and some of the destabilizing phenomena of the affluent society.

All these elements of policy need to be synthesized at the highest level with a number of other fundamental policy objectives discussed previously, in particular: a healthy balance between international involvement and withdrawal and a proper balance between the national and the international interest for each particular stage in the evolution of world society. When so synthesized, "control of technology and its impact" would consist of an articulated, integral element of, and a subordinate entity to, national strategy defined as the aggregate of mutually reinforcing policies aimed at sustaining the nation's vitality and viability.

As we focus our attention on the problem of the control of technology and its impact—which, to repeat, means maximizing technology's benefits and minimizing its harmful effects in the context of social purposes and goals—it becomes apparent that what is involved is not just a matter of articulating a particular desirable policy or a sum total of desirable policies but a continuous development or adjustment of institutions, concepts, and values for this purpose. It is precisely this fundamental activity with regard to institutions, concepts, and values that will determine whether we shall control technology or be controlled by it.

The Rational and the Constituency Method of Decision Making

In an era when horizons for organized human will to mold the future through technology are broader than ever before and in view of the potentially critical changes in the nation and the world through science and technology, it is essential that American institutions and the national decision-making process be effective and responsive to the requirements of the time. This is not the case at present. The single most important reason is the weight given to the "constituency" method in the national decision-making process at the expense of the "rational" method.[3]

The rational method of decision making involves the consideration of all the relevant factors on a given issue or problem and a policy decision based on what is in the best interest of the nation as a whole. The interests of the various groups bearing on the subject would be given due consideration and, to the extent this is justifiable and compatible with the overall national interest, their claims and desires would be met. However, in the last analysis the decision would be based objectively on what is good for the nation as a whole and not on the power, vociferousness, or similar attributes of a particular constituency.

Among the principal advantages of this method might be mentioned its objective rationality and effectiveness in deciding and acting for the society as an entity. It is conducive to farsighted and comprehensive conceptualization and planning for the nation's future. Whereas in theory this method might come close to the ideal, in practice it has certain distinct pitfalls. The centralization of authority normally needed for this method can result in a stultifying system that kills initiative on the lower rungs of the hierarchy. What is in the national or public interest in each major decision is not easy to determine; the decision makers at the top can make mistakes, and the existing system may not provide adequate means of correcting these mistakes in a timely and equitable fashion.

The constituency method is basically a political one in which decisions are made as a result of the interplay among the various constituencies and interests. One advantage is that it tends to stimulate initiative, dynamism, and innovation within and among the various constituencies, since success in having the claims of a constituency satisfied comes with its dynamism and ability to compete. It ensures that the voice of a particular constituency is heard and provides a potential check against arbitrariness and mistakes at the top. Even though political, in practice this method is not completely devoid of elements of rationality; to be effective, a particular group has to justify its claims on objective grounds, and it would normally make an energetic effort to formulate its claim or proposal in a way that at least partially meets national needs.

On the other hand this method makes it possible for an influential constituency to sway a decision in its favor regardless of the merits of its case. It favors compartmentalization and discourages a comprehensive approach in the decision-making process, since each constituency tends to be primarily concerned with its own interest and sphere of activity. The balance of power process that tends to operate with respect to decisions involving allocation of resources between and among the various constituencies militates against an overall national initiative, since decisions can be more readily blocked or slowed down than made and carried out expeditiously. Under these circumstances decisions can be made only if consensus among all concerned is achieved, and consensus usually means deviating little from the status quo or reducing a given decision to the lowest common denominator.

As a result the decision-making process in a predominantly constituency system tends to be incremental, with only one small step taken at a time without necessarily relating such a step to an overall direction of the national effort and without assurance that logical follow-on steps will be made. This is so especially because a later regrouping in the balance of power among the various constituencies and changes in the motivations of some of them may significantly weaken the rationale on which support for the initial decision was based. Given such conditions, partial solutions of major problems and palliatives tend to prevail over real solutions. It takes a major crisis to make the constituency system move or to change the direction of the existing momentum (lethargic as it usually is), but even then there is no guarantee that the response to the crisis would be adequate.

Whereas the constituency system can work reasonably satisfactorily in a period of stability, its built-in static bias makes it highly inadequate in a period of rapid change induced by the swift pace of technological development. Furthermore the constituency method broadens the opportunities for technology to act as an independent variable and impedes the ability of society to subject it to rational control, since each constituency promotes those technologies advantageous to it and tends to overlook or even oppose broader societal interests. Another important limitation of the constituency system is that a certain important interest serving society as a whole may not have an articulate constituency and may thus be neglected, resulting in a detriment to the nation.[4]

Although the discussion above focused on the two types of decision making in juxtaposition, in reality they are combined, in varied degrees, between opposite ends of a spectrum.[5] The task, then, is not to choose one or the other, but to strike the most suitable balance between the two for a particular period of history. Moreover, since each method has its own advantages and disadvantages, a qualitative distinction is appropriate: For a given balance the best elements of each method must receive emphasis.

The theory of the U.S. government combines the rational and the constituency approaches to decision making. The President, who represents the interest of the nation as a whole, is expected to be guided mainly by the rational approach. The various executive agencies are, in part, expected to incorporate the rational method in their internal decision-making processes and in relationship to the Presidency, since they, too, have a responsibility to the nation as a whole. But they also act for their particular segment of the national decision-making process—interior, commerce, foreign affairs—and pertinent vested interests of the nation find their expression through governmental agencies; the activity of the executive departments is thus strongly influenced by their constituency roles. The Congress mainly represents the constituency approach, although individual Congressmen and the Congress as a whole may—and occasionally do—rise above the interests of their particular constituencies.

Because of the strong entrenchment of the constituency system within and among the various executive departments, in reality national decision making reflects most of the negative characteristics of the constituency method and few positive characteristics of the rational. A compartmentalized—rather than integrative—approach to present and future problems prevails, and, in overall policy, the tendency toward inertia significantly prevails over that toward initiative. The problem is further magnified by the sheer size and complexity of the government, with its inevitable bureaucratization and the never completely avoidable temptation to place the protection of the interests of the institution ahead of the purpose for which it was created. Some of the principal agencies concerned with science and technology are briefly discussed in this context.

Institutions and Characteristics of Decision Making

Military Technology: The Department of Defense

The custodian of military technology is the Department of Defense (DOD). In the national decision-making process DOD itself is a constituency that competes for resources for military science and technology with the civilian sector. Under Reorganization Plan No. 1 of 1973 the position of DOD in this regard was strengthened, inasmuch as the Science Adviser to the President no longer had jurisdiction over military technology, and the National Security Council (NSC) did not actually provide a rationalized balance between the military and the civil technological sectors.[6] With the restoration of the Science Adviser to the White House in 1976 he has been granted some restricted authority over military technology, but in reality the situation has changed little. The balance between the two sectors is still predominantly a product of the tug and pull of the various interests vying for resources throughout the U.S. government and of the pluralistic decision-making process (also heavily politicized) within the Executive Office of the President.

Within DOD the constituency system has been mitigated by the strategic and management concepts developed in the last 20 years; the budgetary pressure of recent years, which has forced DOD to make decisions in the overall interest of national security and at the expense of constituency interests (its own and those of individual services); and the increase in the power of the Office of the Director, Defense Research and Engineering, which curtailed some of the constituency interests of the military services.[7] Nevertheless the constituency system still runs strong in the Pentagon; it is reflected principally in the competition among the various groups within the services (which at times spills over into coalitions cutting across service lines) [8] and in interservice rivalry.

The present system in the Pentagon is paradoxical in that it is characterized by both inertia to change and a tendency to be unnecessarily technology-happy. Inertia to change is fostered by the existing balance of power among the various interest groups. It takes a major technological breakthrough—like ICBM or Polaris—to effect a significant change, but then a new constituency is established, and a new balance of power sets in with its inevitable bias for inertia. Given the reality of the self-sustaining balance of power and the reluctance of all concerned to upset it (except, again, in those rare and truly major cases like ICBM and Polaris where the advantages to a particular group or service are so great that it decides to take the risks of upsetting the balance), each group and each service nevertheless realizes that its survival—with regard to other components of the internal balance of power and an external enemy—depends on its competitive strength. It therefore seeks superior performance through improvements in technology. However, operating within the framework and constraints of the existing internal balance among the various interest groups, this pursuit of technology is predominantly incremental. Improvements are undertaken even if they are quite small and very costly and even if it is not clear why they are needed. In this regard the military groups are stimulated by the various manufacturers who attempt to sell their products. At times, however, the deadlocks are broken, or a particularly promising system is given higher priority, by intervention from above.[9]

The existing system could have worked reasonably satisfactorily at a time when resources were adequate to cover the entire range of military technology (and thus satisfy practically every group), but rapidly proliferating technology is vastly expanding that range. The domestic requirements of society and the need to develop nonmilitary technology relevant to national security are likely to curtail even further the resources available to the military. Consequently the Defense establishment will be able in the future to develop and procure only a relatively small segment of the then potentially available military technology. If, under these circumstances, national security is not to be seriously jeopardized, a more rational determination of priorities in the research, development, and procurement of weapons must be established.[10] The prevalence of the constituency system of decision making in the Pentagon is an impediment to this end.

Nonmilitary Technology: The Problem of Criteria

The rational approach appears to be even less prevalent in nonmilitary than in military technology. The characteristics of external threat provide at least some reasonably objective criteria by which the response in military buildup and priorities can be guided, imprecise as the process may be. But there is no single set of criteria to judge the development of, or priorities in, nonmilitary technology. The consideration of profit is one criterion, but it can be highly misleading if relentlessly pursued. Psychological considerations and prestige were factors in the development of some technologies (outer space). National requirements for certain services or products create a stimulus for the development of still other technologies (the mission of the Atomic Energy Commission [AEC] to develop peacetime nuclear energy was established in response to the need for adequate and inexpensive sources of electric power).

Under such conditions of multiple criteria and motives entrenched in institutions, the constituency system is understandably strong and is not readily responsive to change. But the organizational setup for nonmilitary technology requires particular attention because social changes through nonmilitary technology will be fundamental; its early impact—at a time when it is amenable to corrective modification—might not be readily perceived; and its control is likely to be difficult.

Technology and Foreign Relations: The Department of State

The federal agency responsible for America's worldwide interests in the most nearly comprehensive sense is the Department of State, but it has not been successful in providing effective guidance for using technology as an instrument of foreign policy or for coping with the global impact of modern technology as a function of foreign policy. The reason is broader than the case of technology alone; it finds its roots in the Department's nature as a constituency and in its relationship to all functional areas.

The Department of State is not a "natural" constituency but one *designated* to represent the interests of foreign policy in the national decision-making process. Unlike, for example, the Department of Agriculture, which represents the interests of farmers, interacts with them, and changes in response to their needs, the State Department does not have a potentially invigorating grass-roots clientele to be responsible to. Therefore, the Department has a higher-than-usual proclivity to get bureaucratized and ingrown. The conventional requirement to conduct foreign affairs does not appear to provide a sufficient stimulus for timely and constructive change in the Department's role as representing the interest of foreign policy in the national decision-making process.

Moreover, the Department of State does not control resources. This can be illustrated by a comparison of the budgets of State and Defense. For fiscal 1977, the outlays of the Department of State were estimated at $1.0 billion, most of them for administration of foreign affairs. For the same year, the outlays of the Defense Department were estimated at $101.1 billion, of which $10.4 billion alone was for research and development. The lack of control over resources has affected State intellectually and operationally. Intellectually State Department personnel have had limited exposure to, and hence limited appreciation of, the use of functional areas as instruments of policy. This deficiency could have been compensated for by appropriate training and recruiting policies, but such policies were not adequately developed. Thus, traditionally, the Department has had a tendency to view foreign policy as "diplomacy" rather than "national strategy," and its organization and policy followed along country or regional lines. When the changing nature of world politics began to place an increasing emphasis on dealing with resources and employment of resources as an instrument of foreign policy, functional bureaus—for example, the Bureau of Economic and Business Affairs (EB), the Politico-Military Bureau (PM)—were eventually established. Although the Department is gradually changing, the full potential of the functional approach in its outlook and in its activity is far from being realized. Quite apart from the intellectual impact of the lack of control over resources the Department is handicapped operationally in the functional areas inasmuch as they are largely run by other departments and agencies that have power of their own.

But even a fairly high degree of acceptability of a functional bureau in the Department (as, for example, appears to be the case with EB) does not necessarily imply effectiveness in using and applying a given functional field (economic assets or science and technology) as an instrument of policy. It simply means that personnel dealing with the field in question are no longer viewed as outsiders or "second-class citizens" by the rest of State. The progress made in the acceptability of functional bureaus at State being allowed for, regional bureaus and their mentality are still the most influential in the Department, and the bulk of prestige and desirable promotions are associated with them.

The Department of State is especially weak in science and technology. The old Bureau of International Scientific and Technological Affairs (SCI) was one of the youngest functional bureaus in State, and it was at or near the bottom of the acceptability ladder. Staffed largely by technical personnel, SCI focused its attention and energies on winning acceptance in the Department. Progress along these lines was made. However, when, in October 1974, SCI was reorganized into a Bureau of Oceans and International Environmental and Scientific Affairs (OES), the new and enlarged Bureau has inherited a number of problems which SCI did not succeed in solving or simply neglected. These, in particular, included the lack of a policy planning unit; the neglect of fostering appropriate analytic capability, concepts, and initiatives in science and technology policy and the resultant predominantly reactive operational orientation of the Bureau's personnel; the absence of effective leadership and control with regard to foreign activities of the operating technical agencies; and the still not fully resolved problem of acceptance.[11]

The Top Policymaking Machinery in Science and Technology

Although, in the last analysis, decision making at the White House level would be expected to overcome the compartmentalization of the various agencies involved in science and technology and exercise initiative in developing an overall rational policy with respect to the interest of the nation as a whole, in practice this standard of performance has never been achieved. The reasons for this are complex, but an adequate explanation can probably be provided by reducing the complexity to three major interrelated factors: the constituency outlook of the personnel in the White House apparatus for science and technology policy, the nature of its competence, and the relative power of the organization.[12]

The old White House machinery for science and technology policy—which principally consisted of the Science Adviser, his staff embodied in the Office of Science and Technology (OST), and the President's Science Advisory Committee chaired by the Science Adviser [13]—to a considerable degree represented the interests of a constituency, the scientific and technological community (the partiality for science being more strongly pronounced than that for technology).[14] The competence of the White House apparatus was predominantly on the scientific and technological side; it was weak in its understanding of social goals, of the potential role of science and technology in achieving them, and of the impact of science and technology on society. Thus, as a result of the combination of the constituency outlook and the nature of competence of the personnel involved, the advice, studies, and other products emanating from the White House scientific and technological machinery tended to reflect the views and interests of the scientific and technological community and its technical expertise. They fell short of constituting the comprehensive and enlightened science and technology policy the nation needed, a policy that, by its nature, had to go far beyond science and technology itself. When, in the winter of 1972–1973, the Magruder exercise and its aftermath gave some glimpses of the vast dimensions of the problem, the Administration recoiled at the magnitude of the task involved and let the matter rest.[15]

Although the Science Adviser and OST were nominally responsible for the formulation of science and technology policy, the ultimate decisions on policy were left to the White House political arena, and here the science and technology apparatus did not carry a great deal of weight. Both the constituency outlook and its somewhat narrow competence were important contributory, although not decisive, factors in limiting the relative power of the White House science and technology apparatus. The power of the former Science Adviser to the President was strongly constrained by the fact that a number of other agencies or offices in the White House itself—like the Office of Management and Budget (OMB), the NSC, the Council on Environmental Quality, and the Office of Telecommunications Policy—were involved in various aspects of science and technology policy over which he had no control. Above all the OMB had control over the purse strings.

The three principal factors that impeded the development of an overall rational science and technology policy by the White House science and technology apparatus were also largely responsible for precluding it from exercising effective leadership over the operating scientific and technical agencies in such policy matters as the determination of priorities and allocation of resources. Whatever leadership the Science Adviser and his staff exercised, it seldom rose above the function of primarily arbitrating among the positions of the agencies concerned.

Reorganization Act No. 1 of 1973, which transferred the Presidential science advisory machinery from the White House to the National Science

Foundation (NSF), was described in chapter 4. The change shifted power more toward constituency decision making by reducing the influence of the Science Adviser to the President and by giving autonomy to the military sector. In one respect the science advisory apparatus at NSF showed promise. Unlike the old OST, which took an Olympian attitude with respect to the operating agencies, the NSF apparatus adopted a more pragmatic approach to science and technology policymaking. Instead of looking at it from above, it injected itself into it, largely laterally. This approach provided opportunities to get direct access to influential personnel in the operating agencies and to bring desirable change from the inside.

But in doing so, the Science Adviser and his staff lost a clear perception of their role as representing the President and embodying the "rational" approach to decision making. This development, accompanied by the lack of a clear sense of priorities and effective leadership, did not exactly distinguish the period of the President's Science Adviser in NSF as a major historical landmark.[16]

On May 11, 1976, the President signed the National Science and Technology Policy, Organization, and Priorities Act of 1976 reinstating the science advisory machinery in the White House. As compared with the old White House machinery, this Act in some respects broadened the authority of the Science Adviser to the President, now referred to as Director of the Office of Science and Technology Policy (OSTP). At the same time the Act has made distinct concessions to an Administration that was not fully convinced of the utility of scientific and technological advice in the White House and was not ready to give it substantial power.

The appointment of the Director of OSTP and Associate Directors (not more than four) is now subject to advice and consent of the Senate, a requirement that, at least symbolically, strengthens the Office. Other strengthening features include the requirement for the President to submit an annual report to the Congress on science and technology and the stipulation that OSTP produce and update annually a five-year outlook on problems and conditions to be solved with the assistance of science and technology—a potential beginning of systematic planning.

The new White House machinery is distinctly weak in national security. The Direcor of OSTP is authorized to advise the President on scientific and technological considerations involved in national security, but he is not a member of the NSC (which still, in theory, oversees military technology) and can advise the NSC only on its request. In an era when nonmilitary technology is gaining in importance for national security, the Act thus did not provide for a comprehensive machinery overseeing both the military and nonmilitary sectors.

OSTP has not been given new authority over the technical agencies. However, it provided for the establishment of a President's Committee on Science and Technology, to function for two years (unless extended), to examine the entire field of the federal science and technology, and, as appropriate, to recommend organizational reforms. The Committee might thus strengthen OSTP through reorganization of the technical agencies. Moreover, there is nothing in the Act to preclude the Committee from examining the White House science and technology machinery itself. Accordingly, the present organization for science and technology policy has distinct overtones of being provisional.

In summary, the Act has created an organizational framework that is not ideal, but within it improvements over the past are possible. In an operational sense these improvements largely depend on the competence of the Director and his staff, the size of the staff, and its relationship with the President. In a broader sense the improvements will depend on the quality and acceptance of the President's Committee's report, including the organizational changes it may stipulate.[17]

Institutions and Values

In any system of government one can observe certain values associated with a particular institution that, in turn, have an effect on the decision-making process. This phenomenon is particularly pronounced in the United States, largely because of the importance that the constituency approach occupies in the national decision-making process. Constituencies inside the government have power; they view the decision-making process as a phenomenon involving competition with other constituencies; their attitudes and values are reinforced by frequently close associations with related constituencies outside of the governmental hierarchy; the appointive process itself, in part, recognizes the legitimacy of a constituency viewpoint by selecting people from a particular group or interest to fill certain governmental positions.[18] The clusters of shared values thus established and preserved within governmental institutions or their subdivisions or within influential interests outside of the governmental machinery tend to hinder the goal of subjecting technology to effective control by society.

The groups that are of particular relevance to decision making in science and technology policy are scientists, engineers, the military, industrialists, and diplomats. As participants in the national decision-making process, the goal values of these various groups may not differ a great deal, if at all, but their means values and attitudes vary significantly.

Although the values of the scientists vary in degree—depending on the duration and extent of their involvement in governmental decision making—by and large scientists tend to be internationalistic (science is international and serves mankind) and inclined to pursue science and technology for its own sake. Scientists favor free development of science and tend to believe that interchange among scientists will contribute to international understanding and world peace. Accordingly they are likely to support broad international scientific programs and to be reluctant to use science and technology as an instrument of unilateral national policy.[19] Thus, by and large, scientists prefer to see science and technology act as an independent variable; exceptions to this would be cases such as arms control and disarmament and those where technology's harmful effect is direct and obvious.[20]

Engineers are similar to scientists in that they are inclined to promote science and technology and regard this activity as good in its own right. However, having been usually reared in the corporate (and sometimes in the governmental technical) environment, engineers are much more sensitive than scientists to priorities and cost-effectiveness in support and implementation of technical programs. They are also much more instrumentally oriented toward science and technology—in the sense of using them as instruments for achieving a goal—than scientists are. Their appreciation of economic factors, but not necessarily of political and social factors, is better than that of scientists. As a result engineers typically exhibit an understanding of the impact of science and technology that is somewhat broader than that of scientists, but it is still distinctly limited. Many share business values, in particular the concept of *laissez-faire*.[21]

Diplomats are somewhat similar to scientists in that they are internationally minded and, when presented with a choice, are likely to support cooperative scientific programs involving other nations, to share the belief that they might contribute to better relations and peace, and to favor arms control. However, they are largely cautious, do not believe that their problems have a solution, are immersed in short-range palliatives, and are unlikely to appreciate—let alone initiate—bold, large-scale policies and programs in science and technology for social ends. Diplomats are basically not instrumentally oriented with respect to science and technology—in the sense of wanting to use it as an instrument of policy or national strategy—and thus, by default if for no other reason, would let science and technology produce an impact as an independent variable.[22]

Military officers, on the other hand, are instrumentally minded toward science and technology and tend to think in broad strategic terms. Their limitation lies in the fact that, charged with the task of being prepared for the worst possible contingencies, they favor tangible technological capability that they can control but tend to overlook the more subtle impact of technology and to have limited appreciation of, or concern with, nonmilitary technology except when it directly supports military power. Basically the military are nationally oriented and would be suspicious of broad international programs unless they are tied with the existing alliance structure. At their best the military can develop military technology in a timely fashion and apply it, in peace or in war, with measured precision and a keen sense of the requirements of national security. At their worst the military—with the aid of industrialists—can stimulate the spiraling of military technology as an uncontrolled variable.[23]

The values of the industrial world are of significance, not only because they are shared by the businessmen who occupy high positions in government and because industry is influential in Washington, but also because other individuals and groups tend to share business values (e.g., engineers). Although business values have undergone changes in the last 40 years and are being modified further (especially in response to the shortage of resources, ecological concerns, and other social issues), there are certain core values and attitudes that strongly persist.[24]

Business tends to distrust the government, oppose governmental control, and favor *laissez-faire*. It is not, however, opposed to governmental aid or subsidies—especially in science and technology—but would much prefer to see it unaccompanied by control, direction, or regulation. Business is strongly profit motivated, frequently at the expense of other values. Unlike scientists or diplomats, individuals possessing business values are instrumentally oriented with respect to science and technology but view them as potential instruments in a narrower context (e.g., corporate strategy) and not as a part of national strategy involving control of technology on a national or international scale. In general, business values facilitate technology's impact as an independent variable.

The values and attitudes discussed above seldom appear in their "pure" form in any particular individual or governmental office. In the process of interchange between and among the individuals belonging to these various groups a certain degree of modification of values and attitudes inevitably takes place. Some individuals succeed in rising above the values of a particular group and develop a broad and perceptive outlook conducive to the development of comprehensive and effective policy in science and technology.

However, by and large, and for the reasons discussed previously, these values and attitudes—a number of which are clearly obsolete in our scientific–technological age—tend to persist and exert an important influence on the decision-making process. They are frequently found in clusters firmly entrenched in a particular institution or a subdivision thereof that gives them organizational support. They thus tend to reinforce the constituency system and its built-in tendency to resist change.

In summary, the present governmental institutions and the values and attitudes that pervade them strongly militate against effective control of present technology, let alone that of the future. The predominance of the constituency system with weak rational guidance at the top is not conducive to the development of sufficiently comprehensive concepts for this purpose; for that matter it is not conducive to conceptual thinking in general. Even if comprehensive concepts for technology's control were evolved elsewhere—for example, the universities—they could not be implemented within the present institutional framework in the face of the double hurdle of the highly pluralized and compartmentalized power and the values and attitudes of many decision makers whose institutionalized capability for casting a veto is normally so much greater than that for exercising initiative of a sufficiently large scope to implement policies to control technological impact.

Toward a New Balance between the Rational and the Constituency Decision Making

The problem the United States is facing is fundamental; it is one of where to strike a balance between centralization and decentralization, between the rational and the constituency system. Each has its advantages; the question is that of a proper mix. To make our system more responsive both to internal needs of society and to external challenges, we need a shift in the balance in favor of the rational system while still preserving many attributes of the constituency system.[25]

Indeed modern technology permits a measure of centralization essential to move the nation without necessarily creating a Leviathan stifling individual and group dynamism. Once the political system assimilates technology such as computers, advance systems of information, and communications, it would be possible to combine greater centralization of priorities and control of certain critical areas of economy and technology with decentralization on the lower levels nearly instantaneously coordinated on a national scale.[26] In this way the nation will have the benefit of a more effective overall rational guidance and planning while a greater delegation of authority and decentralization on lower echelons of societal structure would further enhance the vitality and dynamism of our pluralistic system.

What measures might be taken to shift the balance toward the rational system of decision making with respect to controlling technology and its impact? Two principal measures can be differentiated: strengthening leadership in science and technology policy at the top and providing that leadership with appropriate conceptual and organizational instruments to ensure that it is exercised with a clear-cut sense of direction and effectively carried out.

The top science and technology machinery needs to be consolidated and given greater power. This can be achieved by giving the Science Adviser and his staff authority in their field at least as great as that of the Council of Economic Advisers (CEA) with regard to the economy and including military science and technology. In recent years proposals have been made that the Presidential science advisory apparatus should, indeed, consist of a three-member Science and Technology Council, on the pattern of CEA, rather than of a single individual.[27] A mere imitation of another organization is seldom successful, but in this instance one can find a rationale that exceeds the ostensibly imitative design:

1. No single individual combines the qualities necessary to provide the requisite high-caliber advice in science and technology policy.

2. As we have seen, the characteristics of scientists and engineers differ significantly. From the point of view of constituency representation, it would be desirable to have a scientist and an engineer on the Council, and not just a single science adviser.

3. Both scientists and engineers have important limitations in appreciating the more complex and indirect impact of science and technology and in designing appropriate means of coping with it. Moreover, scientists and engineers have not always been successful in articulating problems and issues and presenting them with appropriate emphasis to command attention.[28] Accordingly a social scientist with a keen appreciation of science and technology, of society's needs, and of the governmental process—familiarity with which is essential for designing and implementing solutions for societal problems— might be an appropriate addition as a third member. His presence on the Council would help neutralize a constituency image that might still linger around the Presidential science advisory apparatus, and it is likely to strengthen the rational (as distinguished from constituency) elements of the Council's advice.

An effective functioning of the Science and Technology Council would depend heavily on the right kind of staff. Because of the nature of problems involved and the agencies concerned, the staff would have to be both interdisciplinary and interagency. The staff, in part, would be pooled from the various principal organizations—like the State Department, the DOD, the NSF, the Environmental Protection Agency (EPA), the Energy Research and Development Administration (ERDA), and so forth. It would therefore be important that the staff members do not represent their respective agencies (and thus do not politicize their work). They should be selected on the strength of their ability, unbiased judgment, and deep understanding of social processes and goals.

The principal function of the staff would be to initiate or conduct comprehensive mid- to long-range planning for science and technology policy—an activity in which the U.S. government is distinctly inadequate.[29] One can distinguish two types of planning that the science and technology policy staff could profitably pursue: "substantive" planning and "instrumental" planning.[30]

Substantive planning would analyze alternative goals of science and technology policy and alternative courses of action to achieve them. This kind of planning cannot be truly separate from the determination of national goals; if the U.S. government revives the analysis of national goals abandoned in 1970, the two activities could significantly reinforce each other.[31]

Instrumental planning would deal with the means of achieving the ends or goals formulated by the Science and Technology Council. It would handle the implementation of the selected courses of action. Instrumental planning would consist of two types: methodological and administrative. The former would be aimed at the development of appropriate methods or conceptual tools (if they are not adequately developed) to carry out a particular course of action. For example they might include the development of conceptual tools such as the relationship between technology and economic growth or of statistical tools such as the present and prospective figures on the impact of a particular technology—like offshore oil technology—on the gross national product. Administrative planning, subject to a prior approval of a particular policy by the President, would assign specific tasks and steps to be taken to various organizations (e.g., the NSF, ERDA) or their subdivisions and would establish a system of reporting on the various phases of the tasks accomplished.

The science and technology policy staff need not conduct the planning described above all by itself; in fact, if it does, it would court the danger of developing into a bureaucratic monstrosity. Nearly all of the planning could be conducted by existing organizations or by *ad hoc* interagency study and planning groups. Some of the planning (especially methodological) could be farmed out to universities. The staff would, however, have to possess the requisite talent for the planning, the high level of expertise to evaluate it, and the authority to require reworking of inadequate planning. Moreover, to ensure that innovative thinking is not overlooked in the planning process and that the scientific and technological community is given adequate opportunity to contribute to the national decision making in science and technology, it would be important that the staff establish a network of communications with appropriate professional societies, universities, and industrial groups.[32]

An important question remains: the role of OES in international science and technology policy. It is essential for U.S. science and technology policy that its domestic and international elements be closely interrelated; this consideration speaks in favor of primacy of the White House machinery in this area as a whole. The Science and Technology Council would certainly have a better chance for establishing an effective leadership—including that related to foreign affairs—over the operating technical agencies. This does not, however, mean that OES could not play a constructive role in the initiation, planning, and execution of international science and technology policy and in ensuring that it is in basic harmony with overall U.S. foreign policy. Indeed a close relationship between the Science and Technology Council and OES might help in solving the latter's endemic problems within the Department, especially if such relationship succeeds in attracting bright foreign service officers (FSOs) into the Bureau.

Strengthening of rational decision making in science and technology policy at the top of the governmental hierarchy would not be sufficient in the absence of a similar development within the various agencies dealing with science and technology, such as the DOD, ERDA, NASA, and NSF. The development of conceptual instruments and more objective criteria for determination of priorities within each of these agencies would diminish the potentially injurious influence of pockets of obsolescent values and vested interests entrenched in these institutions. One of the principal maladies of our institutions is that programs and priorities on lower tiers of the institutional hierarchy are so entangled and complex as a result of incremental growth or parochial interests that true rationalization of many of our institutions is not possible unless it reaches to some of the lowest tiers of the bureaucratic hierarchy. A large measure of rationalization would thus depend on effective leadership. The head of each agency should be able to overhaul his own organization to ensure high caliber of key personnel, efficiency, and adherence to the basic purpose of his agency and to the guidelines coming from the Science and Technology Council.

If the decision-making process is thus rationalized within each agency, whatever position or claim on resources it develops would be objectively stronger. It would not be marred by headlines revealing cost overruns, waste, and inefficiency. High competence in key positions and objective selection of personnel—phenomena not exactly common in Washington—should allow substitution of quantity by quality, thus cutting down excessive bureaucratization. As the alternatives are brought to the top, they would be more clear-cut and not obfuscated by hundreds of parochial interests within each agency. The consequent strengthening of agencies as constituencies would thus not be necessarily undesirable. Such a development could only strengthen the democratic process where the ultimate decisions made by the President and the Congress are, in the last analysis, political.

The Home Front

12

I SHALL NOW EXAMINE in greater detail the concept of technological strategy and its principal components, and also discuss discontinuities in social and political trends that affect the viability and vitality of the United States.

Technological Strategy

There is a degree of controversy on whether or not the United States has a technological strategy. Some writers argue that it is possible to discern a pattern of practices and objectives that amounts to a strategy.[1] Although, with effort, one could discern a certain pattern, it is doubtful it could qualify for the term *strategy*. Moreover, since the late sixties, America's entire science and technology policy has been in a state of flux, and some of its new directions are but dimly becoming apparent.

In the light of these developments what might be the definition and guidelines for a technological strategy for the future? Technological strategy [2] can be defined as the comprehensive and systematic use of technology to enhance or maintain the nation's viability and vitality. It would involve such activities as the determination of priorities in, and promotion of, technology and technology-related education and skills in the context of domestic and international requirements; domestic and international distribution of raw materials and other productive assets; and political, economic, and technological trends of global dimensions. It would also involve the creation of favorable market or export conditions for products based on new technology. Concurrently, technological strategy would require a timely phasing out of obsolescent technology and modification of the existing technological superstructure to keep abreast with technological advance.

"Technological strategy" implies a certain structure, purposiveness, and a sense of direction. Its major advantage is that, because its components are formulated with a goal in view, they produce a synergistic effect—their total is greater than the sum of their parts. In a situation—such as in the United States at present—where the prevalence of the constituency system tends to deadlock and neutralize the various components of policy rather than give them a stronger sense of direction, the total is probably materially less than the sum of its parts.

The foregoing is not intended to suggest that a tightly structured technological strategy, relentlessly pursuing certain internal and external goals, would be a highly desirable or achievable objective. Although possible in a totalitarian society or a tightly knit one like Japan, it is neither practical nor advantageous in a democratic and pluralistic society like the United States. Indeed a tightly purposeful subordination of science and technology to social objectives might hinder scientific and technological progress. However, a model of technological strategy provides a useful criterion for comparing existing reality with an objective to approximate.

The analysis of preceding chapters suggests some guidelines for the delineation of a technological strategy for the United States. It has been pointed out that the spectrum of science and technology has become very broad and that advanced technology requires immense resources for its development and use. The traditional approach of the United States of developing the broadest possible spectrum of sciences and technologies is no longer tenable. Moreover, although one still hears cries—especially in connection with budgetary constraints of recent years—to the effect that "the United States must maintain its superiority in science and technology," the illusory nature of such an objective will become increasingly evident.

As a part of its technological strategy the United States will have to decide in which sectors of science and technology it will lead the world, which it will develop in cooperation with other nations, which will be borrowed from abroad, and which neglected. In this respect the United States will differ only in degree from other major industrial centers. These differences will nevertheless be important; the United States, if it chooses to do so, would still have the capability of maintaining superiority in selected key areas of science and technology essential to the nation's vitality, whereas such a superiority might not be an achievable goal for other nations. Moreover, as the Japanese demonstrated, in a number of cases effectiveness in the use of technology is more important than leadership in its development.

The foregoing implies that we need a new look at the allocation of our resources aimed at a careful delineation of priorities. The discussion of trends on the three levels of technology—the nuclear umbrella, the subnuclear-umbrella military technology, and the nonmilitary technology level—suggests useful approaches to this problem.[3]

Characterized by mutual neutralization of power, the nuclear umbrella does not represent a profitable area for enhancing America's strength in the future. This has been recognized in our strategic thinking: After a massive allocation of resources to this sector and the achievement of overwhelming superiority over the Soviet Union in earlier decades we have settled for "sufficiency." Insofar as this is consistent with maintenance of effective deterrence, it would be desirable to decrease further the allocation to resources to this area by such means as a disarmament agreement.

Even if further agreements, beyond those of 1972 and 1974 (Vladivostok), are reached in the Strategic Arms Limitation Talks (SALT), the nuclear umbrella is not likely to release truly large resources—the allocation of funds to strategic nuclear weapons constitutes only 7.4 percent of the military budget of $98.3 billion for fiscal year (FY) 1976. With the largest share of our military resources tied up in general purpose forces—32.8 percent in FY 1976—and with this sector displaying increasing symptoms of a stalemate, a rationalization of the allocation of resources to this area might be appropriate.

Unlike the case with the nuclear umbrella, the growing stalemate in the subnuclear military sector has not been generally recognized in our military thinking and in the allocation of resources. But even if it had been, this sector is too complex for seeking a solution by a single formula like sufficiency. It is obvious, however, that it would not be rational or realistic to respond to the growing projection of the Soviet military power in general-purpose forces (in particular the expansion of the Soviet navy) by seeking an overwhelming superiority (as we responded to the Soviet nuclear buildup in the 1950s). A three-pronged approach might be taken.

We may want to accept inferiority, indifference, or noncommittal watchful waiting with regard to certain geographic and technological areas.[4] The salient fact is that it is not really essential for U.S. national security—nor is it possible—to maintain a military advantage everywhere. Furthermore much of the current incremental technological advance in our weapon systems (e.g., the increase in the speed of aircraft or ships by a few knots) is not really essential and is very costly. Accordingly, resources from such geographic and technological areas would be either withheld or reallocated into areas of higher marginal returns.

In still other areas of the general-purpose forces sector we may want to settle for a stalemate as we did in strategic nuclear forces. Here resources would be applied in accordance with the concept of sufficiency. Insofar as consistent with a realistic appraisal of U.S. national security, allies would be introduced more extensively into our strategy and planning to determine how much of our own military power is "sufficient." (A more extensive reliance on allies, in fact, has been initiated by the Nixon Doctrine.) Moreover political and economic factors would be considered to determine how much military power is enough to maintain a reasonably effective stalemate. When a potential enemy initiates a military action, he has to pay a price, not only in military, but also in economic and political terms. Thus it might not be necessary to maintain a matching force with regard to each geographic area to stalemate the enemy; depending on the potential price in political and economic terms, effective deterrence might be provided by a correspondingly smaller military force.[5]

Lastly, there are technological areas of high marginal utility in terms of military power, and here the United States might want to keep or enhance its lead. Through these technologies—developed in part with a geographic context in mind, but at the same time, having broad global functional utility to meet multiple contingencies—the United States would be able to maintain an advantage over the potential enemy and thus not be stalemated in a particular situation, if forced to apply them. Moreover the lead in such technologies might reinforce (or enable us to maintain) stalemate in those areas where we choose to settle for sufficiency. The relevant technologies include electro-optical weapons, highly effective real-time surveillance and detection with global reach, high-speed surface-effect propulsion in ships, deep ocean technology, and certain forms of weather modification that allow operations under adverse meteorological conditions.

A number of technologies on the horizon are promising in both military and civilian application. Nearly all mentioned above have potentially high economic utility. Accordingly a careful appraisal of coming technologies from both the military and civilian points of view and effective cooperation between military and civilian authorities in the development of technologies of mutual interest might be an important step in our technological strategy leading toward a more rational use of available resources. Indeed investment of resources in technologies where civilian and military interests overlap is likely to be especially cost-effective because of the resultant double benefits. Figure 12-1 suggests some desirable directions of reallocation of resources between and within the various sectors of technology.

The sector that will require particular attention in the near future is nonmilitary technology, which is not only growing in importance for national security but also is—and will remain—the principal instrument for increasing the nation's productivity and hence its ability to solve internal economic, social, and environmental problems contingent on the availability of resources.

Figure 12.1. Desirable Direction of Reallocation of Resources

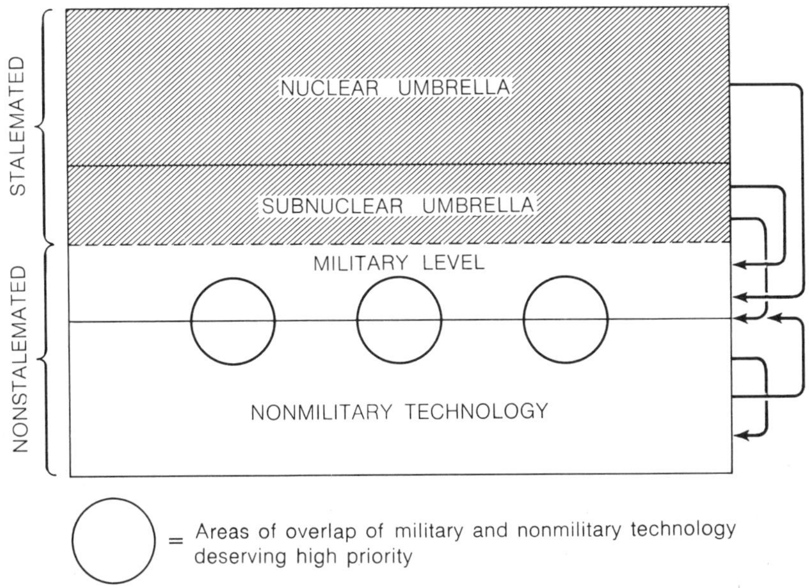

```
        ┌  ┌─────────────────────────────────────┐
STALE-  │  │////////NUCLEAR UMBRELLA/////////////│
MATED   │  │/////////////////////////////////////│
        │  ├─────────────────────────────────────┤
        └  │////////SUBNUCLEAR UMBRELLA//////////│
           ├─────────────────────────────────────┤
           │          MILITARY LEVEL             │
NON-    ┌  │      ◯         ◯         ◯          │
STALE-  │  │                                     │
MATED   │  │       NONMILITARY TECHNOLOGY        │
        └  └─────────────────────────────────────┘
```

◯ = Areas of overlap of military and nonmilitary technology deserving high priority

Nonmilitary Technology and the Question of Priorities

Toward Greater Rationality and Selectivity in National Policy

The question of priorities is more complex in the nonmilitary sector of technology than in the military sector, where the government is both the developer and the consumer of technology; once the government sets priorities, the issue is settled. In the nonmilitary sector private enterprise plays the principal role in determining what is developed and manufactured and what is not. To be sure, the government influences priorities in the development and use of technology in the private sector. However, the role of the government in this regard has been far from ideal.

The professed goal has been "excellence in science and technology," theoretically to be achieved by a division of functions between the government and the private sector. Government is to be responsible for the public sector of technology, for supporting basic research, and for providing incentives to science and technology through means such as the patent system and favorable tax laws with regard to R & D. Private enterprise, operating under the free-market system, is expected to develop technology for its own use. It has been recognized, however, that in certain areas private enterprise may not find it economically justifiable to engage in early stages of technological development. Accordingly the government is to step in selectively in such areas and provide support to technology ultimately intended for the private sector.

In practice, government support did not follow the theory. In the case of agriculture federal support went well beyond basic research into applied research and technology. In the case of medicine the boundaries between the "public" and the "private" sector were also disregarded. The government supports a number of technologies that belong to the private sector and, at the same time and for no good reason, neglects others that need public support. In general the policy is not well thought through.[6] As we shall see later in the case of energy, certain fundamentals in our national decision-making process have something to do with the inconsistent pattern of support rendered to various technologies.

Nor is the situation much better in those areas of activity where support is applied broadly rather than selectively. To be sure, broad federal support to basic science has resulted in U.S. world leadership in this field, which, in the long run, provides a foundation for technological strength. Some duplication and overlapping are healthy in basic science, but their extent in our system suggests that rationalization here, too, is in order. The situation with the favorable tax treatment of all R & D is another case where a change might be needed. Motivated by the prospect of profit, private enterprise takes advantage of the favorable treatment of R & D to generate facilities and products that are useful to society—but also those that are eventually recognized by society as harmful. Aside from the initial waste of resources for the R & D of such products and the harm done to society by their introduction, additional resources are wasted to suppress them.

In the future the private sector of applied R & D will need increasingly greater public attention to ensure that potentially harmful innovations do not enjoy the preferential treatment rendered to R & D by the government. For example it is not clear why research aimed at a more efficient production of cigarettes should have similar tax benefits as that for low-cost electric power. Where innovations under development are clearly harmful to society, they should be identified early, and injunctions should be issued against their development or mass production.

This approach presents political and technical difficulties. Evaluation of private research, development, and production well beyond what is considered acceptable today would be needed. The problem of criteria emerges, along with the danger that an unduly restrictive policy could produce a stifling effect on R & D. One approach that might evolve first would be to limit public appraisal to development alone and not to research. Adequate latitude would have to be ensured to private enterprise, since private initiative itself—not just R & D—could be jeopardized by harsh regulations.

These difficulties notwithstanding, selectivity in regard to what is to be promoted and allowed to be produced in quantity will be necessary, if undifferentiated technological advance, as well as that of society's turning against technology in general, is to be avoided. In the last analysis the issue involved is the one raised in the previous chapter: Should we control technology or be controlled by it? Paradoxically a society-wide opposition to technological advance would not be a case of controlling technology, but rather of technology controlling us; it would, in effect, result in technology's undermining society's rationality.

Responsible social control of technology, free of both antitechnological and protechnological bias, implies that it will be necessary to shift away from the practice of giving the scientists and technologists themselves the single most influential voice in science and technology policy. Rather, a more balanced mix of personnel—among them social scientists and administrators who understand both social goals and the scientific-technological potential as well as those scientists and technologists who have a keen appreciation of social problems and purposes—will have to constitute the group providing guidance for the directions of development of science and technology in the future. Such a restructuring of leadership is a prerequisite to science and technology's effectively serving the nation's vitality and viability, as distinguished from the objective of "excellence in science and technology" as such. If, however, it does not take place and if our leadership in this field is not capable of providing imaginative guidance and priorities in not only meeting but expanding and modifying social goals in response to the growing capability of technology, then we shall fall considerably short of capitalizing on the beneficial potential of technology and shall not meet our objective of "controlling technology."

The Case of Energy: A Belated Priority

President Nixon's announcement on June 23, 1973, of a $10 billion R & D effort for energy, to be completed by 1980, and the subsequent budgetary increases indicating that the actual figure may be as high as $14 billion, suggest that the allocation of funds for this purpose is likely to be adequate. The case of energy, however, illustrates the deficiencies of our present system in selecting priorities. As early as 1964 a major Office of Science and Technology (OST) study, widely considered within the government, urged the long-range development of alternative sources of energy.[7] A number of people and offices in the U.S. Government were aware of the approaching energy crisis a few years before its explosion in 1973. And yet, as late as FY 1973, the many potential means of alleviating the energy problem were underfunded, many very seriously. Discussions with individuals involved in decisions related to energy strongly indicate that accountability for the belated response can largely be ascribed to the prevalence of constituency decision making in Washington and the nation, and to the weakness of the rational approach in decisions.

To return to our earlier example of the 1964 OST study on energy: It was considered on an interagency level, but the existing agencies saw no particular advantage to themselves in implementing the recommendations of the study, and it was quietly shelved.[8] To understand the situation fully, however, one has to go beyond individual examples to some fundamentals. Rational decision making is heavily dependent on forecasting and long-range planning; and in the U.S. government the former is weak, the latter is practically nonexistent. To the extent that some forecasting was done and future developments in energy were anticipated—largely on an *ad hoc* and compartmentalized basis—the absence of effective planning provided no regular avenues through which a desirable course of action could be implemented. In the absence of systematic planning new technology is handicapped because, by virtue of being new, it lacks a constituency. An individual, group, or office that conducted the forecast and foresaw the need for developing new technology usually has to drum up a constituency to obtain support for the necessary R & D. In some cases where a powerful group that would have benefited from a new energy technology did exist, its outlook was focused on the familiar technology that had placed it in a position of power in the first place, and it was not receptive to change. An example is provided by the coal industry, which, until recently, gave virtually no support to the long-standing efforts of the Office of Coal Research to stimulate interest in the development of oil and gas from coal.

There were also strong vested interests in the status quo that could be undermined by new technology. The oil industry was in this category. It certainly had no particular enthusiasm for the rise of alternative sources of energy. There is no evidence that the oil industry actively opposed the development of technologies for hydrogenation of coal, geothermal, or solar energy; it did not have to. In the constituency decision-making process pertaining to energy—in which the petroleum industry occupied a very influential position—mere apathy was enough to kill meaningful action.

By a rational decision to create an agency responsible for the development of energy in general, the U.S. government could have created a constituency for energy in its own midst. This, in turn, would have stimulated the emergence of constituencies for various new sources of energy outside of the government. But such a step was not taken until 1975, when the Energy Research and Development Administration (ERDA) was established.

Before the establishment of ERDA the dominant position of the Atomic Energy Commission (AEC) in the field of energy in the U.S. government tended to overshadow and push to the sidelines nonnuclear energy programs. In comparison with those programs AEC had massive financial support and an influential constituency inside and outside of government. Within AEC money and interest focused on technologies of greater certainty of fruition and relatively close payoff—in a conventional setting, constituencies cannot be developed, and if developed, cannot survive very well if the payoff comes in a distant future. Thus such programs as magnetic fusion had to wait until the push came from the outside—the Soviet success in magnetic fusion and, later, general concern about the energy crisis combined with ecological interests in clean energy.

The rise of environmental concerns in the late 1960s, a time when the relatively clean domestic resources of fossil fuels were being exhausted, also influenced energy policies. It was difficult in the early 1970s to attract serious attention at the White House level to energy. The environment had a constituency—and votes—while decisions favoring energy did not. Moreover, such decisions might, in some of their forms, have conflicted with environmental concerns. There was thus a built-in incentive to delay badly needed vigorous action on energy until the pinch was felt and a broadly based energy constituency could be counted on.

Other Candidates for Priority.

In retrospect the example of energy is a glaring one, but it is not an isolated case. Other important areas enjoy varied degrees of neglect. Perhaps the price the nation is paying for belated decisions on energy can help us to avoid a similar price with regard to other technologies that are not at present receiving adequate attention and the necessary remedial action. Some relevant technologies are highlighted here.

Materials R & D has not been considered a glamor field and has not attracted its share of scientific and technological talent and national attention, although materials create the very foundation of advanced societies.[9] A closely related field deserving a higher priority is that of recycling.[10] Weather and climate modification is a promising area in terms of likely progress, potential breakthroughs, and benefits to society, and yet it is seriously neglected.[11] Oceanology and marine resources development have been depressed in recent years, in spite of the justifiable interest they generated in the late 1960s.[12] America's world leadership in data processing does not necessarily mean that the full potential of computer technology is being taken advantage of. In particular adaptation of computer technology to a systems solution of major problems (e.g., learning process, mass transit, urban sanitation, the processes of bureaucracy) has been relatively neglected.[13]

This enumeration is certainly not exhaustive. Moreover, the expanding range of technology and the changing needs for its application tend to shift priorities to less visible, low-profile areas that require systematic research to be identified. There is still potential for glamor and distinct, publicly recognizable identity in priorities of future technology, especially in areas such as climate and weather modification, marine resources development, and solar energy. However, as we turn to technology to help solve economic, social, and environmental problems, determination of priorities will increasingly require a diffuse, mundane, painstaking activity devoid of imagination-capturing identity: contribution of technology to productivity; the relationship between basic science, applied science, and technology in various technological sectors; the need for cross-fertilization of selected technologies to maximize progress; and so forth.

The problem of priorities could therefore perhaps be best dealt with by institutionalizing their identification in an entity like a National Institute of Technology (NIT)[14] that would apply a systems—as distinguished from a purely technological—approach in its activity. Whereas NIT, through appropriate analysis, could provide the necessary foundation for systematizing (rather than—to use Washington bureaucratese—"*ad hocing*") major decisions on priorities, the decisions themselves, insofar as they involve value judgments, would have to be made on the highest level of the U.S. government. The Science and Technology Council (STC), discussed in chapter 11,[15] would be expected to play an important role in such decisions.

In the last analysis, priorities in science and technology depend on the prevalent social values in a given period of time, be they national security, the preservation of the environment, economic growth and well-being, or social welfare. The development of technology can increase the importance of some values and deemphasize others; to that extent technology plays a role in influencing priorities with regard to itself. At present we do not have adequate intellectual tools to determine priorities with a high degree of certainty. Even if goals, based on social values, are determined, we do not know precisely how to get there, nor do we have an adequate organizational mechanism that could help us make timely decisions on priorities. Given the present state of the art in this area, a hybrid of approaches will have to be used. Some of these would include broad considerations of national strategy and the nature of world political and technological trends. Others would require examinaton of social and economic needs. Intelligent decisions on still other priorities (like technology for productivity) would require systematic analysis and support, preferably embodied in a suitable organizational framework.

Allowing for the necessity of further work in this area, one could nevertheless say that present trends and requirements suggest a major shift in priorities in the allocation of resources. To put the developments in perspective: The 1950s were characterized by emphasis on building up the nuclear umbrella. In the early 1960s priorities were shifted to the subnuclear-umbrella military level and, to a degree, to power-relevant nonmilitary technologies (the outer space effort). The remainder of the 1970s and the 1980s are likely to be characterized by emphasis on those nonmilitary technologies essential to sustaining the nation's viability and vitality.

Modification of the Technological Superstructure

The United States must continue to modify existing institutions and technologies in response to the requirements of the future; this will inevitably encounter resistance. The problem of innovation is not narrowly technical. It is a problem of overcoming the inertia or outright hostility of established interests, of short-term financial sacrifices for the sake of greater longer-term gains, of retraining and reeducating manpower to minimize the pains and personal sacrifices involved in the needed readjustment—and, by doing so, to minimize resistance to change.

In short, the problem is economic, political, and educational, as well as "technological." If we fail to modify the nation's superstructure, our economy will be unable to compete in world markets, capital will be locked into obsolete equipment and high-cost methods of operation, and high prices or slow production will leave broad demand for a given product unsatisfied.

The U.S. housing business provides an illustration of a failure to make a timely readjustment. The U.S. residential housing industry is a $37-billion business.[16] It is highly fragmented and employs largely old methods that, in effect, amount to custom building. The costs are high and growing rapidly, but they do not buy high quality. The business is incapable of meeting the demand for moderate-cost housing.

The apparent solution to the housing problem is to use precut lumber, preassembled components of a house or a building, and, in some cases, factory-built houses. This method, which would amount to systems building, would transform the fragmented housing business into a housing industry. Systems building is employed in Western Europe and the Soviet Union at considerable savings in costs. The potential savings for the United States have been estimated at anywhere between 10 and 20 percent. If systems building is also applied to public and nonresidential construction, many billions of dollars could be saved annually.[17] Besides lower costs, the employment of systems building would result in the construction of houses on a much larger scale and much faster and would thus meet the critical demand for housing in the nation.

However, systems building is resisted by trade unions, since it would largely replace skilled construction workers by relatively unskilled factory workers. In addition, many local building codes preclude the use of certain technological advances, such as plastic pipes, plastic-insulated wiring, and precast concrete. The housing market is largely fragmented; to stimulate the necessary demand for large-scale systems building, the markets in many cities would have to be combined. Systems building technology, in part, would have to be borrowed from abroad (Western Europe) and further advanced to meet American requirements.

Significant progress toward systems building has begun to take place in recent years. Construction business—especially that dealing with individual homes—has turned increasingly to precut lumber, preassembled components, and modular homes. According to the National Association of Building Manufacturers—which represents the fledgling housing industry—the share of the industrialized sector of the housing business amounts to about 15 percent of new residential construction.[18]

In 1969 the Department of Housing and Urban Development initiated Operation Breakthrough, a project intended to stimulate industrialization of the housing business. At a modest investment of about $65 million, Operation Breakthrough undertook to act as a catalyst in modernizing the industry. This was done by subsidizing housing units based on systems approach, encouraging new labor agreements for factory housing production, effecting change in transportation equipment and rates to facilitate the shipping of factory-produced houses and components, and encouraging the establishment of state (as distinguished from municipal or county) building codes and regulations. Some limited results were produced, but for lack of additional funds Operation Breakthrough was phased out in 1973.[19]

Many problems related to technology cannot be effectively coped with unless adjustments and modifications are made in the existing legal, administrative, and political structure. A prime example is presented by our fragmented system of municipal, county, and state boundaries. Appropriate concepts and administrative mechanisms have not been developed, although some steps in this direction have been made: regional councils have been created in some cases to facilitate planning on a broader scale; the Port Authority of New York and New Jersey cuts administratively across two states—operating bridges, tunnels, and airports, among other facilities, in the New York City metropolitan area.

How can policies be designed to meet the requirements of the future and where should they be designed? Change in response to rapidly changing technology cannot emanate from one, or a few, institutions. Society as a whole must recognize that enlightened self-interest requires a continuous adjustment of our technological and institutional superstructure to change. One should be careful to avoid the pitfall of initiating fundamental changes in response to short-range fads, but the magnitude and the tempo of change creates the need to accept the concept of constructive change as a part of the attitudinal and value pattern of society as a whole. Such acceptance would not be a substitute for specific policies, nor would it do away with the social friction and conflict that such policies might generate. But it would facilitate change and help to avoid the convulsive disruptions and instability caused by the failure of a timely adjustment to technologically induced realities.[20]

Private enterprise is becoming increasingly future oriented and, largely under public pressure, socially conscious. It would be in the interest of business, as well as the nation as a whole, if more enlightened business leaders facilitated and accelerated change along those lines. Problems impeding desirable change require greater elucidation and public discussion, so that the positions and activities of various vested interests involved would come to the surface. This would facilitate the making of decisions on the basis of reasonably clear-cut alternatives and would help to clear them from potentially harmful obfuscations. The next decades will require the development of greater cooperation among state, municipal, and local authorities in coping with problems on a regional level. In light of the pluralism of the American politico–administrative system and the fragmentation of authority, the achievement of a desirable degree of cooperation will not be easy.[21]

Largely in response to rapid technological change an increasing number of adults are returning to institutions of higher education on a full- or part-time basis. In many cases these persons are redirecting or completely changing their careers in their late thirties or forties. This phenomenon presents a challenge to the universities to develop the necessary curricula, admissions policies, and placement procedures in order not only to accommodate the mid-career students but also to further encourage and facilitate the needed reeducation of the nation's manpower on the professional and graduate levels.[22]

Whereas many readjustments for the future can be made through the efforts of the various components of the nation—be they state authorities, corporations, or the universities—they cannot suffice without federal activity. The federal government has at its disposal a number of instruments and approaches that it has been using in the past and that it will have to use with greater vigor and purposiveness in the future—acting as a catalyst, regulator, promoter, planner, and enforcer of desirable change.

The full scope of potential federal activity with respect to technology-induced change cannot be given justice here. Since, however, the domain of technology is largely in the hands of business and the federal government, the relationship between the two is briefly discussed.

Government and Business

As we have seen in connection with our examination of Japan (chapter 7), effective use of technology depends, to a considerable extent, on close cooperation between government and business. Support of private enterprise by the government, if conducted in such a way as to facilitate and expand opportunities for industry to take advantage of technology's potential, can go quite far in enhancing the nation's viability and growth, even if the government's science and technology policy itself falls short of being truly exemplary. Ideally, however, a nation would combine energetic, though selective, support of industry as a part of a farsighted technological strategy encompassing a firm delineation of priorities for the future.

The relationship between government and business in the United States consists of a combination of incongruities, anachronisms, and multiple sources of policies that hinder overall policy. The United States government has promoted business enterprise since the late eighteenth century and the philosophies of Alexander Hamilton. At the same time American social mythology has traditionally opposed public ownership of business enterprises and public initiative in business. The sanctity of the system of private enterprise and private initiative is repeatedly asserted by public as well as private sources. However, the U.S. government itself is extensively involved in the ownership and operation of business enterprises, and public intervention into business is substantial. Many federal agencies closely supervise business activity, and the scope of this supervision exceeds, in a number of cases, that of other countries where the sacredness of private enterprise is much less of a fetish than in the United States.[23]

A large share of public policy and regulation with regard to business is exercised by regulatory commissions. Being independent agencies, they have evolved policies of their own with regard to the sectors of the economy they are responsible for. Some of them—like the Civil Aeronautics Board—have been, by and large, conducive to the promotion of technological progress within their particular sector, while others—like the Interstate Commerce Commission—have merely reacted to developments rather than anticipated them, and their policies have tended to impede technological innovation. The existence of independent regulatory commissions, which have developed powerful constituencies of their own, has been an important factor hindering the development of an effective national policy in particular segments of the economy, such as transportation.[24] Largely because of the various vested interests associated with the regulatory commissions, efforts to reform them have not been particularly successful.[25]

Institutional impediments to effective federal policy with regard to business are not the only problem. Perhaps even more important, we do not have enough adequately developed conceptual instruments on how the government can stimulate the growth and vitality of private enterprise on a sustained basis. Productivity is not well understood, and we do not really know how to increase it on a national scale. We do not have enough understanding of technological innovation, and we do not have enough information about its economic impacts.[26] While these and other conceptual gaps need to be filled to provide a solid basis for policy, technological trends and the development of policies in other nations can provide us with some guidelines.

In view of the growing scale and costs of technology, the present antitrust policy would require reexamination. Competition in the future is likely to be increasingly on an international scale, and American enterprise, facing antitrust regulations, is likely to find it difficult to compete against government-supported foreign companies. The need for large-scale capital imposes the requirement for continuous reevaluation of taxation policies, credit regulations, and similar matters to ensure that capital is readily available. The federal government will have to play a central role in facilitating the continuous modification of the nation's technological superstructure. This would involve a number of activities, ranging from the encouragement of business to pool resources for certain types of R & D to financing of prototypes of certain advanced technologies and comprehensive planning of the directions of the nation's technological growth.[27] Along with assisting private enterprise in phasing in new technology, the federal government would have to alleviate the hardships associated with the phasing out of the old—an activity that might encompass large segments of an industry.[28] Whereas a number of these activities are not entirely new to the federal government (in fact, the government is, in varied degrees, engaged in many of them now), to meet the requirements of the future, they will have to be conducted on a more extensive scale and in a more systematic and comprehensive way.

In the last decade or two, it has been recognized that it is a part of the federal government's function to exert its influence to avoid economic fluctuations and to sustain the nation's economic growth. It is doing so mainly through the fiscal and monetary policies developed by the Council of Economic Advisers and the Federal Reserve Board. What has not been adequately recognized—and sufficiently comprehensive instruments for that purpose have not been developed—is that the federal government's function also includes orchestration of the technological composition of the national economy.[29] As we look further into the future, such an orchestration, conducted in a systematic and synoptic manner and closely coordinated with the fiscal and monetary instruments, will be essential for an effective management of the economy. As William Carey pointed out,[30] about 1983–1984, the U.S. gross national product (GNP) will be close to $2 trillion, and the federal budget alone, about $500–600 billion. An economy of this scale will be awkward and require a lot of managing. Like a supertanker it will be much less capable of executing quick, short-term corrections and turns. Running such an economy will require an immense degree of forecasting, anticipation, and new, sophisticated instruments for its steering. Only thus can we minimize the two unwelcome but distinct possibilities that the economy's momentum, combined with poor maneuverability, can carry it to the brink of disaster or, alternatively, lead to an aimless and sluggish drift.

Technological Impact and Sociopolitical Discontinuities

As discussed in chapter 3, the impact of technology produces discontinuities in social and political trends. Not all of these are related to national viability and vitality. The kind that are so related are those producing instability in society or changing society's value system so as to assign little or no importance to pursuits relevant to productive utility.

The technology-induced discontinuities are a relatively new phenomenon. Even the present dimensions of this phenomenon are not very clear, and its future dimensions and forms are yet so much more obscure. The obscure nature of the phenomenon itself creates difficulties in designing the means of coping with it. Thus, perhaps all that can be done at this time is to attempt to differentiate among the various categories or types of discontinuities induced by technology. This, in turn, might suggest different policy approaches toward them.

One way technology produces an impact leading to discontinuities in social and political trends is by opening up new opportunities, giving new freedoms and new instruments of power to those segments of society that, for one reason or another, have not been treated equally or justly. With technology expanding their horizons, the groups thus affected challenge the existing order of society, and, to the extent that they meet resistance, instability results.

One example in this regard is provided by the case of women. Historically women have not been treated equally with men and for a good reason. At a time when physical strength and prowess were highly important for the livelihood and survival of society—whether this was exemplified by hunting, heavy agricultural work, or warfare—the women simply could not compete with the men. In addition they were tied down to the home by the responsibility for bearing and rearing children and by other domestic duties. Thanks largely to technological advance, the importance of physical strength has declined, and since intellectually women are better endowed to compete with men on equal terms new opportunities arose. The typewriter opened new horizons for employment to large numbers of women, at home or in the office; so did the telephone. The washing machine, the dishwasher, and canned food went a long way in freeing women from household chores, while the Pill gave them sexual freedom. But equal opportunities in employment and salaries were lagging behind. The women's liberation movement has largely been a product of this development.

Technological factors were important in the civil rights movement, although it would be a gross oversimplification to explain the movement entirely in these terms. If we go back far enough, industrialization was a significant factor in the abolition of slavery. Whereas the traditional agricultural society was based on the attachment of manpower to land, industrial society required a free market of labor that could provide manpower sufficiently mobile and flexible to meet the flux of industry. Labor had to be available to be hired and laid off on a fairly short notice, without a permanent commitment on the part of a business to support its workers regardless of its own fortunes. In more recent days the modern industrial society—which places emphasis on skills—tends to show the irrelevance of race or religion. Television demonstrates vividly the affluence of the millions of those who have desirable jobs and are not discriminated against and thus stimulates claims for similar treatment. And modern, everyday technology of our affluent society provides readily available means for organization and protest.

A closely related type of social discontinuity—which also may lead to instability and conflict—is that induced by technology's opening up new options in society and thus changing the hierarchy of values. Whereas decades ago a complete eradication of poverty was an idle dream of a few idealists, largely thanks to the high productivity of our industrial society it is a realizable goal now. It is being demanded not only by those who suffer from poverty but also by the millions of others who believe that poverty is a social evil and must be done away with. Technology also changes costs associated with the attainment of certain values; to the extent that the costs are reduced, a particular value is more readily attainable and rises in priority.[31] But the change in the hierarchy of values stimulates claims and counterclaims on public resources, and when the emerging claims are stifled or denied, serious social instabilities may result.

What are the implications for policy stemming from these two types of social discontinuities produced by technological impact? Can social instability be avoided? Some social instability probably cannot be realistically escaped; the problem is to minimize it and to prevent it from reaching proportions consuming a great deal of the nation's energy or being seriously disruptive. Thus legitimate claims for change must be identified early enough and appropriate corrections made before they reach disruptive proportions. The problem with late identification and belated measures is that both the emotionalism and the claims exceed the limits of reasonableness and thus make them difficult, if not impossible, to satisfy.

Technological advance affects the value system of society in a way that may not create instability or be disruptive but might nevertheless undermine the viability and vitality of a nation. It has been noted (chapter 3) that the Protestant ethic and the desire to be productively useful have shown signs of decline among American youth. As society becomes more and more affluent, it is likely that leisure will rise in the hierarchy of values in the United States.

As discussed in chapter 10, the nation-state system, although changing, will still be with us for some time. In this system the viability of society largely depends on the strength of the nation's economy and a certain quantity of military force that the economy must be capable of sustaining. But the case may well be that the decline of productive pursuits as a value is an inherent characteristic of the postindustrial society. Moreover, as the Sprouts suggested in their analysis of the role of Britain in world politics,[32] it is possible that the more advanced a society becomes, the more concerned it is with social programs and welfare, and the less willing to bear sacrifices for its preservation. If so, then it is perhaps inevitable that the most advanced nations, which are the first to enter the era of the postindustrial stage, will decline in their vitality and may seriously jeopardize their viability—just as it was inevitable that the nations that were the first to become industrialized gained in vitality and influence.

If the foregoing assumption is correct, then the United States—the first nation to enter the stage of the postindustrial society—faces a policy problem of how to delay the decline in her vitality so that her security and survival are not seriously jeopardized. This further accentuates the importance of increasing the nation's productivity and instituting the other measures I have delineated. A smaller amount of human effort will simply have to produce correspondingly more to sustain the nation's viability.

A fourth type of actual or potential sociopolitical discontinuity stems from the inability of population to cope, psychologically and physiologically, with the whole aggregate of phenomena introduced by the impact of advanced technology—the very rapid pace of life, the insecurity produced by actual or potential employment dislocation, the general uncertainty about the future, the frustrations resulting from overbureaucratization, impersonality, and rigidity of modern institutions, and so on. This phenomenon, popularized by Alvin Toffler under the name of "future shock,"[33] frequently overlaps with the first two types of social discontinuities and thus raises legitimate injustices to the level of irrationality and greatly exacerbates social instability. It is thus probably the single most serious, potentially very damaging, and the least manageable type of social discontinuity that advanced societies face.

The policy approach to this phenomenon is twofold. First, one must deal

American Policy

with its underlying causes. This might include measures to minimize the impersonality and rigidity of the existing institutions, to anticipate and prevent personal insecurity by retraining personnel, to provide for improved social security coverage where appropriate, and to improve the human environment. Second, one must cope with the symptoms themselves. This might involve an early identification of the alienation of individuals from society, the broadening of counseling and other services for the maladjusted and the alienated, the extension of low-cost psychiatric services, appropriate legislation and adequate personnel for crime prevention and enforcement, and many others.

In general the destabilizing potential of technology forces the issue of social stability to the forefront of policy concerns. The fact that the 1970s appear to be more stable than the late 1960s does not necessarily guarantee America's future stability. The broad scope of the destabilizing impact of technology and its complexity might require that the nation eventually evolve a conscious policy for social stability, with a coordinating body and an integrated conceptual framework for this purpose. Even before such a development all major policy decisions, whether they pertain to foreign or domestic affairs, will have to be carefully evaluated with regard to their potential effect on social stability. The anticipated instability, if any, will then have to be considered, along with other factors, as the price to be paid for such a decision. The case of U.S. involvement in Indochina suggests that it may not be enough to compute the price for a particular decision only in terms of such criteria as the size of the troops mobilized, money spent, or lives lost.

The impact of technology on the home front and the ability of the United States to control that impact hold the key to the future viability of the nation. America's policies with respect to the outside world are important and can enhance the nation's growth and vitality, but only insofar as they emanate from a strong and healthy home base.

The Outside World

13

IT MAY BE USEFUL to introduce some conceptual considerations on the role of technology in foreign policy. Doing so will help us understand both its potential and limitations in expanding the effectiveness of foreign policy.[1]

As a factor in American policy technology can be either a dependent or independent variable, or a mixture of both. As a dependent variable, it is clearly an instrument of policy. For example, the United States uses various technologies in assisting developing countries. As an independent variable, technology is a force to which policy must adjust itself. For example, the existence of nuclear weapons influences American foreign policy not only with respect to the Soviet Union but also to a host of other states.

A particular development of technology need not necessarily impose constraints on foreign policy or run counter to its interests or objectives; in fact it may facilitate their attainment. To give a hypothetical example: If appropriation of the continental shelf beyond the 200-meter depth were an initial objective of American foreign policy, then the growing technological capability to exploit marine resources at greater depth and their use by private enterprise would, under the terms of the 1958 Convention on the Continental Shelf, effectively assist the attainment of this objective (see chapter 9). Insofar as the policymaker becomes aware of such development of technology and incorporates it, along with other instruments, into the nation's foreign policy, it then becomes an instrument of policy—that is, a dependent variable.

From the point of view of the policymaker the ideal situation would be if technology were always a dependent variable. This is, however, an impossibility. To the extent technology is developed and applied by foreign nations, its impact on American policy is primarily as an independent variable, although, under certain circumstances, we might have a measure of control over it or steer it in a certain direction. The impact of technology cannot always be foreseen; if so, it affects foreign policy as an independent variable—regardless of whether the effect is beneficial or harmful. Even if foreseen, known to be harmful, and physically within our own borders, the impact of technology cannot always be prevented—for political and other reasons. In any case we do not have a sufficiently developed conceptual framework to effectively use technology (or functional elements in general, of which technology is but one) as a dependent variable and as an instrument of foreign policy.

In view of these limitations, present-day American foreign policy can, at best, strive to achieve three goals in an effort to improve its effectiveness in using technology (or functionalism in general) as an instrument:

1. To refine, conceptually, the distinction between the impact of technology as a dependent and an independent variable; in particular, to determine where and to what extent developments in technology produce an impact of concern to foreign policy without such an impact's being a conscious product of the policy.

2. To monitor the impact of technological factors as an independent variable and, if this impact is desirable, to let it continue or to reinforce it; if it is undesirable, then, to the extent possible, to stop, redirect, or neutralize it.

3. To subject to careful scrutiny the extent to which technological factors and forces can be employed as instruments of policy—ranging from specific short-term objectives in a particular region to the achievement of a desirable international system.[2]

The following discussion of American foreign policy with particular reference to technological factors should be viewed in the light of both the policy's present limitations and its potentially broader scope.

U.S. Policy and the Soviet Union

Any discussion of technology in United States–Soviet relations inevitably touches on the fundamentals of our postwar policy toward the USSR, since, in one form or another, technological considerations have always loomed large in that policy. Insofar as it leans on technological factors and instruments, our Soviet policy contains a number of dichotomies and crosscurrents that are likely to increase rather than diminish. The reasons lie principally in that the impact of technology contains a number of imponderable elements and that the effect of technology has assumed new dimensions not clearly perceived at the time the policy was formulated.

Background

Our post-World War II policy toward the USSR, as largely developed in the late 1940s, had two principal objectives: (1) To stop Soviet territorial expansion, mainly by building up the military power of the West and by assisting the economic recovery of nations bordering on the Communist bloc and thus minimizing their susceptibility to Soviet influence. Later, when the Soviet Union developed nuclear weapons, a strategy of deterrence was added that, in essence, supported the original policy. (2) By thus "containing" the Soviet Union, several results were expected to be achieved. Realizing the futility of their efforts at spreading Communism beyond the existing borders of the Soviet bloc, the Soviet leadership would reevaluate their expectation of the "inevitability" of the victory of Communism and lose dynamism in their pressure outward. It was also expected—or, at least, hoped—that the Soviets would decrease or abandon their reliance on military power to promote their interests. If the rivalry were thus shifted to the nonmilitary sector (economic and technological), this would be to our advantage, since the outlook for war would be greatly diminished, and, just as important, this was the sector in which the United States traditionally excelled and was, therefore, confident in holding more than its own.

Furthermore failure on the part of the Soviets to extend their power would weaken the force of ideology among their population. "Pluralizing tendencies" would develop in the Soviet society. These would be facilitated by internal technologic–economic factors: a higher level of industrialization and concomitant desires for a higher standard of living and for more consumer goods, all leading toward the decrease in Soviet aggressiveness and the reduction of their military power. This "mellowing" of the Soviet Union—and of the Soviet bloc as a whole—was to be expedited by an eventual expansion of economic relations between the United States and the Soviet bloc and by joint cooperative projects in technological and other spheres. These measures would also promote decentralization in the Soviet bloc and a leaning of the Eastern European satellites toward the West.[3]

The U.S. Soviet policy achieved a measure of success, and a number of its objectives have actually materialized, at least partially. The single most important result of American policy has been the success in stopping Soviet territorial expansion. Since 1949, in a quarter of a century of containment, the only Communist gain in the Free World was Cuba—a stroke of luck rather than the result of a planned Soviet strategy. The Sino–Soviet split—an unanticipated development and, in effect, a windfall for the West—gave impetus toward pluralization in the Communist bloc, which American policy had attempted to promote. Rivalry in the military sector was not abandoned, but in the late 1960s the USSR began to emphasize the nonmilitary economic–technological sector as the focal point of the long-range struggle between the Communist and the capitalist systems.

Whereas U.S. military containment on the ground was reasonably successful, Soviet military power moved into new dimensions made possible by modern technology. These were primarily strategic nuclear weapons and seapower, including potential amphibious capability. This made it possible for the Soviets to achieve a global reach in military power and, at least in nuclear weaponry, develop a capability comparable to that of the United States. In part these new dimensions spared the Soviets from some of the frustrations envisioned by the original containment policy and reduced the need for their seriously questioning the success of Communism. American military withdrawal after Vietnam, the decline in the U.S. competitive position in the world market vis-à-vis its own allies, and difficulties at home were other developments that tended to bolster Soviet confidence and stimulated the view of America as a declining capitalist power.

Thus the success of U.S. policy toward the USSR in the first 25 years after World War II was qualified at best. Then Soviet economic difficulties and the resultant decision of the Soviet leadership to enter into an extensive exchange with the West in general and the United States in particular to obtain badly needed technology broadened horizons for American policy. Some limited cooperation with the USSR in science and technology and related areas had developed in earlier years, but it was the Agreement on Cooperation in the Fields of Science and Technology signed at the Moscow summit of May 1972, as well as other related agreements signed at that meeting, that marked the opening of U.S.–Soviet cooperation on a large scale.[4]

A noteworthy feature of U.S.–Soviet cooperation after May 1972 is that it focuses on the establishment of new, joint U.S.–Soviet committees or commissions for cooperation in their respective fields and thus attempts to insulate scientific and technological cooperation from the stresses and strains of international politics by placing authority directly into the hands of appropriate government agencies rather than dealing through foreign ministries. Cooperation in the past was primarily confined to the exchange of scientific information and scientists, but the new agreements place emphasis on joint projects.

Current Policy

By the second half of the 1970s three categories of cooperation with the Soviet Union crystallized.

One category consists of bilateral programs developed under the U.S.–USSR Agreement on Cooperation in the Fields of Science and Technology

The Outside World 239

by the US–USSR Joint Commission on Scientific and Technical Cooperation. Specific projects in this category are managed by appropriate technical agencies of the government, although they also include private entities (research institutions and companies). The range of topics covered under these programs and under discussion as candidates for programs is very broad, including fields such as cancer research, civil aviation, and fusion energy. A secretariat in the Department of State is responsible for overall coordination of this form of cooperation with the USSR.

A second category of cooperation with the USSR in science and technology consists of industrial cooperation agreements concluded between private U.S. firms and Soviet agencies, principally the State Committee for Science and Technology. Potentially this category can be by far the most significant, amounting to many billions of dollars, although not all of the agreements concluded have been finalized and implemented.

The signature of such agreements is considered to be proprietary information by many firms. The agreements are not controlled or systematically monitored by the U.S. government, but insofar as they result in specific transfers of technology, such transfers are subject to approval by the Department of Commerce in accordance with the Export Administration Act of 1969. The principal purpose of this requirement for approval—which also applies to technology transferred within the first category of cooperative programs with the USSR—is to preclude the transfer of technology with potential military applications. The range of technologies in this category is also very broad, covering, among others, computer systems, agricultural equipment, measurement instrumentation, commercial aircraft, and construction and mining equipment.[5]

A third category of cooperation is exemplified by, and principally embodied in, the International Institute of Applied Systems Analysis (IIASA), a nongovernmental organization established in Vienna late in 1972. Although scientific institutions of some 14 countries from the Soviet bloc and the West participate in the work of the Institute, it has strong overtones of bilateral U.S.–Soviet relations inasmuch as the superpowers organized IIASA, they control its key positions, and are by far the principal contributors to its budget.

The Institute's avowed function is to apply systems analysis and other advanced techniques to complex problems of highly developed nations, but, as suggested by the research strategy of IIASA, its political impact could be more far-reaching than the problem solving itself, for IIASA aspires to world leadership in the development of techniques, concepts, and approaches to problem solving that could be applied to many problem situations. It aims to supplement, coordinate, and improve activities of existing research organizations of member nations. It thus has the potential for influencing Soviet research and thinking along the lines of pragmatic, technologically conditioned solutions, in which ideological considerations will recede into the background or disappear altogether. Insofar as IIASA's research may lead to concrete, large-scale cooperative projects in such fields of the Institute's present concern as energy systems, water resources, the environment, urban and regional systems, and computer application, it may significantly influence the two other categories of U.S.–USSR cooperation mentioned above.[6]

The current objectives of U.S. national strategy with regard to the Soviet Union can be summarized as follows:

1. Through cooperation in science and technology and related areas with the USSR, the United States hopes to draw the Soviet Union into a web of relationships with the outside world and to acquire a stake in its stability that would result in Soviet support of an international system based on peaceful change. This component of U.S. national strategy leans on the non-zero-sum elements in the international political process, but it recognizes the existence and validity of the zero-sum elements insofar as it acknowledges competition and rivalry with the USSR within an international system that both superpowers would be expected to uphold.[7]

2. The United States hopes that its policy will further contribute to institutional and group pluralization in the USSR and will result in the decrease of Soviet distrust of the West and a decline in the allocation of resources to the military. This aspect of policy finds its roots in the old containment concept and has overtones of the zero-sum game.

3. The United States expects to benefit in material terms from the policy of cooperation, whether this benefit comes from transfer of certain technologies from the USSR, savings of U.S. resources through cooperative projects in science and technology, or imports of raw materials and sources of energy in return for U.S. technology. This aspect of policy is rooted in certain non-zero-sum assumptions in U.S.–Soviet relations.

4. A fourth aspect of U.S. policy is closely related to the first but requires separate consideration. Insofar as the Soviet Union derives benefits from cooperation or other relations with the United States, Soviet resort to the use of military force or explicit military pressure is likely to deprive the USSR of those benefits. Thus nonmilitary elements—be they Soviet imports of technology from the United States, joint projects for the development of Siberian resources, or imports of grain from the United States—impose constraints on Soviet use of military power and thus play a role in deterrence. The benefits

from resorting to force will have to be very substantial to justify its use, and although force may still be used, the Soviet leadership will have to resolve some serious questions before the decision is made. Would, for example, a resort to military force in the Middle East against the United States justify the military risk and the loss of the infusion of nonmilitary technology from the West? Thus, even if the USSR does not decrease its military power, certain non-zero-sum elements in the international political process will impose constraints on its use. Although this objective of U.S. strategy was not clearly perceived at the time of its formulation, its implications have begun to gain recognition.

The Outlook

In considering the outlook for success of U.S. policy, we must keep in mind that all four objectives of U.S.–Soviet policy are closely interrelated; success in one is dependent on at least a degree of success in the others.

The USSR is at present entering into a number of "linkages" with the West, and it will enter into more. However, both the extent of this development and the degree to which the Soviet Union will support the existing system (rather than overturn it) would depend on internal changes in Soviet society. As we have seen in chapter 5, the Soviet Union reveals signs of pluralization; cooperation with the outside world is likely to increase it further. But if the USSR solves its internal economic problems and the Soviet leadership arrests the process of pluralization and reasserts its control over society—which could happen—many of the linkages with the West can be used as instruments for projecting Soviet power rather than vehicles for pluralization of the Soviet society. Thus internal developments in the USSR deserve close attention of American policy.

Internal changes in the Soviet society principally depend on three elements: (1) The impact of the cooperation and broadened contacts with the external world: (2) internal dynamics within the Soviet society, a highly important element of which is technological backwardness and shortage of resources; (3) the vitality and viability of the United States and other major non-Communist nations. At this point our discussion focuses on the first two.

Cooperation with the USSR is an important leverage for influencing change in the Soviet society but alone is not likely to be adequate to bring about the desired change. Probably most important in effecting such a change is economic difficulties and resultant shortages of resources. However, if cooperation with the United States in science and technology enables the Soviets to generate major resources internally, then the Soviet leadership will no longer have the inducement to pursue policies with major systemic implications. To put it differently: If the cooperation with the USSR and the impelling nature of Soviet resource-technology difficulties are properly balanced, they can go far in stimulating social change in the Soviet Union; but if they are not so balanced, then their influence is likely to be greatly diminished.

Thus the magnitude of the flow of technology and resources to the USSR is critical. The present U.S. policy does not control, nor even systematically monitor, the magnitude of the flow. The only restrictions imposed on the transfer of technology to the Soviet Union are those on defense-related technology, which are not really important in the context of our concern with social change. Whereas, in policy terms, the extent of the flow of technology and resources is virtually free of restrictions, in practice it is restricted by, for example, the availability of requisite resources and technologies in the United States, the Soviet ability to pay, the success in reaching mutually satisfactory agreements, and occasional politically stimulated restrictions that flare up (e.g., the reluctance of the Congress to approve the most-favored nation clause because of Soviet restrictions on emigration of Jews).

The question arises: Can the United States, as a matter of policy, regulate the volume of resources and technology flowing into the USSR so as to optimize their role in systemic change? Several considerations appear to make this course of action impractical:

1. A policy deliberately designed to regulate the flow for systemic change is likely to be counterproductive. It will probably strengthen the influence of those groups and individuals in the Soviet ruling circles who oppose cooperation with the West and may result in a reversal of policy.

2. Internal organization in the United States is not suitable for this kind of regulation and, given the free enterprise system, would be difficult to establish.

3. Even if the United States could restrict the flow of its resources and technologies to the USSR, the latter still would have alternative sources in Western Europe and Japan.

The United States thus seems to have no choice but to proceed with what amounts to an "unleashing" of technology as an independent variable on the USSR. In doing so, the United States can only hope that whatever the Soviets obtain from the United States and other advanced nations will not be completely adequate to solve all their problems and thus induce them to take steps leading to desirable (from the U.S. point of view) social change. At the same time the United States must recognize that, insofar as the impact of the imported technology on Soviet society is to be controlled, this function will be left largely to the Soviet government.

Many in the West have expected that, once détente gets underway, the USSR would reduce its military power. Indeed the Soviet military budget has come to be viewed as a barometer of both the success of détente and of social change in the Soviet Union. This view is somewhat simplistic.

Internal economic difficulties may force a major diversion of resources from the military sector. However, if the Soviet Union is reasonably successful in solving its economic problems, it woud be premature to expect an early reduction of military power—even if pluralization in the USSR continues—for at least three reasons:

1. The USSR has lost its ideological appeal, and it needs power to boost the nation's morale at home and its image abroad.

2. Since military technology favors stalemate, the marginal utility of the ruble in stalemating American military power, especially in the "conventional" sector, is reasonably attractive. The Soviet leadership may thus be tempted to continue to allocate large resources to military power.

3. Even given a substantial degree of institutional pluralization in the Soviet Union, it would take a strong leader to reverse the momentum of internal military influence. It might take time for such a leader to emerge; and, in the absence of such a leader, it may take even longer for the internal balance of power to shift against the military.

Therefore, at least for a while, constraints on the development and application of Soviet military power will have to be external. One such constraint is the cost associated with disrupting the advantages derived from cooperation with the West. A second constraint is the military capability of the United States and its allies. This suggests that, détente notwithstanding, the United States will have to maintain a military capability commensurate with the requirements of deterrence. Since, in recent years, military power has declined in relative importance, there is a potential danger that, especially in the climate of détente, its role in maintaining international stability can be overlooked. Military power has declined in importance because of technological trends in the military sector favoring mutual neutralization of power and because of its relative success in deterrence. If military power is unilaterally reduced by the United States below a certain threshold, there is a distinct possibility that it may not be adequate to maintain stability, and the resultant destabilization may again push military power, at substantial costs, to the fore, while political and economic factors decline in relative importance.

This point should not, however, lead us to an oversimplification in the other direction, namely, a narrow obsession with Soviet military capability. The impact of modern technology is so variegated and far-reaching that one can no longer afford to base one's policy on a proximate analysis of the Soviet capability (which may or may not be potentially dangerous) and the Soviet intent (which may or may not be benevolent) in a given period of time. The analysis has to be expanded to encompass the likely constraints that a given cooperation might impose on the Soviets in the future and thus modify their intent several years later. To provide a hypothetical example (on the assumption, for the sake of argument, of its technical feasibility): A joint project between the United States and the Soviet Union to dam the Bering Strait—and thus to modify the climate of Northern Siberia and Alaska [8]—would, besides producing major material benefits, impose important political constraints on both the United States and the Soviet Union. Would the United States be a relative beneficiary from such constraints? How might such constraints modify future Soviet intentions? An answer to these questions cannot be separated from larger considerations of world stability and American policy with regard to a future international system. Thus, imponderable elements and a degree of calculated risk are unavoidable components of American policy toward the USSR.

As to the economic and technological benefits from cooperation with the Soviet Union: They, no doubt, exist and, in the long run, could be very important. However, in the near- to mid-term, the USSR would by far be the greater beneficiary. The extent to which present cooperation with the Soviet Union will be beneficial to the United States will also depend on the evolution of Soviet society.

In the next five years or so the transfer of technology to the Soviet Union is not likely to generate significant competition for the United States in the world market. The Soviet Union is not likely to sufficiently upgrade its industry to compete with the United States, and it is likely to repay the United States largely with exports of raw materials and semifinished products. If pluralization expands in the USSR, followed by liberalization of the politico-economic system and accompanied by emphasis on the needs of consumers, the growing Soviet market may open increasing opportunities to American firms. There will also be mutual advantage in joint large-scale scientific and technological projects. To be sure, even a pluralistic USSR will in time become competitive with the United States, and the United States must be prepared for this development.

If, however, the Soviet Union successfully proceeds on the road toward an ideologically constrained, technology-oriented model, then competition from the USSR will come considerably sooner, perhaps by the mid-1980s. This is so because, under these circumstances, the Soviet Union is likely to place emphasis not on domestic consumption, but on the production of manufactured goods for exports to obtain more hard currency for imports of technology and for other power-oriented pursuits.

In conclusion, present U.S. policy with regard to the USSR has two distinct weaknesses:

1. It needs a stronger rational foundation. In the public debate on the policy and in actual attitudes and activities of public officials, businessmen, the Congress, and other individuals directly and indirectly involved, it tends to manifest either euphoria or skepticism. Business circles are largely euphoric (although lately this euphoria appears to be showing signs of decline) and the rush to deal with the USSR has overtaxed the ability of the U.S. government to monitor the development and assess its meaning. Skeptics are concerned about the absence of an early arms control agreement with the USSR, the prospect of "subsidizing the Soviet military–industrial complex," the tightening of internal controls by the Soviet leadership, and such developments as interference in Angola. A more balanced position, accompanied by a careful and systematic intellectual scrutiny of developments inside and outside of the USSR, is necessary.

2. The present large-scale transfer of technology to the USSR makes sense only if the United States puts its own science and technology in order, strengthens its competitive position in high-technology products, and otherwise enhances itself domestically. Quite apart from meeting the future commercial competition of the Soviet Union, the substance and the image of a vital America are essential to help catalyze internal change in the Soviet society.

Western Europe

U.S. interests with regard to Western Europe have certain conflicting elements. Western Europe has emerged in recent years as a serious commerical competitor of the United States, but the United States has an interest in cooperating with Western Europe in advanced science and technology. In the long run America's own security depends on a stable and economically sound Western Europe. Thus the preponderance of U.S. interests lies overwhelmingly in supporting Western Europe, although competitive elements will be unavoidable in the relationship.

It may be recalled that, in the context of technological impact and depending on the policies of Western European countries, three different "pure" models were envisioned in the next decades: a technological-trends-resistant Western Europe, a functionally oriented Western Europe, and a unified Western Europe.[9] None is likely to materialize in its pure form, but the models provide us with a convenient yardstick to measure the course of developments in Western Europe and of American policy in influencing them. For the simple fact is that Western Europe's future depends to a large degree on what the United States chooses to do. Because of her resources, technological capability, and important stake in European security, America has both the potential for and a vital interest in steering Western Europe into one of the alternative futures—or at the very least in steering it away from the least desirable alternative.

It is obvious that a technological-trends-resistant Europe is an undesirable alternative from the point of view of the American national interest. It would create instability in Western Europe, weaken its defenses, and provide a temptation for the Soviet Union to apply political pressure and resort to other means to capitalize on the new opportunities available on the European scene. This alternative might rekindle the American–Soviet conflict. The United States would have to compensate for the declining Western European capability for defense by strengthening its own military power, with a consequent curtailment of the resources available for the nonmilitary technology so important to America's national security and vitality. Whatever chance for regional disarmament in Europe had existed would disappear. It is conceivable that an intensified rivalry between the United States and the Soviet Union over a weakened Europe would result in an arms race in conventional forces, which might spill over into strategic nuclear weapons. As a result of this rivalry (say, in the mid-1980s), the superpowers might feel impelled to invoke the escape clauses of strategic arms limitation agreements then in force.

Whereas the need for preventing technological-trends-resistant Western Europe is clear, the questions of which of the other alternatives is most desirable and whether or not it could be realistically implemented with the assistance of an appropriate American policy are more complex. In general it can be said that both a functionally oriented and a unified Western Europe would be desirable alternatives; an exception would be a Western Europe unified by a dictatorial or totalitarian regime as a result of a period of instability generated by our technological-trends-resistant scenario. However, in the next decade or two, the likelihood of a unified Western Europe (except, perhaps, for the dictatorial variant just mentioned) is small. Therefore, an American policy that opts for a unified Western Europe as a direct or primary goal is not likely to be productive and would involve a waste of effort.[10] America should therefore focus on supporting a functionally oriented Western Europe—but which one?

There are two functionally oriented outcomes the United States could choose to promote: a functionally oriented Western Europe engaged in extensive cooperation in science and technology with the United States or a nationalistic Western Europe mindful of the United States as a competitor in the world market and within Western Europe itself. It is possible to envision a "mix" of the two but, as a deliberate product of policy, such a mix would not be desirable inasmuch as it would carry the danger of half-measures and a relapse into a variant of a technological-trends-resistant Western Europe. Accordingly a definite stress on one or the other will be necessary, although some elements of the mix will be unavoidable.

The emergence of a Western Europe imbued by technological nationalism has a certain appeal in that the American policy designed to bring it about might be relatively simple. The United States could strengthen the cohesiveness of European nations by adopting a narrowly nationalistic science and technology policy that might involve, among other things, strong support of American multinational companies in Western Europe. Such a policy is likely to trigger a strong European reaction, leading to the pooling of technological capabilities and resources. Although this policy might be effective in achieving a functionally oriented, nationalistic Western Europe, strengthened technologically and successful in avoiding internal instability in the 1980s, it would carry a number of disadvantages: undesirable political repercussions on American–European relations, no sharing of the costs of technological development and its use with Western Europe, and no guarantee of Western European success in meeting the requirements of technology beyond 1985.

It thus appears that the most desirable alternative is an American policy designed to support a functionally oriented Western Europe through the establishment of cooperative scientific and technological programs. This policy has the disadvantage of being difficult to implement. If it is not carefully thought out, not purposeful and energetic, or not backed by an adequate allocation of resources by the United States, it may result in Western Europe's ending up in an approximation of the technological-trends-resistant model.

On the other hand, if this policy is successfully implemented, it would meet both the medium- and long-range objectives of European security. A dynamic, functionally oriented Western Europe might attract Eastern European countries (and perhaps even the Soviet Union) in their own search for growth. A suitable climate might thus be created for cooperative programs in science and technology with Eastern Europe in which the United States would have an opportunity to participate. Such cooperation could help break down further the divisions between East and West and prove helpful in solving some internal European problems that hinge on East–West relationships.

Present American scientific and technological policy with regard to Western Europe does not appear to have a strong sense of direction; it mainly pursues targets of opportunity. With a few exceptions the United States has been drifting in the direction of bilateral cooperation with European nations. This course of action does not merely fail to provide an adequate solution but may indeed be harmful to the longer range interest of both Western Europe and the United States. To meet the requirements of future technology and of the huge costs associated with it, Western Europe must develop a mass market and internal institutions capable of coping with technology on a sufficiently large scale. Compartmentalized, bilateral relationships between the United States and individual Western European nations help to meet some short-range needs and create an appearance of progress, but they tend to delay, and may eventually inhibit, the achievement of this objective.

Thus the task of U.S. policy is complex. On the one hand the United States cannot afford to be insensitive to the interests of its companies in those areas of technology where Western Europe is strong. On the other hand it has an interest in inducing Europeans to cooperate among themselves as well as with the United States and to formulate this cooperation in a way that would be conducive to meeting the challenges of technology in coming years. There is no single, foolproof way to achieve this, but the formulation of mutually attractive, large-scale, multilateral programs (as distinguished from small-scale, bilateral programs with individual European nations) is one means toward this goal.

The U.S. initiative in establishing "post-Apollo" cooperation with Western Europe and Canada on the National Space Transportation System ("the Space Shuttle"), involving about $10 billion over 10 years, provides an example of a program in this category that is a going concern. Another major area of potential cooperation is in the field of energy. Both the United States and Western European nations responded to the energy crisis by separate R & D programs intended to develop new sources of energy. Duplication of effort between the two regions is likely to be substantial. Major cooperative programs in this field could both save money and promote the objective of European technological integration. The United States did not propose large R & D programs in the preliminary planning for the International Energy Agency (IEA) in the spring of 1974, and negotiations in the next two years showed little improvement.[11] There is still room for major American initiatives in this area, but strong leadership on the part of top U.S. officials is essential to produce results. ERDA program managers, who largely determine the nature of international programs in energy R & D, are not interested in major cooperative projects and are not likely to initiate or support them energetically, inasmuch as such projects are more complex to manage and might cut their own budget by reducing duplication of research on an international scale.[12]

The IEA offers a suitable forum for major tripartite cooperation also involving Japan. A way may thus be opened for Japan to participate in shaping a functionally oriented Western Europe, and through such cooperation Japan herself would have the benefit of sustaining her viability in coming years.[13]

A third major undertaking might be a U.S.–European program in marine research and development, focused around the exploration and exploitation of resources of the North Atlantic Ocean. There are at present some joint marine programs—primarily bilateral, involving exploration—with Western European nations,[14] but these barely scratch the surface of the full potential for trans-Atlantic cooperation. The discovery of hydrothermal mineral deposits in the Mid-Atlantic Ridge in 1973 (see chapter 10) may be only the beginning of other major discoveries of important raw materials in the largely unexplored Atlantic, possibly including large deposits of oil. American ocean development programs have been in abeyance in recent years for lack of funds; cooperation with Western Europe could revitalize them. Potential resources from the Atlantic Ocean could, in the long run, transform the economies of resources-poor Western Europe. In addition to direct benefits an early, well-planned cooperative effort between the United States and Western Europe in the Atlantic could also be instrumental in the avoidance of potential conflict and rivalry over resources as they are eventually discovered and become exploitable.[15]

Japan

As we have seen in chapter 7, Japanese society is in transition. Japan's national goals of recent years are being questioned, but new goals have not yet crystallized. The transition may continue for a number of years, with old values clashing with new, possibly resulting in serious instability and continuous lack of a clear external sense of direction.

In the immediate post-World War II years, the United States assumed direct responsibility for transforming Japanese society, principally through the inoculation of democratic values. With the implied consent of the Japanese the United States, in effect, assumed tutelage over Japanese foreign policy for some 25 years after the war. Remarkable progress has been achieved in making democracy acceptable in Japan. However, neither Japan's decision-making process nor the outlook of her leadership makes the nation quite ready to assume the responsibility in world affairs corresponding to her economic and technological achievements. Since this lack of preparedness has deep cultural roots, it is a moot question whether the blame can be placed at the door of America or not. The situation is further complicated by the uncertainty of internal societal change; the interplay between internal and external factors may go far enough to erase America's principal achievement in the Far East in recent decades—a stable and democratic Japan.

The leverage the United States had over Japanese society in the early postwar years no longer exists. Japan herself will have to resolve the problems of internal social change and find new directions. However, as a superpower and a friendly nation, the United States is in a position to influence Japan's behavior in foreign affairs. Since Japan's involvement and activities abroad are closely linked with, or readily find repercussions in, developments at home, the United States, to a degree, would also have influence over Japanese internal societal change.

A long-range policy that the United States might pursue with regard to Japan—and that would be in the interest of both nations—is to foster Japan's understanding of, and active involvement and acceptance of responsibility in, foreign affairs on a global scale commensurate with Japan's capability and interests. This policy would be designed to change the outlook of the Japanese elite from insular and self-centered to global—somewhat similar to the responsible attitude the British elite had, except that it should be oriented to contemporary trends in world affairs and devoid of imperial connotations. In this role of responsibility Japan would be expected to be an active participant in the molding of a future international system, an activity in which both U.S. and Japanese interests largely coincide.

U.S. officials might point out, with a considerable degree of truth, that this was precisely what the United States has been trying to do for the past 10 years, if not longer. Japan has been urged to contribute more to foreign aid and thus enhance international stability. She was encouraged to be a more active participant in the regional affairs of Southeast Asia and thus assume greater responsibility in that area. The United States has been strongly in favor of greater military responsibility of Japan for her own and regional defense.

Whereas these activities have no doubt tended to give Japan a greater role and responsibilities in international affairs, in many instances they also intended to free the United States from some of her own burdens and responsibilities or to induce the Japanese to do what the United States would do anyway if Japan did not. Moreover, they have been mostly compartmented, concrete actions into which the Japanese were being nudged; they did not foster initiative or broad responsibilities on the part of Japan, and they did not usually have global reach. Since U.S. self-interest was directly involved, American policy in broadening Japan's participation and responsibility in world affairs suffered somewhat in its ability to inspire.

What is needed is a more farsighted policy in which the payoff may not be immediate and in which Japan is given opportunity for initiative and contributions of her own so that, eventually, the Japanese would shed their insecurity and timorousness. One example might be to obtain a permanent seat for Japan in the U.N. Security Council and help Japan develop her own position of responsibility there, free of American nudging. Another example would be to invite Japan into the diplomacy and decision-making process with regard to the Middle East and to do this early enough so that she could be a full-fledged participant in meaningful decisions, and not somebody invited after the fact. The United States could foster an attitude of responsibility on the part of Japan with regard to Western Europe and its future stability and cultivate a view among responsible Japanese officials and businessmen that technological and economic cooperation with the Soviet Union and China is not just an opportunity to obtain raw materials or other economic benefits but a momentous development whose goal, in part, is to change social values of millions of people and to mold the development of world society. Given these goals, the payoff cannot be measured in purely economic terms.

As Japan is drawn into these various activities and made a responsible participant through concrete major programs, the Japanese may lose their prejudices regarding certain nations, shed their narrow market orientation, evolve a rational and realistic national strategy in world affairs, and gain confidence in themselves while constructively contributing to the molding of an international system that would safeguard their long-range well-being. This kind of attitude and activity is likely to have a beneficial feedback on the domestic scene and contribute to internal stability.

Concrete policies and programs in science and technology can play an important role in helping to buttress Japan's economic vulnerability and the limitations of her resources. Timely cooperative programs in advanced technology could help to solve the problem of limited resources for both Japan and the United States and become important pillars for strengthened relations between them. In some of these technologies where mutual benefits of such cooperation clearly exist and for which Japan is especially suitable—for example, marine resources development—it might be advantageous to the United States to take the initiative, to stimulate an early interest by Japan in this area, and to broaden the framework of cooperation to other nations.[16]

The United States and a Future International System

A policy directed toward a future international system would involve the appraisal, monitoring, and orchestrating of a number of variables. These include world distribution of power, the prevailing international system of values, international law, international institutions, and a number of factors (including technology) affecting international stability. As discussed in part 3, these variables are basically interdependent; changes in one affect changes in all or some of the others. Some further illustrations can be provided.

The growth of international institutions—the United Nations, its specialized agencies, and multinational corporations (MNCs)—is likely to create greater interdependence in the world, strengthen international law, curtail the power of nation-states, enhance property- and peace-oriented values and increase international stability. Conversely the spread of militant ideologies (as, for example, was the emergence of right-wing nationalism as the prevalent value system in both Europe and Asia in the 1930s) is likely to result in instability, weaken international law and institutions, and affect the existing world distribution of power. A different variant of the interrelationship of the various components of an international system began to evolve in the 1960s: The impact of nuclear weapons created a concern for international stability, inducing the United States and the Soviet Union to a degree of cooperation (and thus ending the Cold War, although not international rivalry). This, in turn, has begun to be reflected in the strengthening of certain areas of international law and international institutions (e.g., the Outer Space Treaty; the Non-Proliferation Treaty; the beginning support of international institutions by the Soviets). On the other hand a major shift in world power in favor of the Soviet Union in the next decades could rekindle Soviet militancy, weaken international institutions, and result in a prevalent value system leaning in the direction of the "socialist" camp.

The entry of Communist China into the United Nations in November 1971 created the possibility that one of two divergent developments might take place, with major implications for a future international system. It is possible that, by joining the United Nations, participating in its multiple activities, and thus expanding contacts with other nations in matters of mutual concern, China will be drawn into the mainstream of activity of the world community of nations. She may become less suspicious of the "capitalist" states and less power oriented and, in general, may eventually "mellow." This development is likely to strengthen the kind of a future international system that the United States might find desirable.

However, a discontinuity might be on the horizon in the evolution of the existing international system. The power gap between the less-developed nations and the advanced world is narrowing. Moreover, the United States—as all other nations—can no longer solve all of its problems unilaterally or on a subsystem level, and a number of them require solutions on a universal scale. If the Chinese make substantial progress—as they profess to attempt— in establishing their leadership of the LDCs through the United Nations, the United States will be facing new and not readily predictable challenges in an international system increasingly polarized along North–South lines.

It is thus paradoxical that the United States, which has a major stake in the outcome, has neither a firm, carefully developed policy for a future international system nor an institutional setup to foster it. Our universities have not yet developed a sufficiently comprehensive conceptual framework that would encompass and integrate the various elements of the international system and that could provide a solid foundation for policy. At a time when, largely owing to technological developments, the pace of life is very rapid and when changes that used to take generations occur in a single decade, the need for a sophisticated and purposeful policy with regard to a future international system is greater than ever. To be sure, the Secretary of State and the President himself occasionally assert the nation's concern with the building of a peaceful, stable world where mankind and the United States would live in freedom and prosperity. But the fact remains that there is no office in the U.S. government that has a full-time and continuous responsibility with regard to the future international system in its many ramifications.

It would be unnecessary and counterproductive to build a new and large bureaucratic mechanism for a future international system. A number of existing offices are concerned with various aspects of this subject. The problems are that the concern of these organizations with a future international system is highly compartmentalized and that not all of them are conscious of the relevance of their activities to a future international system, although this does not necessarily mean that their actions run against the objective of building a desirable one.[17]

The establishment of an Adviser to the President for a Future International System would help significantly. His or her function would be to formulate general guidelines of policy for a future international system, to monitor the activities of the various agencies that affect its development, and to advise the President on a possible redirection of policies of these agencies. Such an adviser would also watch the balance between national and international interests, and advise the President when the balance is shifting too much in one or the other direction (at present, the balance between these two interests is largely determined by the tug and pull of the various agencies). The very existence of such an adviser would stimulate the consciousness of the agencies concerned of the relationship of their activities to a future international system and would have a catalytic effect in promoting their thinking and, as needed, reorienting policies in that direction.[18]

It is not enough to introduce improvements in governmental machinery alone to provide a balanced and comprehensive policy for a future international system. Such a policy would require a coherent conceptual framework and a fountain of expertise and of intellectual support. For the lack of a better term one might say that a future international system needs a "constituency," on which governmental policy could lean, that would be capable of invigorating policy by fresh thinking. To be sure, there is such a constituency at present, but, as with governmental institutions, it consists largely of scholars focusing on international organization and law. As a result their product is not truly comprehensive and is frequently one-sided, with overtones of proselytizing for a stronger universal organization. Although a limited number of studies do provide a broader political approach,[19] it still remains the task of the universities to develop, perhaps in cooperation with certain business and governmental circles, a stronger and well-balanced intellectual community for a systematic analysis and support of a desirable international system.[20]

A major problem with important implications for the international system which requires immediate attention and a careful follow-through in coming years is that of foreign assistance. The Agency for International Development is nearly exclusively involved in *concessional* assistance, and the focus of its development activity is overwhelmingly on food, population problems, health, and education, and not on industrial development. But availability of petrodollars and progress in economic development of a number of LDCs not blessed with abundance of oil are shifting requirements to *reimbursable* technical assistance and *industrial* development. Opportunities for molding a future international system through science and technology are significantly broadening, since most of science and technology lies in the industrial sector. The U.S. government has not, however, adequately faced up to this issue in both policy and organizational terms.[21]

In the formulation of policy with regard to a future international system the direct and indirect impact of technology as an independent variable on the system requires close attention. Concurrently an examination is needed of how technological instruments could be effectively used to implement the nation's strategy for a future international system. Lastly, the totality of the nation's capabilities and the extent to which the United States is willing to allocate resources toward shaping a future international system will be an important consideration in the next decades.[22] It is precisely this willingness to allocate or not to allocate resources that will determine which among the several principal variables involved—world distribution of power, international organization and law, the prevailing system of values on the international arena, and the state of international stability—must be viewed as dependent and which as independent variables. If the nation chooses to withdraw in the future and allocate limited resources to international pursuits, more of these variables will, from the point of view of the United States, become independent. But even with limited willingness to allocate resources for this purpose, the requirement for a strategy for a future international system will not disappear. If Switzerland and Sweden have carved out a role for themselves—and they have—in influencing the development of a future international system, a superpower can afford to do no less.

Over the Horizon: Technology and Directions of Societal Development

part 5

The Changing Nature of Advanced Societies

14

WE SHALL NOW TURN to a discussion of the next century, and of how the remainder of this century will be affected by the potential developments and requirements of the next one. I shall cover a larger span of time and entities of larger magnitude, and will thus have to resort to a higher level of abstraction than previously. The issues analyzed include the nature of evolving society, the directions of its evolution, and the scope for human will to influence these directions. I raise no claim to being exhaustive or definitive in addressing them; in fact a good part of the remainder of this book is largely impressionistic and speculative. These caveats notwithstanding, a reexamination of issues is warranted for at least two reasons: (1) My analysis may give a new perspective useful to other researchers in their more systematic scrutiny of the subject. (2) Since these issues are brought within the context of international relations, I hope that my analysis may expand the growing concerns and frontiers of that field.

I shall go beyond science and technology proper in my analysis. Any serious attempt to examine the impact of technology and the question of how it can be controlled must place it in a broader social context and look beyond technology if it is to succeed.

The Postindustrial Society: Where Is It Taking Us?

The idea of the postindustrial society was developed by Daniel Bell. It has been accepted by other analysts of the future (notably, Herman Kahn and his colleagues at Hudson Institute), although with some variations in its meanings. Other writers have used a different term to describe the evolving society of the future, but in its substance it has remained close to Bell's postindustrial society.[1] It would therefore be appropriate to begin by examining the postindustrial society to determine where, in the view of Bell and Kahn, we are going.

The Postindustrial Society of Daniel Bell

The single most important characteristic of Professor Bell's postindustrial society [2] is structural. He believes it has been achieved when more than half of its labor force is employed in services rather than the production of goods. This transition is the third (after the agricultural and the industrial) stage in the evolution of society. In the United States the goods-producing and the service-producing sectors were about evenly balanced in 1950; by 1970 more than 65 percent of the U.S. labor force was employed in the service sector. The United States is thus the first nation to enter the early stages of the postindustrial society, with other advanced nations following, in varied degrees.

Largely as a result of the structural change in the economy the professional and technical class rises to preeminence. The base of new power shifts from property and political criteria to knowledge—not any knowledge, but theoretical knowledge—and thus signifies the primacy of theory over the empiricism that characterized industrial society. Consequently, asserts Professor Bell, "the university—or some other form of knowledge institute—will become the central institution of the next hundred years because of its role as the new source of innovation and knowledge." [3]

The primary resource of the postindustrial society is human capital. Whereas the "design" of industrial society was "game against fabricated nature," in the postindustrial society it is "game between persons." Information has supplanted energy as the "technology" of the postindustrial society. The chief problem of the new society is the organization of science. Its "axial principle" is "centrality of and codification of theoretical knowledge," which has supplanted "economic growth: state or private control of investment decisions," which was the axial principle of industrial society. [4]

Like any social system the postindustrial society must have an ethos of its own—"the values enshrined in creeds, the justifications established for rewards, and the norms of behavior embodied in character structure." Thus, we learn, just as the Protestant ethic was the ethos of capitalism, "the ethos of science is the emerging ethos of post-industrial society." A society permeated by the ethos of science "comes closest to the ideal of the Greek *polis*, a republic of free men and women united by a common quest for truth." To be sure, the community of science has its own hierarchies and prestige rankings, but they are based on achievement confirmed by peers, and not inheritance, contrived manipulation, or brute force. [5] The ethos of science thus comes close to another principle central to Professor Bell's postindustrial society: meritocracy. Technical competence determines access to power in the postindustrial society, a society that is "the logical extension of the meritocracy; it is the codification of a new social order based, in principle, on the priority of the educated talent." [6]

This, in brief, is the concept of the postindustrial society, a concept of universal application, provided that a given society has reached the requisite stage. However, after presenting a formidable theoretical structure, Professor Bell goes rather far in minimizing its practical importance. We are told that "there is no specific determinism between a 'base' and a 'superstructure'; on the contrary, the initiative in organizing a society these days comes largely from the political system. Just as various industrial societies—the United States, Great Britain, Nazi Germany, the Soviet Union, post-World War II Japan—have distinctively different political and cultural features, so it is likely that the various societies that are entering a post-industrial phase will have different configurations." The importance of the concept appears to be reduced even further as Professor Bell informs us that all it suggests is "that there is a common core of problems, hinging largely on the relation of science to public policy, which will have to be solved by these societies; but these can be solved in different ways and for different purposes." [7]

The foregoing suggests a number of questions not explicitly raised or answered by Bell's analysis. What is the magnitude of constraints that the tenets of the postindustrial society will impose on different political and cultural structures? Does this society favor *some* political systems or cultures, while running against others? If there is no determinism between the change in the social structure as embodied in this concept and the politico–cultural superstructure, what is the price for disregarding implied requirements of the postindustrial society? Are different societies truly free in the postindustrial phase of their development to mold their future as they see fit?

This latter view would certainly be consistent with the highly important role that planning and forecasting play in Bell's society and in his emphasis on "a process of direct and deliberate contrivance," which, allegedly, characterizes our time and includes conscious transformation of societies.[8] However, a reader who would cheerfully anticipate the American society's remolding itself for a brighter future as it proceeds further into its postindustrial stage is in for a rude shock. Indeed, something truly funny happens to this theoretical concept on its way toward fulfilling its potential in the United States.

To follow Professor Bell's analysis further: In its industrial stage the United States was governed by a market economy and market decision making. There was an "unspoken consensus" on the utility of this system, which coincided with the major consumer values of the society. In moving into the postindustrial stage, the United States has become a national society in which critical decisions affecting all parts of society simultaneously are made by the government, and not the market. While the market disperses responsibility, governmental decisions and planning accompanying them make the points of decisions highly visible and thus attract antagonistic interests and viewpoints toward them. As a result "government becomes a cockpit" of political struggle among the various interests involved. Moreover, there is greater participation in political life, there are more groups and interests to check each other, and this leads to an impasse and frustration. The situation is further aggravated by the lack of consensus. The old consensus based on the premises of individualism and market rationality is gone; according to Bell the United States is moving to a communal ethic, but without the community's being as yet wholly defined. Decision making has therefore become highly politicized, and "the conception of a rational organization of society stands confounded. . . . Rationality . . . finds itself confronted by the cantankerousness of politics, the politics of interest and the politics of passion." [9]

While politics is thus perhaps the single most important barrier to evolution of the "ideal" postindustrial society, another one is "culture." Capitalism, to sustain mass production, undertook to promote a hedonistic way of life. Acquisition of material possessions as a status symbol and pursuit of pleasure were encouraged. The new hedonistic culture destroyed the Protestant ethic, and we are thus facing a growing disjunction "between the social structure (the economy, technology, and occupational system) and the culture (the symbolic expression of meanings), each of which is ruled by a different axial principle." [10] The new culture not only fails to sustain the social structure in the postindustrial society but also is antagonistic to it and contributes to the politicization of decision making and more group conflict: "cultural issues are transformed into political issues as women press for the repeal of anti-abortion laws, young people for the legalization of marijuana, and sexual deviants for the end of discrimination." [11]

If the theoretical concept of the postindustrial society did not have enough determinism to assert itself, then the forces stimulated or evoked by this society do appear to have enough to deadlock society in a crisis.[12] Will the United States succeed in getting out of this deadlock? Professor Bell proposes no specific course of action, and he sounds very pessimistic about the ability of some of the key variables to assert themselves so as to get us out of the woods. In fact one detects notes of despair in his analysis: "Politically, there may be a communal society coming into being; but is there a communal ethic? And is one possible?"[13] The question is left unanswered. Elsewhere he strikes an even more gloomy note:

> But a technocratic society is not ennobling. Material goods provide only transient satisfaction or an invidious superiority over those with less. Yet one of the deepest human impulses is to *sanctify* their institutions and beliefs in order to find a meaningful purpose in their lives and to deny the meaninglessness of death. A post-industrial society cannot provide a transcendent ethic—except for the few who devote themselves to the temple of science. . . . The lack of a rooted moral belief system is the cultural contradiction of the society, the deepest challenge to its survival.[14]

At one point Professor Bell offers a prediction:

> The politics of the future . . . will not be quarrels between functional economic-interest groups for distributive share of the national product, but the concerns of communal society. . . . They will turn on the issues of instilling a responsible social ethos in our leaders, the demand for more amenities, for greater beauty and a better quality of life in the arrangement of our cities, a more differentiated and intellectual educational system, and an improvement in the character of our culture.[15]

But this statement does not sound very convincing in the face of the odds—which Professor Bell himself pointed out—against this turn of events and in the absence of concrete suggestions on his part about how to get these issues high priority on the societal agenda and, even more importantly, how to ensure their favorable resolution.

The Postindustrial Society: Herman Kahn's Version

Herman Kahn's analysis of the future [16] leans heavily on Bell's concept of the postindustrial society, but in some of his assumptions and in his projections of future developments Kahn differs significantly from Bell. Unlike Bell he is not concerned with a theoretical concept. In practical terms Kahn's postindustrial society is open-ended, with multiple developments possible. These developments include both the continuation of trends and reactions to them—"many different branching points," as he puts it.[17]

This open-endedness notwithstanding there is a certain core of developments to which Herman Kahn subscribes, or, at least, he considers them more likely than others. The postindustrial society will generate more and more affluence—and there will be growing leisure. Affluence is more likely than not to increase the average person's feeling of alienation—he will be, to a large degree, freed from economic constraints but will increasingly feel his inability to influence decisions in society. The individual will thus tend to escape into a private world of his own, some of which will border on the pathological (drugs, neuroses) and some not. Although the postindustrial society is likely to be characterized by a mixture of values—for example, there will be people who will still pursue the old Protestant work ethic and there will be others embracing humanistic values—the general trend is toward hedonism, the seeking of pleasure defined in material terms, with some admixture of humanistic pursuits.[18]

Unlike Bell, Herman Kahn does not subscribe to the view that the postindustrial society will be a knowledge or a learning society. "A learning society would probably be more indicative of the last stages of an industrial society than of a post-industrial one." Education will be a useful tool, but in itself it will not play the central role: "in a post-industrial society technocrats will become routine."[19] Also, unlike Bell, Kahn is not worried that the counterculture may deadlock society or "win." In the Hegelian tradition, countercultural values are most likely to synthesize with those of traditional society. Indeed Kahn believes that such a synthesis could work quite well: "In fact, most of the historically successful societies were characterized by both hard work and orgies, virility and vice."[20] Alternatively, a counterreformation might arise—fascist, populist, or religious—and wipe out the counterculture. Nor is Kahn concerned that, with affluence and leisure, the postindustrial society might be characterized by boredom. It is the affluent literati who get bored; most people have no problems filling their time with their families, personal property, and hobbies. Thus the postindustrial society will not be a purposeful society. When tired with trivia, people may search for meaning and purpose by occasional flares of religiosity, but a true religious revival that would reverse the march of the sensate culture is doubtful.[21]

Is there an "ideal" or a "desirable" society that Kahn would like to see evolve? Indeed there is, and it is likely to materialize. It is a "Westernistic" society, resembling Hellenistic Greece. The difference between Hellenistic Greece and the Westernistic variant of the postindustrial society will be that, in the latter, everybody will be rich and will lead the life of a Hellenistic gentleman. Unlike the aristocracy of later periods Hellenistic gentlemen were not especially sophisticated, genteel, or chivalrous. They were primarily interested in their personal lives and in having a good time rather than in their philosophical justification. The Westernistic society will be mosaic, characterized by coexistence of various groups adhering to different values—managers clinging to a neostoicism, convinced that it is their duty to keep society running; others pursuing neoepicureanism; periodic religious revival movements; and so on. Material horizons will be expanding, but Westernistic society will lack spiritual content. There will be a widespread feeling that the Westernistic civilization represents a culmination of history, but at the same time, it will be characterized by a sense of decadence and a degree of alienation. To those historians who described the Hellenistic period as one of decay, Kahn retorts:

Perhaps so, but it was a period of decay that lasted some seven hundred years, in which people lived relatively well and did some remarkable things. If a Westernistic syncretic culture and society can grow out of the Western experience the way the Hellenistic did from the Hellenic, we should be more than satisfied.[22]

Whereas Kahn's postindustrial society is not deterministically stalemated by forces and counterforces basically generated by technological impact, it is permeated by technological determinism in a different sense. The evolution of society is basically propelled by technological influence with its strong materialistic and sensate overtones. "People," as Kahn says, "are amazingly adaptable." [23] They adapt to various influences and elements, they synthesize them; in Kahn's postindustrial society they do not seem to control or run them. To be sure, individuals and groups may rebel against a particular manifestation of the technological impact or try to escape it in private worlds of their own. But these actions are usually on a fragmented scale and do not really change the somewhat amorphous, pluralistic flow of social evolution; in fact they are part and parcel of it. Whether in rebellion to the impact of technology or in cooperative response to that impact, the end product has strong deterministic overtones.

A "Westernistic" society—whose sights are set much lower than those of Daniel Bell's theoretical postindustrial society—in part reflects Kahn's value preferences. But there is little doubt that the concept of a Westernistic society, to a large degree, is all that Kahn considers to be realistically attainable, given his implicit assumptions regarding the constraints that exist in contemporary social evolution. At one point Kahn asserts that man is an important term in an equation of history—if you vary him, you change the outcome. Thus, "it is possible to alter future normatively." [24] But he provides no clues about how this could be done, and in the total context of his analysis, this assertion is not persuasive.[25]

Some Questions about the Postindustrial Society

One can disagree with this or that assertion of Bell or Kahn, and one can express reservations about some of their premises, but it is difficult to deny that their analyses have articulated certain fundamental trends and developments that have already appeared or may develop in the future. A more constructive approach, therefore, would be not to criticize individual points, but to go beyond their analysis and raise the question of human ability to steer a postindustrial society. Do we really want the predictions of Daniel Bell and Herman Kahn to come true? What sort of alternatives do we have? What sort of constraints does a postindustrial society impose on individuals in their effort to mold their future? What are the leverages for steering society in a certain desirable direction? Is a postindustrial society the foreseeable form of society for the next century, or is it a mere phase in societal development to be supplanted in a not-so-distant future by a new society, with distinct characteristics of its own?

I do not use the term postindustrial society to imply adherence to a tightly structured theoretical framework such as advanced by Daniel Bell, but simply to refer to a society with certain general characteristics of the postindustrial stage, it being recognized at the same time that many of these characteristics have originated in the industrial society and simply continued.[26] My analysis is essentially from the political point of view.

The Postindustrial Society, Political Authority, and Collective Human Will

Technology, by fostering an immensely complex, pluralistic society, imposes definite constraints on certain forms of political authority in the postindustrial stage—constraints that can be so compelling as to make some forms of political authority nearly unworkable.

One can distinguish four distinct types of authority in the modern world: (1) democracy, in which power genuinely lies in the hands of the electorate; (2) oligopoly (to borrow the term from economics), in which power basically lies in the hands of a few large institutions or interest groups; (3) autocracy, where power lies in a single individual or a small group that tolerates some dispersion of authority but is basically effective in running society from the top; and (4) totalitarianism, which tightly controls every aspect of society— the economy, the government, and the population, including its public opinion and ideology.

The postindustrial society, because of its high complexity, and, hence, the need to rely on experts and large organizations, very seriously impedes the role of the electorate in running democratic societies. It tends to convert a "pure" democratic state into oligopoly, with some democratic overtones. More specifically, democratic elements are impeded on two levels: (1) The effectiveness of the electorate is curtailed by its inability to pass judgment on highly complex issues requiring expert analysis, and (2) the ability of the elected chief executive (who, in theory, can command the needed expertise) to give a direction to the nation is limited by the power and amorphous nature of large organizations and interests, which he can control only partially.

The postindustrial society's need to rely on experts, the rise of large organizations with a degree of autonomy of their own, and the generally highly skilled and educated population characteristic of advanced economies pose a formidable threat to totalitarianism. Indeed, these factors make impossible the extent of control normally associated with totalitarian regimes. The Soviet Union discovered this long before reaching the postindustrial stage. Stalin's rule was the closest to totalitarianism the USSR knew; even so, the society was much too complex to be effectively run by totalitarian methods in the last years of his life, and totalitarianism eroded after his death in 1953. The present autocratic rule in the USSR is under pressure to turn increasingly oligopolistic, although it still manages to preserve its basically autocratic nature. Thus, to an extent, we can see some convergence in the form of government between the United States and the USSR—a certain gravitation toward oligopoly and pluralism, although political differences remain important.

The impact of technology in the postindustrial stage is thus paradoxical. On the one hand it enables a given advanced society to mold its future unhindered by nature and physical environment and to do so on a vast scale. On the other hand, by promoting complexity and pluralization, technology inhibits the mustering of the collective human will essential for an effective molding of society's future. This latter tendency is stronger than the former, and it exhibits distinct deterministic overtones.[27] In this sense both Bell and Kahn are right. For a democratic society like the United States there are two typical models of determinism: either a deadlock among the various forces involved or a free, amorphous, and aimless interplay among them, seeking synthesis or coexistence to avoid deadlock.

"Pure" models of determinism aside, in practice the deterministic aspects of technological impact combine and tend to produce a semiparalysis of national will, a phenomenon observable in both the United States and the Soviet Union. As compared with democratic societies autocratic regimes have certain advantages in mitigating the deterministic impact of technology: Usually they have a stronger central government and can resort to force. Advanced democratic societies, however, are not devoid of means for molding their own future. In recent years this has been demonstrated by Japan.

Japan has become an effective and dynamic society principally because of her communal spirit and strong sense of national purpose. The problem is that Japan's national purpose—economic aggrandizement—is somewhat narrow and is predominantly dependent on the very tool that has the propensity to undermine the nation's communal spirit and pluralize her society, namely, technology. Signs of change are already detectable in Japan and, especially when triggered by an economic setback or other adverse developments, can accelerate rapidly. Nevertheless the case of Japan does demonstrate that a democratic society imbued by communal spirit and a sense of purpose can go far in molding its destiny.

It is possible that, in an advanced stage of the postindustrial society, both democratic and autocratic regimes might regain some of their "pure" characteristics and give a stronger sense of direction to their societies. As pointed out in chapter 11, with the widespread assimilation of computers and advanced systems of information and communications, it would be possible to combine a greater degree of control at the top with the decentralization on the lower levels that could then be coordinated on a national scale. When this capability is accompanied by a reorganization of institutions at the top to strengthen the chief executive, democratic governments can develop a greater sensitivity to local problems, give greater degree of participation to local population, and provide more effective leadership by the chief executive over large organizations. A similar capability in the hands of autocratic regimes would sensitize the leadership to local problems and stimulate incentives (and, hence, efficiency) on the lower level while strengthening overall control over governmental organizations and the economy. Thus, whereas the technological instruments will be similar, the critical difference between the democratic and the autocratic regimes would continue to be a free election of leadership in the former, and hence a fundamental difference in the purpose and extent of responsibility to the public with which the available technological instruments will be used.

The Postindustrial Society: Whither Power?

The question of influencing the direction of societal development is inseparable from the question of where power lies in the postindustrial society. After all, it is only by finding the locus of power that we can exert leverage on society.

In Bell's "ideal" postindustrial society theoretical knowledge will become the base of new power, and eventually "the entire complex of prestige and status will be routed in the intellectual and scientific communities." However, even in this ideal society, government would still play an important role—it would control economic forces, and all the crucial decisions regarding the balance of the economy and its growth would come from it.[28]

Considering the complexity of future advanced societies and the magnitude of the task of steering them, one can agree in principle that the main instrumentalities of power will be high-grade intellectual instruments and effective organization to apply these instruments—most likely, a government. Nevertheless one can raise a number of reservations about Bell's delineation of the sources of power.

One reservation pertains to the nature of knowledge. Although theoretical knowledge has been gaining in importance in recent years, it may be that with regard to power, applied knowledge may be more important in the future. Theoretical knowledge is relatively free, it is transferable, it usually does not recognize boundaries, and by itself it may be but remotely relevant to power or to the steering of society. It is how the theory is adapted to a particular environment and actually applied that is important in power terms. Thus, as we look into the coming decades, when a number of nations enter the postindustrial stage, it is conceivable that one or two regions will generate most theoretical knowledge, while the rest of the world will focus on applying it. And the bulk of power is likely to go to those who will be the most effective in the development of applied knowledge and its reduction to practice.[29]

A second reservation arises with regard to the relative importance of government—when compared with the intellectual and scientific communities—as a locus of power. Even in the most critical periods of the nation's reliance on science, scientists have been on tap rather than on top. This is likely to be so in our "ideal" model of the postindustrial society. Scientists and experts are primarily specialists, but what is necessary to run society is syncretic knowledge combined with political acumen. Ideally a government can develop a corps of capable and responsible civil servants and administrators who lean on universities and research institutes for the necessary specialized knowledge. The prestige of the universities and their faculty—and other sources of such knowledge—will increase, but the bulk of power will probably rest with the users of knowledge—the government—rather than its source.

Given our "ideal" conditions, perhaps one should assume more participation by professors in governmental offices, but it is not at all clear that the generators of theoretical and specialized knowledge will have the requisite qualities to make decisions and exercise power, as distinguished from doing high-grade research. With few exceptions the record of academicians in government in the United States has not been encouraging in this regard.[30]

Moreover, our hypothetical "ideal" postindustrial society may never materialize, and we should consider circumstances short of ideal. In this connection note that our institutions of learning have not been particularly successful in providing a solid body of knowledge to guide society and help to solve its fundamental problems. There is no guarantee that they will do so in this century or beyond.

The "hard" sciences have been more successful than others in demonstrating solid achievement, but their successes have been used without necessarily giving the scientists true power. Whereas the social sciences would be expected to provide a body of knowledge to help guide society, the situation here is not encouraging at all. To cite a few examples:

Economics has probably been more successful than other social sciences in developing a theoretical foundation and specific conceptual tools for influencing economic stability and growth, but even here deficiencies are glaring. Economists have never been successful in incorporating technology into their theoretical constructs; in fact they have been known to resist efforts to remedy this shortcoming. In a number of instances their failure to do so has significantly weakened the utility of their analyses. Nor have economists been especially successful in providing effective prescriptions for the economy in the various phases of its development. An illustration is provided by the December 1973 meeting of the American Economic Association, which held a post mortem on the accomplishments of the expiring year. As reported by *The New York Times*, the economists "did not know what the Federal Government's controls program had done to hold down prices and wages. They were not sure what economists could most usefully do for policymakers. And they were confused about what questions their profession ought to be ready to answer in the years ahead." [31]

Political science, which more than any other field is responsible for providing a body of knowledge essential for steering society,[32] has been far from successful along these lines. In defense of the discipline one can say that its task is more difficult than that of the "hard" sciences, but a good part of the blame can be placed at the door of the political scientists themselves. There is a great deal of stress on theory in political science, but this stress largely takes the form of attempts to imitate the "hard" sciences in focusing on scientific method and scientific thinking. As a result, analysis of goals and elucidation of values have suffered. Models and constructs have been built, many of which have remote relationship to reality. Policy science has been largely neglected. As Evron Kirkpatrick has noted, because of the influence of science on the discipline, it is predominantly oriented to the past and not the future, while the policymaker, the decision maker, and the politician are future oriented.[33]

Stimulated by the upsurge of social and environmental problems of recent years, many political scientists have reacted by calling for "relevance." [34] As viewed by the reformists, "relevance" mainly means the application of the intellectual and human resources of the discipline to immediate social and political problems, including political action. Perhaps in response to the shallowness of "relevance," there has been recently a resurgence of interest in policy studies, but its viability remains to be proved. As yet the fundamentals essential for a judicious steering of society into a not-so-clear future remain largely unexamined.

In short, even if there are certain objective factors in the postindustrial society that favor the rise of the university in relative influence—although, contrary to Bell, it is not at all likely that the university will become "the central institution of the next hundred years"—this can happen only if the universities bring themselves up to this role through a thorough reexamination and redirection of their scholarship.[35]

A question must be raised about the relative power of business in the postindustrial society. According to Bell, in the postindustrial society business is increasingly subordinated to the government, and the business corporation increasingly shifts its function from the "economizing mode" (production and profit) to the "sociologizing mode" (taking care of employees, developing social responsibility). While the power of business declines, Bell notes "a paradox"—the rise of the multinational corporation (MNC), "the spread of the world capitalist economy," which appears to escape subordination to political decision. Characteristically Bell sees another potential deadlock in this development: At the end of the twentieth century the world may end up divided in a "class" or "color" struggle between the rich and the poor nations.[36]

To Kahn the role of the MNCs is less complicated. They will be the carriers of the "capitalist" values to the Third World and will help to develop it. Though important and powerful, most MNCs will have to adapt to the native political environment. However, they will have a strong penetrative power—even if the local government is antagonistic to MNCs, it cannot afford to discriminate against them for long, since the nation will fall behind in economic development. Kahn goes further to say that historians of the distant future are likely to view the MNC of the late twentieth century as "the engine of progress," and "a uniquely dynamic institution which imparted a good deal of its dynamism to the society as a whole."[37] Presumably all this dynamism would be displayed by the corporations essentially in their "economizing mode" (to borrow Bell's terminology).

Probably, in the later stages of the postindustrial society, the business corporation, at least in its "economizing mode," will markedly decline in influence. However, the power of business, whether nationally or internationally, cannot be readily dismissed in the earlier stages. Business commands huge resources and is likely to use them to protect its interests. Moreover, if a combination of technological impact and American policy succeeds in pluralizing the "socialist" camp, we may see powerful quasi-multinational corporations emanating from Eastern Europe and the USSR and spreading throughout the globe. Although largely state owned, they may possess a certain autonomy of power and behave much like their capitalist counterparts.

Depending on the policy of governments and international organizations, a somewhat different turn in the development of business corporations may be seen. The U.S. government, along with other governments and international organizations, could formulate systems-oriented projects for coping with major public problems in the United States and the Third World. These might include improved police protection based on alarm systems geared to computers; computer-based educational techniques that may revolutionize the learning process; integrated, large-scale approach to sanitation using advanced automation techniques; advanced methods for eliminating illiteracy in the less developed countries (LDCs); mass-production techniques for providing low-cost housing in advanced and less developed nations; solutions to the problem of bureaucracy; and so on. If bids for such projects were offered to major corporations, this could eventually result in a large, highly advanced, and efficient public-service-oriented industry with global operations. Although operating for profit,[38] this industry might attract idealistic, highly intelligent young people and develop a strong sense of responsibility for social improvement. Moreover, it might develop high-grade "think tanks" of its own, capable not only of providing solutions of specific problems but also of scanning the intellectual horizons in search of social directions and potential future problems. Thus we may see two types of business corporations existing side by side—the more traditional product- and profit-oriented corporations and the expanding, newer, public-service-oriented corporations, with a stronger sense of social responsibility. The new corporations need not be subservient to the government, passively waiting for whatever project the government chooses to develop. Like the leading nonprofit corporations of today—for example, the Ford or the Rockefeller Foundations—these corporations may seek out promising areas for public service and eventually get local or federal government interested in supporting them. They are likely to have influence and power in their own right, but insofar as this power is exercised within the framework of a growing communal ethic of the postindustrial society, it need not be at the expense of, or in conflict with, other principal loci of power—such as the government. It is conceivable, however, that if, in some nations, public responsibility in the government declines and its bureaucracy degenerates, such corporations may preserve a sense of responsibility and effectiveness and thus provide a potential source of remedial influence. Given public support, these corporations may play a constructive role in restoring governmental responsibility and effectiveness.

The foregoing example brings us to another point in appraising the loci of power in the postindustrial stage, namely, that not all societies will be equally successful in solving their problems. The horizons for organized human will to mold society's future are expanding vastly, but the minimum threshold for societies to govern themselves is higher and more demanding to surmount. The loci of power in the postindustrial society will vary, depending on the degree of success in social organization and its ability to solve the society's problems.

In dynamic, vital societies having a strong sense of purpose, the locus of power may rest with the government, assisted—but not necessarily shared in terms of power—by the universities. Alternatively, if the universities fail to provide the necessary knowledge and guidance to society, the government may develop its own internal sources of knowledge. A different pattern may develop: The locus of power may reside primarily in the government and secondarily in large, public-service-oriented corporations with a strong in-house intellectual and scientific capability, perhaps assisted by independent organizations like the RAND Corporation.

We can go further down the line and envision a "Westernistic" society, with its mosaic culture. Here power would be largely pluralized, even diffused, so as to lack a strong center of gravity. Some of it would reside with the government; some with conventional business corporations; some with public-service-oriented corporations; some with religious leaders who would enjoy periods of revival, stimulated by individual alienation and a search for a purpose; and still some with the leadership of cultural, racial, or ethnic groupings, which would seek security, a sense of meaning, and identity in their respective components of the mosaic culture. This society would be reasonably successful in providing the material needs for its population, it would be affluent, it would resort to periodic technological spectaculars to provide a psychological uplift to its people, but it would be largely characterized by drift, hedonistic pursuits, and a sense of decadence and rootlessness.

We can go down a few more steps and envision a society less successful in solving problems of its postindustrial stage. Here power would be even more diffused, to a degree not readily identifiable. Sprawling bureaucracies try to cope with problems as they can, never really hopeful of solving them and condescendingly disdainful of those idealistic young people who demand the standards of justice and performance that the bureaucrats know they cannot meet. The level of technology is fairly high, and there is considerable affluence in society, but it is accompanied by antagonistic claims on resources by various groups, resulting in frequent deadlocks in the decision-making process and resort to palliatives rather than true solutions of problems. Alienation is considerable and finds expression in escapism of people into private worlds of their own or occasional flares of violence triggered by such events as a breakdown of the central computer distributing pension and unemployment checks.

In a number of its forms the impact of technology on future advanced societies is potentially destabilizing. Some societies in the postindustrial stage may be unable to prevent or cope with this destabilizing impact. Thus the locus of power may differ considerably, depending on their ability to cope with the problem of stability.

If a particular society is ineffective in coping with social instability triggered by the impact of technology, its locus of power—for a while, at least—may reside on the streets of its principal cities rather than in government. The value system of the population of postindustrial societies will be significantly affected, perhaps determined, by the prevalent degree of stability. It is conceivable that offbeat ideologies, mostly of the autocratic variety, may appear to squelch insecurity generated by persistent instability. Cycles in values and forms of government may appear. Autocratic regimes, accompanied by security-conscious values, may appear in the aftermath of social instability, followed by eventual erosion of autocracy and by pluralistic values. If the society does not establish an appropriate equilibrium between pluralism and governmental authority so as to ensure the society's effective functioning, another period of instability may ensue, to be followed by a different form of autocracy.

The Semistationary Society: A New Direction in Societal Evolution?

Our discussion so far has focused on the postindustrial society as an entity with certain clearly defined characteristics—recognizing, however, that considerable variations may exist, depending on the stage of the society's development, political differences among various postindustrial societies, and their relative success in solving the problems associated with this stage of societal advance. It might be appropriate to ask whether, as we survey the horizon, we cannot envision a new society emerging that may have characteristics significantly different from those of the postindustrial society. In this connection one must consider the outlook for a stationary-state society.

The appearance of the MIT study on *The Limits of Growth*[39] triggered a major debate on the merits and future of economic growth. The study concluded that, largely because of the growth of population and the exhaustion of resources, the world economy is heading for collapse in the twenty-first century. Accordingly it advocated a series of policies leading to "the state of global equilibrium" or a "no-growth" world society. The study's various weaknesses (in particular its gross underestimation of the potential of technology for solving economic problems) have been pointed out by its numerous critics. Suffice it to say that world economic collapse in about a hundred years is not a strong or even a realistic possibility.

Although the conclusions of the MIT study leave much to be desired, one of its aspects cannot be readily dismissed. It is a mere question of time when the world will have to face the prospect of a "no-growth" society—or, as it has been called by others, "the stationary-state society." The time of the advent of such a society appears to be quite remote. However, for reasons other than those envisioned by *The Limits of Growth*, a variant of the stationary-state society—referrred to here as "the semistationary society"—is likely to emerge in the early decades of the next century.

From the point of view of the flow of materials, our present world economy is "linear." It is a voracious, wasteful mechanism that, in the long run, carries seeds of its own destruction. The economy consumes a vast amount of raw materials from the environment and thus leads to depletion. At the same time it returns to the environment an almost equal amount of materials in the form of waste and thus causes pollution. In addition to this primary cause of pollution there is also the secondary cause, the energy sources used to drive the system. This economic system cannot continue indefinitely—eventually it will deplete virgin raw materials supplies and clutter the earth with waste.

An ultimate solution lies in converting the present "linear" economies into "closed-materials economies." Here the economic process would be circular, with nearly all raw materials coming from waste. The principal "external" input—that is, an input from outside of the closed, self-regenerating system—would be the energy used for the production of goods and recycling of waste. This energy would also be the principal pollutant, since, according to the second law of thermodynamics, energy cannot be recycled. Eventually it will appear in the form of waste heat. The energy to drive production and the recycling process need not come from exhaustible resources on earth—it could come from the sun, which would have the virtue of a minimal polluting effect.[40] A graphic illustration of the present-day "linear" economy and the potential closed-materials economy is presented in figures 14-1 and 14-2.

A closed-materials economy would provide a foundation for two types of potential societies: the semistationary society and the stationary-state society. The semistationary society would have an almost stationary stock of materials, continuously recycled, and a stationary population. The stationary-state society would have a stationary stock of materials and wealth and a stationary population. The critical difference between them would be that, largely because of technology, the stock of wealth of the semistationary society would still expand, while the stationary-state society would be, by definition, a no-growth society. The semistationary society is likely to occur considerably before the stationary-state society.

The principal pressure for a semistationary society is likely to come from the pollution end of the production cycle, not from the resource-availability end. Even now the problem of pollution and waste—especially solid waste—is serious in advanced countries. The oceans can no longer be viewed as a limitless dumping ground, and the advanced nations are running out of land for disposal of waste. Even where, physically, the land is still available, social and political difficulties impose almost prohibitive restrictions.[41] Another important factor is the growing economic attractiveness of solid-waste reuse, stimulated by high cost of virgin raw materials and energy as well as the costs of environmental protection. It tends to take two forms: recycling, resulting in recovery of materials (in most instances recycling requires much less energy than extraction and processing of virgin materials) and conversion, applied principally to organic raw materials, which can be converted to a derived product (compost, fiberboard building material) or energy (by direct burning or as storable fuel).[42]

Figure 14-1. Present-day "Linear" Economy

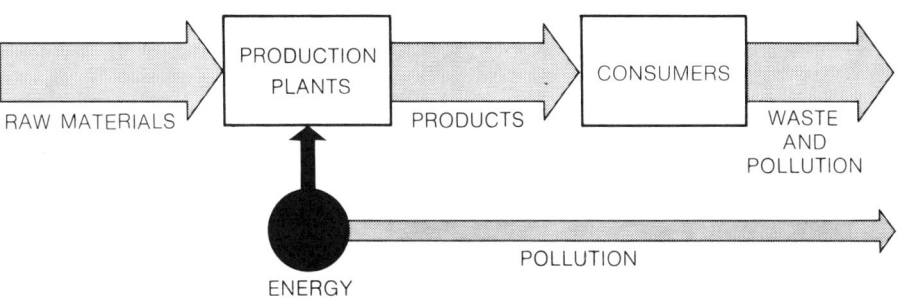

Figure 14-2. "Closed-Materials" Economy

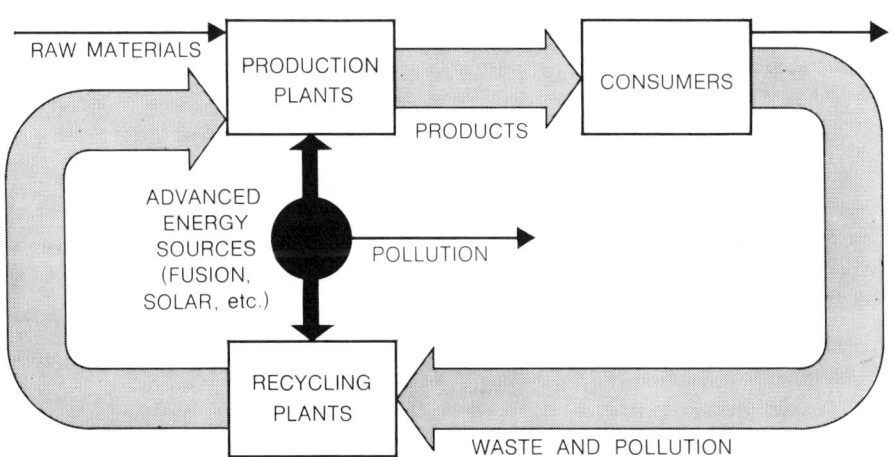

Adapted from *Scientific American*, Vol. 224 (Feb. 1971), p. 61 (From article by William C. Gough and Bernard J. Eastlund, "The Prospects of Fusion Power.")

Pollution considerations will increasingly influence foreign investment policies of advanced nations. So far the principal impetus for establishing subsidiaries in LDCs was provided by the availability of raw materials and cheap labor. However, Japan is now planning to locate systematically pollution-intensive industries abroad because of their impact at home. Taking the present situation as a point of departure, the following scenario can be envisaged for the coming decades.

Gradually pollution-intensive enterprises of advanced nations begin to follow the Japanese example and try to escape the mounting environmental costs by moving plants into the LDCs. This trend, however, begins to have adverse effects on the internal growth of advanced nations and, in particular, on their balance of trade. The United States, being more strongly affected than other nations, attempts to gain uniform environmental regulations throughout the world and thus neutralize the economic advantage of the poor countries. Most LDCs oppose this move, however, apparently convinced that their capacity for absorbing waste is a resource to be capitalized on in international economic competition.[43] As a result the United States, followed by other advanced nations, imposes restrictions on the relocation of pollution-intensive enterprises to the LDCs.[44] Concurrently the United States undertakes measures to stimulate the recycling of solid waste.

Progress is being made in eliminating certain types of cardiovascular diseases and cancer. The numbers of aged, yet infirm, people are gradually rising. Both senior citizens and working population—mostly for different reasons—favor large allocation of resources to gerontology. By 1985 the U.S. government undertakes a crash program in this field. Around the year 2000, a number of breakthroughs are achieved in gerontology as well as in cardiovascular diseases and cancer. Early in the next century the average life expectancy in the United States and in other advanced nations rises to 92.[45]

Since the increase in the average age is not offset by a corresponding drop in birthrates,[46] the rate of population growth in the United States rises from a low 0.3 percent to more than 1 percent, adding about 2.7 million people annually to the U.S. population.[47] The substantially larger numbers of adult, fairly vigorous people engaged in normal consumption patterns of affluent societies impose a major burden on the already badly strained waste disposal situation.[48] Recycling appears to be the principal remedy, but effective implementation of a recycling system requires lead times of 10–20 years.[49] The United States is thus caught poorly prepared by the crisis. The existing relatively limited recycling capability cannot meet the demand.[50]

The situation promises to get much worse before it gets better—perhaps sometime in the second decade of the next century. In addition to depressing the economy by fiscal and monetary instruments, the U.S. government is considering imposition of far-reaching restrictions on production and drastic lowering of the standard of living of the population. The U.S. press notes that, in terms of governmental preparedness, the situation is similar to the energy crisis of 1973, except that the present crisis is of much greater magnitude.

Advanced Societies 279

This scenario provides an illustration of some potential developments likely to impel advanced nations to resort to reuse of solid waste and thus lead to a closed-materials economy. What about the actual feasibility of large-scale recycling—in particular the requisite technologies? Will they be available and would they solve the problem of solid waste at acceptable costs?

Development of technology for recycling waste has been limited, but even the present technology could go rather far in providing much larger amounts of secondary materials at acceptable costs. The main obstacles to recovery of resources at present are institutional and economic, not technological. The entire economy is geared to the consumption of virgin materials. Producers of them enjoy tax privileges and do not pay full costs for environmental degradation created by their activity. By comparison, producers of secondary materials get no credit or special consideration for the benefits to society they bring, which include conservation of natural resources, favorable effect on the balance of trade, and reduced pollution. There is also a certain lethargy in resorting to recycling when its economic benefits are clear-cut, even under the existing regulations favoring virgin materials. Largely because of these factors, only a small fraction of materials used by industry are derived from recycling.[51]

Advances in recycling technology are likely to facilitate reuse of solid waste and to reduce its cost.[52] A technological development that has the potential for revolutionizing the recycling of waste is the fusion torch, a concept originated by William C. Gough and Bernard J. Eastlund (then at the AEC) and now in an early stage of research. The fusion torch would use a portion of the ultra-high temperature of a fusion power plant to vaporize, dissociate, and ionize waste (either solid or liquid) and thus reduce it to its basic elements. These elements could then be separated and reused. To make the process more economical, the exhaust energy could be put to further use, such as production of electric power. The advantages of the fusion torch include its low pollution effect; the convenience of recovering the recycled materials in their elemental form, which broadens options for reuse; and the fact that it would not lead to further material residue as most present recycling technologies do. On the other hand, to capitalize on the torch's full potential, one would have to wait until fusion power becomes a reality—perhaps in the mid-1990s.[53] This would make it possible for the fusion torch to produce a significant economic impact in the first decade of the next century.[54]

In short, from a technological point of view, the crisis envisioned in our scenario is not inevitable. Given timely initiation of appropriate governmental policies that, among other things, would promote the recycling of materials and increase materials effectiveness and the durability of products in general so as to minimize the amount of waste, it can be avoided. However, these developments would mean that, some time in the early decades of the twenty-first century, the world is likely to witness the emergence of semistationary societies. Initially these societies would not comprise "pure" models. They would emerge in certain highly advanced nations; there will not be, at that time, a worldwide semistationary society. Their population would still be growing. They would not be completely closed-materials economies; considerable amounts of virgin raw materials will still be used, although a large majority of the materials will be recycled. Some use of materials from outside the cycle of the closed-materials economy is likely to come in the form of imported finished products, manufactured in the LDCs and introduced into the recycling stream as they wear out. (These imported manufactured products will be paid for by exports of technologies; therefore, there will be no corresponding exports of products to keep the flow of materials perfectly balanced.) Gradually more nations would become semistationary societies. At the same time the most advanced semistationary societies would increasingly approach their "pure" model, in which almost all materials are recycled; relatively little virgin raw materials are brought into the economy; [55] and the population is constant.[56] Eventually the semistationary society would encompass the entire world, while some of the most advanced nations might increasingly assume the attributes of the stationary-state (no-growth) society.[57]

Sociopolitical Implications of the Advent of the Semistationary Society

When viewed in the context of the technological potential, the coming of the semistationary society does not lead us to an economic catastrophe or present unsolvable economic problems. It does seem, however, to have some interesting sociopolitical implications. These arise largely from the combination of the growth in population resulting from the rapidly declining mortality rate with a general slowdown in the rate of economic growth.

In the incipient stages of the semistationary society the economy will be expanding, but its rate of growth will be considerably lower than in the old linear economy: In the process of conversion from the linear to the closed-materials economy, the emphasis will be placed on maximizing use of solid waste and minimizing extraction of virgin raw materials. In an early phase of the conversion process the upsurge of investment in recycling plants might compensate for the curtailment of economic activity in the virgin-materials sector. However, with time the situation will change. Increasingly larger amounts of materials will be recycled, and although the total stock of materials in the economy will still be growing, its rate of growth will be decreasing. The focus of R & D will not be on new products and growth but on reducing the cost of recycling and improving the durability of goods. Durability decreases the flow of materials through recycling and manufacturing and thus reduces both the economic and environmental costs of energy used to drive the system.[58] It also reduces the depletion of virgin resources and the pollution that this depletion originates. Thus durability and a slower rate of what economists call the "throughput"—the total flow of matter and energy—will be sought to minimize economic and environmental costs. These developments are likely to dampen the rate of economic growth.

Another factor contributing to slower economic growth will be the large share of services and the correspondingly smaller share of the production sector in the economy. Since the productivity of the service sector is generally lower than that of the production sector, the large size of the service sector will adversely affect the overall rate of growth.

Finally, another dampening effect on economic growth will be produced by the large and growing number of old people. The vast majority of them will likely retire long before exhausting their production potential. Thus they will not contribute to productivity and economic growth but will claim sizable shares of its wealth and output. Especially since their sheer numbers will carry political weight, the older people are likely to be successful in obtaining legislation favoring their interests, and this legislation will probably not be conducive to economic growth.

With population growing at a fairly high rate and the rate of economic growth declining, per capita income is likely to decline or, perhaps, to become stationary. This development has important political implications.

A growing society, with a rising per capita income, has a major political advantage: It opens up avenues for upward mobility and thus diminishes the potential for pent-up frustrations and instability in society. Talented and ambitious young people can rise in social status and income without necessarily depriving others of these assets. In the absence of growth, however, upward mobility can be achieved only at the expense of somebody else; economic and status competition thus becomes a zero-sum game.[59] If, in addition to limited growth, society faces increased longevity and expanding numbers of older people, the most successful of whom tend to stay on their jobs rather than retire, the situation for the younger and the middle-aged people may turn out to be intolerable. It may be so even if we assume that there will be no significant ideological differences between the expanded number of the separate generations living at the same time.[60]

A question can be raised: Is this likely to be a condition of relatively short duration through which advanced societies could somehow "muddle through" or meet by palliatives rather than correct with fundamental policy adjustments? The answer does not appear to provide reasons for optimism. If advanced societies are caught unprepared by the increase in the growth of population and do not have an adequate recycling capability, then the need to depress the economy while such capability is being developed—which may take as long as 10 to 20 years—would require fundamental adjustments in society to make this condition acceptable in the face of the growing population, blocked promotion for bright young and middle-aged people, and potential ideological conflicts between several coexisting generations. But even if a timely development of recycling does take place and no need to depress the economy would arise, the other factors would still be powerful enough to necessitate fundamental adjustments in society. The population problem would not readily go away. Since the potential increase in life expectancy is assumed to be about 20 years within a relatively short period of time, beginning early in the next century the population of advanced societies is likely to grow at a rate higher than in the late 1990s for some 20 years and thus depress per capita income. Toward the end of that period its rate of growth will probably taper off. Perhaps in another 10 or 20 years, the population will stop growing or become stabilized at a very low rate of growth.[61]

In this nearly "pure" semistationary society the principal and nearly the only factor in economic growth will be technological advance. Technology will continue to improve materials effectiveness, and, from a given amount of materials, more products will be made. Moreover, technological innovations will result in new, more sophisticated products. In fact, since there will be but very limited need to invest capital and human resources in extraction of virgin raw materials and since recycling technology will reach a point of low marginal returns, more capital and manpower will be available for technological advance capable of increasing the available stock of wealth.

Also contributing to economic growth will be the use of virgin raw materials that will still be expanding the total materials stock of the economy. However, the use of virgin raw materials is likely to be small, if one discounts a mere replacement for losses of materials in the production and recycling process.

Thus the advanced semistationary societies will have to rely predominantly on technology to provide economic growth, but technology's capability in this regard will be quite limited. With the other factors contributing to growth being constant or nearly constant, technology can contribute to growth primarily through increases in productivity. Because of the predominance of the service sector with its low productivity and because of the greater spread of leisure, economic growth might not exceed 1 percent per year.[62] Unlike the situation in the early stages of the semistationary society, per capita income will be increasing, but this increase will be small. In a more pessimistic variant of the advanced semistationary society, per capita income might be stationary, perhaps as a result of growth in population.

Eventually—perhaps in another hundred years—our most advanced semistationary societies will begin entering the stationary state. This development is likely to take place, not because the virgin raw materials will be exhausted (their consumption will be significantly curtailed by advanced nations beginning with the early stages of the semistationary society), but because the total stock (physical wealth and population) of a society combined with its rate of throughput will reach a point of becoming unduly burdensome to the natural ecological processes of a particular society. This point can be postponed by lowering the rate of throughput, but sooner or later it will be reached. At that time economic growth will stop, and so will per capita income, unless population begins to decrease.

The phenomenon of providing rewards and incentives by material means—one of the foundations of industrial and postindustrial societies—is running into a serious problem. The margin of the growing wealth necessary for a smooth functioning of this mechanism will begin to shrink in a not-so-distant future. Its worst period is likely to be in the early stages of the semistationary society, when per capita income is likely to decline. In a couple of decades the situation will ease up somewhat, but the old mechanism can never be quite the same. In the more distant stationary-state society, the situation will worsen again, although not quite so much as in the early stages of the semistationary society.

To be sure, when the incipient semistationary societies begin to face this problem, they will be considerably richer than they are now. The problem will be, however, not the total size of wealth, but the effectiveness of its role as a means of reward and a symbol of success. Here what matters is the *relative* wealth as allocated to individual members of society. Unless the issue is squarely faced well ahead of its brunt and appropriately dealt with, it has the potential for disrupting societies. Those societies—like the United States—where materialism is especially strong and where material rewards have a particularly high status value are likely to be most seriously affected.

The problem probably can be solved, but it would require the opening up of new avenues for mobility and status in society—avenues that require limited allocations of material wealth. The task is not a mean one—it will necessitate a fundamental rethinking and restructuring of the traditional concern of political science: Who gets what, when, and how.[63] Among the three principal reward-related values in society—deference, security, and income—the first two will have to receive greater emphasis, and the last will be deemphasized and partially disassociated from the other two. Once society reaches a dead end in being able to reward its members by material wealth, it will have little choice but to find alternative means of reward—especially since there should be more than enough wealth in society to provide a reasonably decent and comfortable living for everybody.[64]

What are the avenues into which the aspirations of the ambitious young talent might be reoriented? They might involve the intellectual and cultural pursuits, the arts and the humanities. However, unlike the present situation in the performing arts, where an important avenue for gaining a high income is an appearance on a magazine's cover (or, better yet, centerfold), only serious and deserving talent would receive recognition, and it need not be expressed in high material rewards. In addition to the high status associated with success in the intellectual and cultural fields, the individuals concerned will also be provided sufficient security in their careers. In turn this would decrease the potential pressure on them to acquire material wealth as a means of security. Once the value orientation of society is shifted toward intellectual and cultural fields and commercialism is no longer the principal road to success, leisure is likely to acquire a new meaning and attraction. Intellectual and cultural activities can be pursued by many people during their leisure time. Aside from the inherent satisfaction these activities provide, the individuals involved would also be rewarded by the associated prestige.

A number of analysts have noted that highly advanced, densely populated technocratic societies deprive man of certain human, "civilized" aspects of life that he might rightfully be entitled to enjoy. It might be appropriate to review briefly some such effects of technologically advanced societies [65] and to raise the questions: In what way, if any, might the semistationary society be different in this regard? Can we expect an improvement?

One aspect of highly advanced societies is the reduced probability for enduring relationships. The likelihood of encountering a person for a second time or dealing frequently with him or her is small. The urban cocktail bar and the department store do not compare with the village pub and the country store for generating warmth and enduring friendships. If highly advanced societies greatly reduce chances for durable, nonsuperficial relationships, it is a considerable loss in itself, for, as one analyst put it, "much of life is almost a pathetic search for such relationships."[66]

The flux and the impersonality of advanced societies have contributed to the decline of customs and rules of behavior that once helped to sustain relatively more agreeable and considerate human relationships. In the course of centuries man has developed ethical rules, enforced by the threats of retaliation or ostracism, and by religious beliefs. The influence of religion has declined in highly developed societies, and their impersonality has produced a serious "free-rider" problem. Why should one be courteous or honest to a person, if to be rude or deceitful is more advantageous under a given set of circumstances and if one is not likely to see that person again? Why should one refrain from occasional pollution or noise, if it is only a drop in the bucket, and other people are unlikely to retaliate? In impersonal, affluent societies it costs more and is worth less to serve as a witness or to interfere with muggings. As a result the implied social contracts expected to sustain the agreeability of society seriously weaken.[67]

As Linder pointed out,[68] the high productivity of advanced societies and high real income result in high value of time. This value is not limited to the time an individual spends at the regular job, but "spills over" into leisure time.

One way to increase one's income is to get promoted. But in highly competitive advanced societies with a high level of education, promotion for many aspiring people does not depend on job performance alone. It might depend on attendance at parties where contacts are made or where a man's social attributes (and those of his wife) can be demonstrated to superiors; on playing bridge or golf with the boss; or lunches or dinners to obtain contracts; and so forth. A more direct way to increase the financial value of leisure is to engage in moonlighting—to work overtime. Still another way to raise the value of leisure is to do for oneself the services normally done by service personnel, like painting the boat, repairing the garage, fixing the basement, and washing the car. Incentives for doing so are considerable: It is expensive to hire services, and the loss of time associated with obtaining the necessary services also makes it much more economical to do the job by oneself.

A less direct, but fundamental way to derive greater benefits from leisure time is to increase the amount of consumer goods to be enjoyed per each unit of time. Thus leisure activities tend to revolve around television sets, boats, snowmobiles, cameras, and other expensive consumer goods. Once these various goods are acquired, not to use them is in itself not economical—indeed a waste. The demand on time thus tends to snowball.

The upshot of these developments is paradoxical—in societies where labor-saving devices and high productivity make leisure increasingly available, the high value of time increases its scarcity. To the extent that leisure is available, it is harried and consumer-goods intensive. A number of values are lost in the process. There is little time available to spend with one's parents. Time has to be spent efficiently; in the pursuit of efficiency, affection and love are shortcircuited for the sake of sex. Moreover, sex itself gets shortchanged and substituted for the more readily available pornographic literature or simply by involvement in other activities. It takes time to cultivate enduring friendships; hence such friendships are scarce. There is also little time for thinking and for cultural pursuits.

There is no clear-cut indication that the advent of the semistationary society would do away with, or even diminish, these undesirable effects of highly developed societies. The size of the population in the twenty-first century will be considerably larger, and impersonality of society is likely to increase even more. The free-rider problem is likely to remain, if not increase. With the decline of per capita income in the early stages of the semistationary society, the value of time is likely to decrease and thus alleviate somewhat the pressure on leisure time, but by that time, productivity will be considerably higher than in the 1970s. Hence, relative to the present, the value of time will not be smaller.

There are, however, other reasons why demands on one's time might decrease: (1) Because of higher productivity, working hours would be shorter and thus leisure time would be increased. (2) With successful older people continuing to work, the incentive to devote leisure hours to activities that might help in promotions will decrease, since for the large majority of ambitious young and middle-aged people, promotions will be blocked for many years. (3) Aside from performing services with regard to their own property, opportunities to translate leisure time into materially rewarding work through moonlighting and similar activities will significantly decrease. These factors are likely to reduce the value of leisure time and make it less harried, but they will not necessarily produce a more agreeable and happier society. If materialistic values remain strong, the reverse may be true. Stifled in promotions and having considerable leisure time, the young and many middle-aged people may engage in activities detrimental to social stability.[69]

Some other characteristics of the semistationary society might, however, facilitate development of a more agreeable social climate. A large part of the population will consist of older yet vigorous retired people living on fixed incomes. They will have a great deal of time on their hands that will be of relatively low value. These people will not be harried; they are likely to develop lasting friendships and warm relationships among themselves. Because of their large numbers the free-rider problem will not disappear among them, but it is likely to be mitigated by closer interpersonal relations. Because of the senior citizens' large share in the total population it is conceivable that their warmer and more considerate attitudes may have a spillover effect on the rest of society. On the other hand, it is entirely possible that cleavages along age lines may develop. Thus the older people may constitute a happier but self-contained community.

Whereas the outlook for a matter-of-course disappearance or even decrease of the undesirable effects of economic advance does not appear to be especially promising in the semistationary society, the situation becomes significantly different if a particular semistationary society undertakes, *as a matter of deliberate policy*, to reorient values and rewards away from material goods. The value of time, in economic terms, will decrease. More time will be spent in a stimulating conversation and artistic and cultural pursuits, which are conducive to improvement in human relations and a general agreeableness of society. Alienation of individuals from society is likely to lessen. Ethical standards will probably rise, and this, along with a generally greater degree of agreeableness in society, is likely to diminish (although, probably, not completely eliminate) the free-rider problem. Indeed these developments might help to bridge the potential gap or, as the case may be, decrease antagonism between the older retired people and the rest of the population and thus further contribute to social harmony and stability. In a society that is less harried and pressured, more agreeable, personal, and with fewer people feeling alienated, the rate of crime, neuroses, and a general feeling of malaise are likely to recede. This would not only contribute to a happier and healthier society but would also diminish the demand on resources allocated to combat these phenomena and thus release these resources for other purposes.

International Implications of the Advent of the Semistationary Society

The transition of highly advanced societies into their semistationary stage is likely to affect their conduct internationally. The internal demands on resources of the incipient semistationary societies will probably diminish their willingness to allocate resources to external pursuits. Moreover, if advanced nations experience internal instability in their transition to a semistationary society—which is very likely—their vitality, and perhaps viability, on the international arena could be jeopardized.

The involvement of most incipient semistationary societies in world affairs is likely to decrease. This phenomenon will be facilitated by their declining dependence on raw materials from abroad. On the other hand interdependence among the semistationary societies (as distinguished from interdependence between the semistationary societies and the rest of the world) might increase. The early semistationary societies will feel the pressure for reducing the costs (economic and environmental) of energy necessary for the production of goods and the recycling of materials. They will probably attempt to take advantage of the time zone differences and seasonal diversities and resort to huge power grids—such as those suggested by Glenn Seaborg and discussed in chapter 3—cutting across national frontiers. The establishment of such grids is likely to proceed cautiously and selectively—the semistationary societies will still be mindful of the implications of the grids for national security. However, this development is likely to strengthen the non-zero-sum aspects of international politics.

The emergence of semistationary societies might cause a considerable friction between them and the raw-materials-exporting nations. Many of the latter will no longer be "less developed countries" in the present sense of the word—in the early twenty-first century such nations as Argentina, Brazil, Venezuela, Mexico, Uruguay, Chile, Peru, Jamaica, Egypt, Iran, Iraq, and China are likely to be extensively industrialized, although they will not yet be postindustrial societies and will be many years away from becoming semistationary societies. For our purposes we may call them "less advanced countries" (LACs). The friction between the early semistationary societies on the one hand and the LACs and the LDCs on the other would stem, not from the resentment of the LACs and the LDCs to what they considered "domination" by the MNCs of the most advanced nations—that period will have long passed. Instead a conflict of interest would arise from just the opposite development—the restrictions imposed by the semistationary societies on investments abroad (accompanied, in some cases, by liquidation of such investments) and the curtailment of imports of virgin raw materials. To the extent that imports of virgin raw materials into the semistationary society will continue, they will tend to come largely from international areas—the deep ocean seabeds, which will supply oil and manganese nodules with high metal content. For the semistationary societies these imports are the more desirable since they do not aggravate the balance of payments problem.

As the relations between the semistationary societies and the raw-material-producing LACs and LDCs polarizes, the latter nations go through periods of economic depressions and social instability. The situation is further aggravated by the availability of medical means to extend the longevity of population substantially. A number of LACs and LDCs find it impossible to deny these medical innovations to their people, and the resultant upsurge of population growth heightens social instability.

In short, in the early decades of the next century, there will likely be a considerable degree of international instability, finding its roots in the social instability plaguing the highly advanced and the not-so-advanced nations alike, although not always for identical reasons. It is possible—although not inevitable—that this instability may result in wars—undertaken, in some cases, to divert attention of the population from internal problems. These wars might rip the fabric of international institutions and provide a setback to the trend toward the gain of the non-zero-sum elements in the international political process.

As an incipient semistationary society the USSR is more likely than any other power to gain from the developments in the early decades of the next century. Having a large territory relative to its population, it is likely to be in a better position to absorb solid waste short of resorting to large-scale recycling. Thus it would be able to push economic growth, based on virgin raw materials, for a longer period of time than other advanced nations. The Soviets would resort to the recycling of solid waste on a partial basis, only in highly congested areas, and where high transportation costs do not justify remote landfill. Moreover, the USSR has never been a significant importer of raw materials from abroad; thus it will not antagonize raw-materials-producing nations by stopping imports. In fact the opposite is likely—the Soviets would be in a position to take advantage of the drastic slump in world prices in virgin raw materials and step up imports. They might also increase their investments abroad to take advantage of cheap, unemployed labor and the low prices for plant and equipment available for sale from the "capitalist" MNCs curtailing their operations. Thus the USSR is likely to benefit both economically and politically, since it would gain friends in raw-materials-producing nations and extend its physical presence in them.

As the semistationary societies move to a more advanced stage, their international behavior will largely depend on how successfully they will have solved their internal problems in the earlier transitional phase. If a particular society is still strongly imbued by materialistic values, it is likely to be preoccupied with competition for smaller resources for the size of its population and will probably have fewer resources and energies to spare for external pursuits and for national security. It may still be plagued by social instability. If, however, it will have reoriented the values of a large part of its population toward nonmaterialistic pursuits and thus released pressure on its resources, the society under consideration will strengthen its position to influence world affairs and protect its interests and security. The intellectual and cultural pursuits of the majority of its population are likely to strengthen the society's world outlook and interests. Thus an early reorientation of values away from narrowly material pursuits would be important for semistationary societies. Conversely, a failure of such early reorientation may jeopardize their position in foreign affairs, including their security, in a world environment that might still not enjoy a high degree of stability.[70]

An interesting question arises with regard to "mature" semistationary societies: What is likely to be their capability and propensity to wage wars? The picture is not entirely clear. However, raising a few questions sheds some light on the subject.

A divergence is likely to emerge in the semistationary society between the requirements of the civilian and the military sector. Whereas the civilian economy will be geared to the production of high-quality, very durable goods at a slow rate of throughput, wars will require high-speed mass production of military goods that will be rapidly consumed (i.e., destroyed). How much materials could a closed-materials economy afford to have consumed in war and still be functional? Could civilian plants be readily converted for military production? Or would the semistationary societies rely on a separate skeleton war industry, idle in peacetime? Could they afford such industries, in view of their relatively slow growth and the shortages of resources? On the assumption that a semistationary society had to produce an all-out effort of military production for a conventional war, when would it reach a danger point of "overheating" the economy?[71] Presumably the more advanced the semistationary society, the less room it would have for increasing the rate of its throughput. Thus the most advanced semistationary societies might be seriously handicapped in waging wars.

A few more questions could be asked. To what extent could the stockpiling of munitions compensate for limitations on wartime production? Since the economy will have limited capability to sustain a conventional war with another major power, would there be a greater incentive to escalate to nuclear exchange? Will conventional wars be fought with striking power in being, while war potential will lose much of its significance—as it has now for nuclear wars? What about the composition of population of the semistationary societies, heavily loaded at the top with older people? How will it affect the nation's decision making with regard to war-related activities? Will it be the LDCs alone that will be able to afford the luxury of a protracted war? Or will the interdependence among nations be such as to impose a different and a more compelling set of restrictions on warfare? Will all these restrictions on the ability to wage war provide a sufficient incentive for major disarmament agreements?

Many of these questions cannot be answered with any certainty. However, it does seem that the entry of the most advanced nations into the semistationary stage imposes serious obstacles to warfare, if not to the maintenance of armed forces. Perhaps one can discern a pattern in the impact of technology with regard to war and related activities. Nuclear weapons have produced a stalemate, which has begun to descend gradually into the conventional sector and has thus further narrowed the effectiveness of military power for waging wars, if not for the maintenance of international stability. The advent of the semistationary societies might impose even greater constraints on warfare by the highly advanced nations. Perhaps wars between these nations, if fought at all, will be fought only by proxy through the LACs. And, unlike the case of Vietnam, the highly advanced nations will be very careful not to get directly involved.

There is of course no such thing as certainty about a particular future, and the developments envisioned above may not necessarily materialize. I do suggest, however, that we cannot afford to rely on the concept of the postindustrial society as envisioned by its founding fathers to provide us with guidance for an indefinite future. Discontinuities in social trends may occur, leading us to a society with a different set of problems, challenges, and opportunities.

Moreover, our future does not necessarily consist of either the determinism of a deadlock of various forces among themselves or the determinism of a diffuse, almost amorphous, overpluralized society devoid of a sense of direction. One has to be aware of the trends and the constraints involved at each stage of societal development, but room does exist for human will, expressed through society's institutions, to assert itself, to introduce change, and to give society a sense of direction.

The ability to move society in a desirable direction may simply be a matter of opportunity perceived, as would be the case if governments stimulated the development of public-service-oriented corporations that could become important vehicles of constructive change and carriers of social responsibility. This ability may receive a stimulus from certain social trends that might make an otherwise desirable move compelling because the alternative to it is less than acceptable. This might be the case with the reorientation of human values away from material pursuits in the incipient stage of the semistationary society. The question of human ability to steer society in a desirable direction, with particular reference to the United States, will be explored further in the next chapter.

The National Purpose in an International Context: Whither America?

15

THERE ARE AT LEAST five reasons why, when examining America's future, we should bother to discuss her national purpose.

1. Each society, to be viable and healthy, must have a purpose with which its population can identify. Depending on historical circumstances and conditions in a particular society, this purpose may not necessarily be explicitly and strongly promoted by the government. It may proceed toward its fulfillment through interplay of social forces within the governmental framework and a system of beliefs and values originally set up or developed by society.[1] In either case the existence of a national purpose, its cohesiveness and strength, the extent of the people's identification with it, the relative success of the process of its fulfillment, and the degree of its compatibility with the trends in a given period of history provide an important barometer of the viability of a particular society, now and in the future.

2. A related but elusive and less clearly understood reason stems from the human desire, detectable throughout history, to find meaning in life transcending the finiteness of temporal existence. Some individuals find meaning in religion, still others in personal creativity. For most, however, such meaning can seldom be found apart from their society, the faith in its institutions and beliefs, and the purposes it was set to accomplish. In short, society's purpose and meaning are related to the individual's own purpose and meaning in life and, hence, to human happiness.[2]

3. A national purpose is important as an instrumentality in overcoming the inertia of society's excessive pluralization by the impact of technology and thus in making possible its movement in a desirable direction. We have seen that, through national purpose, Japan has succeeded in moving her society. Perhaps this can be done elsewhere.

4. The lack of a consensus on values in society impedes our ability to subject technology to control. Perhaps an examination of the national purpose might help us find values around which society's collective human will could be mustered. If so, we would overcome another deterministic impact of technological advance—the tendency toward a deadlock among the various countervailing interests.

5. Historians and philosophers of civilizations—from Spengler to Toynbee and Sorokin—provide us with substantial evidence that civilizations never stand still; they grow, reach a peak, and then tend to decay.[3] In recent years some voices have expressed concern that America, after going through a meteoric rise as a civilization of world significance, may be experiencing an even more meteoric decline.[4] An inquiry into America's national purpose might help to shed some light on the future of the United States.

The following section takes a closer look at America's national purposes in historical perspective. At this point a clarification on terminology is in order. As used here, the term *national purpose* encompasses a major objective that a society attempts to achieve. The term *national goal* is narrower, usually signifying an instrumentality of the government designed to promote a particular aspect or a phase of the national purpose. As we shall see later, "national goals" had to be resorted to on a number of occasions when, for one reason or another, the expected automatic realization of a particular national purpose was impeded or proved to be nonfunctional.

America's Purposes and the Extent of Their Achievement

Our socioeconomic system was expected to accomplish essentially three purposes: (1) to improve continuously the material conditions of its population, (2) to achieve a high degree of individual freedom for its citizens, and (3) to bring, in the wake of mass prosperity and liberty, the flourishing of spiritual values aggregating into a virtuous and enlightened society.[5]

In an effort to accomplish the first purpose the socioeconomic system was leaning on the philosophy of Adam Smith. In his philosophy the continuously rising production of goods and their spread among the population at large were to be accomplished by several instruments (all of them founded on a free market economy and operating automatically), the principal ones being competition, the division of labor, the law of accumulation, and the law of population.

Competition would regulate individual and group self-interests, would convert them to public good, and would ensure dynamism in society geared toward innovation and productivity. The division of labor would ensure high productivity. As Adam Smith put it, if each worker applied himself separately to production of pins, he would do well to produce 20 a day; but 10 men, each performing a separate function—drawing out the wire, straightening it, cutting it, and so on—could make 48,000 pins a day. The law of accumulation, by plowing profits into the means of production rather than into charity or consumption, would be important in the continuous growth of the economy. It would also raise wages as the capitalists bid for workers and thus increase their standard of living and expand consumption, which, in turn, would fuel economic growth. The law of population—the expansion of labor force in response to higher wages, and its eventual contraction as economic conditions become unfavorable—created another mechanism in a *laissez-faire*, self-regulating system, which ensured, among other things, that, because of shortage of labor, wages do not run up to unreasonable levels and thus do not consume an inordinate share of capital.[6]

Not all of Adam Smith's concepts proved to be completely valid, and none of them functioned perfectly. Correctives had to be introduced into our socioeconomic system to ensure an effective functioning of some of its instruments, especially of competition. Indeed, under Theodore Roosevelt, "trust busting" as measure to safeguard competition was practically elevated to the status of a national goal. However, the basic fact remains that, leaning largely on the principles of Adam Smith—in part modified by other instruments such as the "new economics"—we are close to achieving the first goal of our socioeconomic system, namely, a broadly based prosperity.

Much has been done in the United States to expand individual freedom. The principles on which our economic system is based were conducive to this goal. To work successfully, the economic system needed free, energetic individuals to establish enterprises and to compete freely in the market place. It needed free, mobile labor; accordingly, economic forces joined hands with humanitarian factors in opposing slavery. However, as society became more integrated and interdependent, the concept of expansion of human freedom was redefined to encompass also the promotion of conditions conducive to human freedom. Can individuals, living in overcrowded, substandard conditions or reaching a point of starvation, be considered truly "free," no matter what equal rights they might enjoy under the law?

It is precisely when human freedom began to be increasingly dependent on governmental decision making in creating conditions suitable for its enjoyment that questions began to arise about whether our socioeconomic system were adequate to the task. Our predominantly constituency decision making is slow, incremental, and susceptible to being readily blocked by antagonistic interests and is not responsive to rapidly changing conditions. The amount of change necessary to ensure conditions conducive to human freedom will be vast, which casts a serious doubt upon the ability of our system to cope with it.[7]

Although varied degrees of progress have been achieved in the pursuit of the first two purposes of our socioeconomic system, the situation is much less than ideal with regard to achieving the third—the flourishing of spiritual values.

Our socioeconomic system—again, in tune with the theory of Adam Smith—has rejected the normative, elitist approach of philosophers like Plato and Aristotle, who were interested in creating a high and memorable civilization, even if it were shared only by a small minority. Instead our system chose more modest goals for its civilization, but its benefits were to be shared by the entire citizenry.[8] It would have been difficult enough to bring everyone to the level of a virtuous and enlightened society, but the task became next to impossible. Our system provided no suitable instrumentalities for the development and implementation of the kind of civilization it aspired to. Certain forces in our system favored the development of affluence and freedom in a predominantly *laissez-faire* environment. But this was not so with regard to the development of an enlightened and virtuous society. It was assumed that spiritual values and a moral society would prosper with the rise of affluence and freedom, but this assumption was based on little more than hope.

In addition, the instrumentalities and values developed by our system in pursuit of economic growth and affluence (a purpose that tended to dominate the American scene) proved, in some respects, antagonistic to the development of a virtuous and enlightened society. Probably the most important instrumentality of our economic system is competition. It is predicated on the assumption of the existence of antagonistic self-interests that, in competing among themselves, produce better and cheaper goods and thus benefit society. There is little doubt that competition and its corollary—the acceptance of legitimacy of self-interests—have been immensely valuable to the economic growth of the United States, but they have become so deeply entrenched in our socioeconomic system as to result in the subordination of broader social values, such as ethics.

In competition the goal is to win, and in our economic system winning in itself carries the stamp of moral approval. After all, is not the best product—or the best person—supposed to win and thus benefit society? Conversely a person or a firm that has lost is, by implication, not good enough to benefit society. The pressure of competition and its emphasis on the morality of the end—success—leads to the tendency to overlook the ethics of the means.

The degree to which ethical considerations are disregarded where their violation cannot be easily detected is illustrated by the example of radio and TV repair shops. In up to 90 percent of the cases, customers are billed for nonexistent repairs or are overcharged for the repairs that are made. Watergate is another example of winning at any cost. In various degrees and in a more subtle form the disregard of ethics has penetrated the executive suites of our corporations and our governmental bureaucracy.

Competition is not the only instrumentality transferred from our economic system to society at large that impedes social improvement. In some respects the principle of *laissez-faire* has produced a similar effect. When, in the early 1920s, radio broadcasting began to spread rapidly in the United States, the position of the broadcasting stations was that, in broadcasting, as in other sectors of the economy, the government should interfere as little as possible. As a result (and at variance with European democratic nations, where governments have played a key role in broadcasting), the privately operated stations, the advertisers, and the listening public became the principal determinants of what went on the air. The same principle was followed with television when it was commercially introduced in the late 1940s.

The system has some advantages; especially in radio, it has facilitated innovation and the development of variety in musical styles. However, an opportunity was lost for using such powerful instruments as radio and television as a means for systematically raising the cultural and esthetic standards of the American people. Programs tend to cater to a fairly low common denominator, and although one can detect improvement over decades, it is slow. Some of the best TV programs have a hard time surviving on the national networks, which carefully watch the predominantly mediocre entertainment standards reflected in the Nielsen ratings.[9]

A third reason why the attainment of an enlightened and virtuous society was impeded can be traced to our success in the pursuit of affluence. Affluence has developed certain entrapping attributes that prevent society from striving for higher goals beyond materialism. Imbued by materialistic values, society has translated material goods into a status symbol. Once material goods became much more than a means of satisfying economic needs of the people, their acquisition turned into a self-perpetuating process. However, while affluence does have certain entrapping qualities, their strength is not unequivocally clear. It is possible that Linder overstated his case when he argued that high productivity, affluence, and enjoyment of expensive consumer goods have largely driven meditation and time for cultural activities off the market.[10] The emergence of first-rate orchestras in a number of American cities, the rise of interest in ballet in recent years, and the growth of public TV indicate that there may be a grass-roots movement attempting to break through this materialistic entrapment.

298   Directions of Societal Development

In short, whereas the fulfillment of the first two purposes of our socioeconomic system was, to a considerable degree, mutually supporting and had certain self-executing characteristics requiring only a partial governmental intervention and support, the case with the third purpose is much more complex. It can be fulfilled by an act of societal will that would have to be strong enough to overcome, tone down, or neutralize some of the instrumentalities and effects of the accomplishment of the first purpose of our socioeconomic system—material affluence.

Some Perceived National Goals Reexamined

It might be worthwhile to take a look at national goals as recognized and pursued by the U.S. government as well as those proposed by observers of American society. In turn this examination may help to clarify desirable directions for societal development and the means of achieving them.

The U.S. Government and National Goals

Some of the principal national goals pursued by the U.S. government can be identified (not necessarily in the order of importance) as follows: [11]

1. Economic growth
2. Full employment
3. Eradication of poverty
4. Advancement of health
5. Improvement in the well-being of the senior citizens
6. Equality of opportunity for minorities
7. Equal rights for women
8. Protection of privacy
9. Reduction of crime
10. Solution of urban problems
11. Improvement of the environment
12. Reforming and improvement in education
13. Protection of the consumer
14. Encouragement of the arts and the humanities

The first purpose of our socioeconomic system—to provide affluence for the nation's population—has largely been achieved. Therefore, Goals 1 through 3 constitute what may be called a "mopping-up operation." As we look into the future, economic growth will have to be modified to fit the requirements of a society whose linear economy will be increasingly transformed into a closed-materials economy. The maintenance of full employment has become recognized as a largely routine function of the government, and is hardly a national goal that carries meaningful implications for the future. Eradication of poverty merely reflects a resort to political means to speed up what our system should have accomplished anyway. Goal 4, advancement of health, is a corollary to the achievement of widespread prosperity.

Goals 5 through 7 attempt to spread the benefits of affluence to those who may not have previously shared fully in them. But even to a larger degree the last two of these goals, 6 and 7, are intended to fulfill the second purpose of our society—the expansion of individual freedom by removing discrimination. Goal 8, protection of privacy, attempts to safeguard the freedom of the individual against encroachments by modern technology. Goals 9 through 11 are aimed at creating conditions necessary for the enjoyment of freedom. A partial exception to this is Goal 11, improvement in the environment. To the extent it is motivated by esthetic considerations, it spills over into the third purpose of our socioeconomic system—that of creating a spiritually rich society.

Goals 12 through 14 carry the promise that we may finally address ourselves to the task of creating an enlightened and virtuous society in which spiritual values are pursued. However, as one looks more closely at what the first two of these goals attempt to accomplish, one need not get unduly excited about the prospect of entering such a society, since this is not quite what these goals have in mind.

The educational goals deal with such problems as a more equitable financing of schools, student assistance programs to ensure a broad-based admission of qualified students, the establishment of a National Institute of Education, and so on.[12] These efforts appear to fall more appropriately under the rubric of expanding equality of opportunity, affluence, and conditions for the enjoyment of freedom, that is, the more conventional pursuits of our socioeconomic system. Unless a deliberate and concerted effort is made to introduce ethical, esthetic, and intellectual (as distinguished from materialistic) values into our educational system, these values are not likely to emerge from the spread of education alone. And there is nothing in our present pursuit of education as a national goal that indicates the federal government is about to make a commitment to such an effort.

One might think that the protection of the consumer was motivated by the desire to upgrade ethical standards in the nation and thus move it in the direction of a virtuous society. A closer look at the background of the governmental concern with the consumer reveals a somewhat different motivation.

Governmental action for the protection of the consumer was stimulated by "consumerism"—a modern breed of consumer movement. As a response to a demand—however justifiable—of a constituency (the consumers), this national goal is thus well within the tradition of our political process and is hardly novel. Moreover, present-day consumerism claims that technical complexity of products and the existence of many varieties of essentially the same product make it difficult to choose intelligently. As a result the consumer's ability to police the marketplace is eroded.[13] When viewed in this light, protection of the consumer is but another corrective designed to uphold one of the principal instrumentalities of our economic system—consumer sovereignty—and thus make sure that the efficiency of the economy is maintained.

Among the present national goals, encouragement of the arts and the humanities comes nearest to the promotion of an enlightened and spiritually rich society. However, its purpose is narrower than the promotion of such a society. We are told that the young people lost faith in their country in the 1960s and that this faith can in part be renewed by "a deeper understanding of who we are, where we have come from, and where we are going—an understanding to which the arts and the humanities can make a great contribution." We are further told that the government has a limited function in this field—"that of reinforcing local initiatives and helping key institutions to help themselves." [14] And the request for appropriations is correspondingly modest, when compared with other national goals.[15]

In short, however worthwhile and important some of the national goals currently pursued by the U.S. government are, they are not very imaginative and, when viewed in the context of the future, not truly adequate. They are overwhelmingly focused on expansion of material well-being and conditions conducive to the freedom of the individual, and do little, if anything, to accomplish the third purpose of our system—a virtuous, enlightened society in which spiritual values flourish.

One might say that national goals as actually pursued by government at a given time must always be addressed to the most pressing concerns, while the more farsighted national goals should be looked for in other government documents, those dealing with the future. However, the latest U.S. government report on national goals, that of the National Goals Research Staff, *Towards Balanced Growth: Quantity with Quality* (July 1970) is a disappointing document. It focuses on "balanced growth" and expresses nothing but hope that "America's talents and wealth can be used to provide a quality of life commensurate with its potential." Moreover, the report emphasizes the role of the American people in formulating their goals and thus, in effect, expects the goals to emerge almost deterministically in the process of societal evolution.[16]

Independent Analysts and National Goals

Independent observers of American society are more imaginative in suggesting goals for the future. The problems with their societal goals are usually of two types. They may suggest societal goals but provide no guidance on how they can be achieved—it being assumed that the goals are truly desirable. Both Daniel Bell and Herman Kahn fall into this category. Alternatively, other analysts may propose means of achieving certain desirable societal goals, but these means are not sufficiently practical to produce the expected results. An example is John Kenneth Galbraith and his *The New Industrial State* (1971). Following are the highlights of his thesis.

Galbraith's principal misgiving about the state of affairs in America is the hold that "the industrial system" has over society. For the sake of its own growth the industrial system manages and expands material wants of the population well beyond truly rational needs of human consumption. It seriously curtails liberty by manipulating human belief and subordinating it to the system's needs. In its symbiotic relationship with the state the industrial system influences the state's activity. The system favors those services of the state that meet its needs (like defense and technological development), while services not needed by the system—like public clinics, assistance to the poor, parks and recreation areas—are neglected or discriminated against. Moreover, the industrial system not only fails to support esthetic aspects of life but also endangers them, as evidenced by strip-mining over virgin mountainsides, the pollution of streams, and the strings of highways over urban open spaces.

The societal goals that Galbraith calls on the United States to pursue fall principally within the framework of the last two purposes of the American socioeconomic system—the expansion of the freedom of the individual and the achievement of an enlightened, spiritually rich society.[17] But how does one achieve these goals, considering the professed power and interests of the industrial system? Professor Galbraith's answer is less than convincing.

A part of his solution is somewhat Marxian. The industrial system has to support education for its own purposes, but in doing so it supports general enlightenment. The growing educational and scientific estate will eventually liberate society from the holds of the industrial system.

Like Marx, Galbraith is not satisfied with leaving change to favorable deterministic forces alone. He calls for action by the educational and scientific estate to free society from the pervasive influence of the industrial system. The action is to be exercised on two levels—educational and political. On the educational level the university must become aware of its power, must assert its autonomy and authority, and must develop the proper educational policy aimed at the cultivation of intellectual and esthetic values not subservient to the industrial system. On the political level the intellectual community must strive to achieve cohesiveness, resort to direct political action, and gain political power to counter that of the industrial system.

But, contrary to Galbraith, the problem of being able to steer society into a desirable direction is much more complex than countering the influence of "the industrial system." In part this problem finds its roots in the pluralizing influence of technology, which affects, directly or indirectly, governmental institutions, the scientific and educational estate, values, and society at large. In part the problem lies in the effect of some of the instrumentalities to which we have resorted to achieve broadly based affluence. In part it lies in the entrapping consequences of success in achieving affluence, raising the economic value of time and thus inhibiting our ability to go beyond materialism.

Given these formidable obstacles, the prospect for unifying the university community for a concerted educational policy aimed at the cultivation of intellectual and esthetic values is slim indeed. Our educational system is too decentralized to provide a realistic leverage for change. Can one truly expect to persuade the many thousands of university and college presidents, deans, chairmen of the departments, and influential individual professors to turn against the values of what Galbraith calls "the industrial system," while the real culprit is even more nebulous, the pluralizing impact of technology, with its multiple interests?

The prospect is equally slim for unifying the scientific and educational estate for concerted political action. But even if such a unity were achieved, this approach would hardly provide a solution. Because the size of "the educational and scientific estate" is not very large, its unity for a political purpose would amount to the addition of another constituency, with an interest of its own, to our political process in which politics based on interests lead more to deadlock than progress.[18]

The Crisis of National Purpose

In the past 15 years or so the United States has been going through a crisis of its national purpose. The symptoms of this crisis are multiple, and many appeared in such a form that, on the surface, they bear little or no relationship to each other or to the nation's purpose. Fundamentally, however, their relationship is unmistakable, and they are worth bringing together.

One symptom is the serious concern about the national purpose and national goals. In 1960 President Eisenhower's Commission on National Goals produced a report on *Goals for Americans*. In the same year *Life* magazine ran a series on the national purpose. In 1970 the National Goals Research Staff of the Executive Office of the President came out with its report on national goals. A number of independent analysts raised the question of national goals. Despite all these efforts, we have gotten nowhere in resolving the national purpose.

Even before the impact of Watergate and the energy crisis, surveys of the attitudes of the American people showed a high degree of frustration, distrust of American institutions, aimlessness, moodiness, a degree of despair, and inward orientation. More than ever before, people seem to have lost belief in their ability to change things, and so they tend to be cynical, indifferent, and withdrawing.[19] Watergate and the energy crisis accentuated many of these symptoms.

A study conducted by Louis Harris and Associates, Inc., for the Senate Subcommittee on Intergovernmental Relations in September 1973 showed that, for the first time in more than a decade of opinion sampling, the feeling of powerlessness, cynicism, and alienation among the American people had reached a majority (55 percent).[20] Other studies by the same organization revealed that public confidence in key American institutions had been dropping since 1966, and hit a record low in 1976.[21] Having experienced severe inflation, long gasoline lines, and shortages of many materials, the American public began to question the old-style American dream. We have been told by some writers that the historical dynamism and drive of Americans to new frontiers—any frontiers—should be abandoned. We should adjust ourselves psychologically to a limited future and embrace a "postachievement ethic."[22]

Nor is the international situation much more encouraging. In contrast to the situation in earlier decades the United States is much less a source of inspiration to other nations. The LDCs are not enthusiastic about our foreign policy, and the energy crisis seems to have papered over the fact that the

United States has more problems than accord with its friends in Western Europe and Japan. If the Soviet Union and China seem to be reasonably pleased with the United States, they are so for their special reasons, and not exactly because of our vitality and purposiveness, which they admire.

Perhaps the present situation is not a dead end. Perhaps the crisis of the national purpose can be resolved, the nation can regain a sense of direction, find new frontiers to explore, restore its confidence and vitality, and proceed toward building a brighter future, at home and abroad. To accomplish these objectives, the nation's purpose and its goals will have to meet a number of criteria.

Criteria for a National Purpose and Goals

The crisis of the national purpose consists of two components: finding desirable goals and implementing them in our socioeconomic system. The latter is more difficult than the former, but the ability to succeed on both counts is necessary to solve the crisis. Thus one criterion of a chosen national purpose might be that its objectives, formulated as national goals, be reasonably attainable.

Furthermore, we shall probably do our society a disservice if, in the search for national goals, we narrow our sights and confine them to the fetishes of the day, whatever they might be—ecology, equality of opportunity, or the energy crisis. These issues may be extremely important, their resolution may be a prerequisite for the attainment of the society's higher purpose, but it is more important to recognize their transient nature and not to lose sight of the need to look deeper into the processes and stages of societal development and thus help to crystallize fundamental purposes and goals that may not be quite so readily perceived as, say, pollution. In highly complex times the search for national purpose and societal goals is a highly demanding intellectual process that involves a careful scrutiny of both the ends and the means. The shrillness of voices predicting doom is hardly an adequate substitute for high-caliber insight and intellectual integrity of analysis.

If the search for national goals is recognized as primarily an intellectual process based on objective analysis and a sincere desire to find directions for societal development (rather than on the traditional political process based on the interplay of interests), the goals and directions revealed are likely to be more convincing and hence easier to implement. However, the delineation of goals alone (convincing as they might be) is not enough to ensure their implementation. The delineation of intermediate steps and of the obstacles to be overcome before we can reach desirable goals would be extremely important. In short, it is necessary that the search for goals and the means for their implementation proceed hand in hand. Something akin to a "strategy" of societal goals needs to be developed, in which both means goals and end goals are carefully considered and arranged in the proper order. The implementation phase would thus be a combination of both an intellectual and a political process. It will be the ultimate deciding point for what the national purpose and goals are; it is at this point where the political process would be expected to introduce, as necessary, correctives into the rationally derived goals and where the rationality of the goals would—it is hoped—improve the political process.

An important criterion for a national purpose related to those mentioned above is its ability to meet requirements of the future. In an era of rapid change largely induced by technology, a national purpose can hardly have depth and durability unless it is formulated with a maximum possible awareness of the challenges and opportunities of coming decades and, perhaps, centuries. Insofar as the formulation of the national purpose and national goals is an intellectual process, it must be future oriented.

A national purpose cannot succeed unless it is in harmony with certain fundamental characteristics of the people and contains a promise for a better future. Such characteristics of the American people as a high degree of idealism (reflected, for example, in the Declaration of Independence), optimism, dynamism, and the propensity to seek and explore new frontiers are fully compatible with the basic component of a viable national purpose—an achievable promise.[23] However, while geared to the needs of the American people, it is important that America's purpose have a meaning and an appeal of universal dimensions. In a world becoming increasingly interdependent owing to the impact of science and technology, America's future cannot be effectively molded in isolation from other nations. Furthermore, America's destiny as a civilization heavily depends on its being able to offer an inspiring example to other peoples.

Last, the national purpose and its social goals, to be meaningful and inspiring to society as a whole, must be such as to benefit every member of society. This does not mean that the national purpose would have to be reduced to the lowest common denominator. On the contrary it should be sufficiently high to challenge the imagination and energies of those who have attained a great deal in life and who might think that there is nothing else to be attained as well as of those who are less fortunate than that. Just as in the case of Adam Smith, who promised affluence to society, not everybody will be able to partake in the benefits of the new national purpose at once or to the same degree, but if a succession of clear, worthwhile, and attainable means goals is provided, then the sense of participation and the benefits derived will be society wide.

Toward an Articulated National Purpose and Societal Goals

Given the above criteria, what national purpose and societal goals might the United States pursue? A purpose the nation might do well to turn its attention to is not very novel but is unfortunately neglected: the development of man's potential for the best there is in him and the projection of this best to the entire society for its benefits. This would include man's spiritual values, standards of his conduct, integrity, esthetic values and pursuits, intellectual growth, and responsibility to society. The reader will recognize that these objectives are very close—in some respects almost identical—with the third purpose of American society discussed earlier—to bring about an enlightened and virtuous society in which spiritual values would flourish. However, the national purpose proposed here differs in three important respects from that discussed earlier.

First, unlike the case with the traditional third purpose of our socioeconomic system—which was not very clearly defined and was more like a dream, a hoped-for consequence of the achievement of the first two purposes of the American society—the present purpose will be more specifically articulated, with its objectives set in the context of present and future needs of our society. Second, these objectives will not be expected to be achieved automatically but will be specifically sought after by a deliberate and sustained societal effort. We got trapped in the pursuit of economic goals, and we can move to a higher level of societal development only by a combination of intellect and human will working in concert. Third, in response to changed conditions and to expected further changes in American society and the world, the proposed national purpose will differ in some substantive characteristics from the traditional third purpose of American society.

The goals for an enlightened, model society suggested here are not only the more modest goals of Adam Smith but also the achievement of the highest standards of a memorable civilization to which the earlier philosophers, including Plato and Aristotle, aspired. Unlike the case of earlier times, when the benefits of such civilization were confined to a few, in our society they will be shared by as many as possible at a given stage of societal development, and their benefits will be continuously spreading as the ethical, esthetic, and intellectual standards of our society grow.

The achievement of a memorable civilization as a part of America's national purpose meets a number of criteria discussed previously. It is a new frontier and a challenge, and historically Americans have always risen to the occasion when facing worthwhile challenges and promising new frontiers. A memorable civilization can be a powerful symbol, which is important for the implementation of the new national purpose. Effective symbols are essential if the nation is to be moved away from preoccupation with material pursuits to other values and rewards and thus become a more agreeable, happier society, capable of avoiding the perils of transition into a semistationary society of the twenty-first century. As we shall see later, the goal of memorable civilization meets the criterion of a universal appeal and will be important for America's role in the international forum. Lastly, the goal of a high civilization is indispensable if man's full potential for growth is to be realized. This point requires elaboration.

Biologists point out that human evolution is both biological and cultural, with the two aspects proceeding simultaneously. The interdependence between the biological and the cultural evolution was sustained by the development of a positive feedback system between the cultural pattern, the body, and the brain, a system in which each of the three shaped the progress of the other. In the early stages of evolution, the biological aspect—the extension of the arm, the opposable thumb—was perhaps the more important. However, in the later phases of man's evolution, as the central nervous system reached a high stage of development and extensively interacted with social organization, the cultural aspect gained in relative importance. As Clifford Geertz puts it, "there is no such thing as a human nature independent of culture." [24] Just as men are incomplete or unfinished animals who complete or finish themselves through culture, striving toward a higher civilization is the spearhead of improving and perfecting culture and, hence, man himself.[25]

A next question to be addressed is what, in the context of our national purpose, is a "virtuous society"? It is a complex question that cannot be adequately discussed here; however, some tentative criteria for such a society can be suggested.

A virtuous society, almost by definition, implies a society adhering to high ethical standards. These standards will be a product and a part of the nation's goal of creating a high civilization whose benefits are shared by all. However, the acceptance of high standards of ethics in the nation's value system is not enough to create a virtuous society. It must also be governed by a concept of justice that, in a given stage of societal development, can be accepted by society as supporting and enhancing its characteristics as a virtuous society. For discussion purposes, we might, perhaps, accept a modification of John Rawls's concept of justice [26] as suitable to meet the requirements of the national purpose suggested here and those of society, now and in coming decades.

This concept of justice could be defined as follows: "All social primary goods—liberty and opportunity, income and wealth, and the bases of self-respect—are to be distributed equally unless an unequal distribution of any or all of these goods is to the advantage of society as a whole, with due regard to those of its members who are the least favored." [27] This definition of justice would recognize inequality of rewards in society but would base inequality on the individual's contribution to society. This concept of justice would help to promote a communal ethic, since an individual's privileges would be derived from his contribution to society, and he would be indebted to society for them. By the same token it rejects the primacy of self-interest that pervades society and constitutes the principal determining factor in the allocation of rewards. Whereas, in economics, self-interest did play a role beneficial to society, in this stage of society's development, and when projected to all spheres of social activity, it does more harm to society than good by a wide margin.

The concept of "the advantage of society as a whole" depends on the hierarchy of values in society at a given time. It could include the achievement of a memorable civilization or the enhancement of the economic well-being of the nation. It could include both, with due consideration to priority of one or the other, in the event of conflict.

Thus the part of our definition, "the advantage of society as a whole," would include a perfectionist element—the striving to excellence, to new frontiers—including the intellectual, esthetic, and spiritual. This perfectionist element, of course, will constitute but a part of what is to the advantage of society. The phrase "with due regard to those of its members who are the least favored" advances the principle of spreading the achievements and benefits of society to all. These need not necessarily be material; one can enjoy a reasonably high standard of living but be among "the least favored" in the esthetic, ethical, or spiritual sense. This component of our definition would also tend to foster responsibility to society and a communal ethic.

One might raise an objection to the national purpose of the development of man's potential through an enlightened and virtuous society and a memorable civilization on the grounds that there are still poverty, discrimination, and other more immediate problems in our society. Why not first take care of these problems and then address ourselves to higher goals? There are a number of reasons for not waiting. Only the more fundamental ones will be mentioned here.

It is true that we have a number of loose ends from the other purposes of our socioeconomic system that have not been adequately taken care of, but we would be doing injustice to those who are being liberated from poverty and discrimination by not showing them anything else to strive for. Full justice is hardly being rendered to the millions of Americans who are reasonably well off materially but who feel alienated, powerless, and cynical in a society that seems to have lost a sense of direction. Perhaps a hierarchy of goals could be established in accordance with which those who are still trapped in poverty and other shortcomings of our society could, along with the efforts to improve their predicament, be exposed also to higher goals and values and thus short-circuit the stages of societal development and facilitate their emancipation.

No less important is the consideration to the effect that many of the "more immediate" problems our society is trying to solve are actually symptoms of more fundamental social phenomena. A number of these phenomena can be either removed or ameliorated in the process of achieving the higher national purpose, and thus the symptoms can be made to disappear, diminish in number, or become more tractable. In a society such as ours, strongly imbued by materialistic values, poverty is more difficult to eradicate than would have been the case if society's values were reoriented away from materialism. Because personal achievement is largely expressed in material terms, distribution of income and wealth is mainly a zero-sum game on which the attention of millions is keenly focused and in which sacrifices in actual or potential income or wealth for the sake of one's fellowman do not receive ready acceptability. Moreover, since the material level of individuals is closely linked with status and self-respect and is systematically fanned by advertising, the concept of poverty is not a static one. The material standard considered as "poverty" tends to increase over the years and thus make the eradication of poverty a receding target.[28]

Developments such as crime, drug addiction, and the "counterculture" can, to a considerable extent, be traced to the impersonality of our advanced society and the high economic value of time, which inhibit durable, nonsuperficial relationships, human warmth, and the general "agreeableness" of society. A reorientation of values and rewards away from material goods can go rather far in removing or ameliorating the fundamentals on which our social ills tend to breed. It is precisely such a reorientation that the national purpose of an enlightened, virtuous society and a high civilization would be expected to accomplish.

Indeed it is entirely possible that if we merely continue fighting the "immediate problems," the symptoms, and do not address ourselves to the fundamentals, we may be fighting a never-ending battle that will sap our resources and patience and leave society more frustrated, exasperated, and disillusioned than ever before. This danger can become very serious, especially given the narrowing material horizons of the next century.

A Road to Implementation

It is very difficult to introduce major innovations in our socioeconomic system, which is not readily adaptable to change. The reason is the highly pluralistic nature of our society, its decentralized governmental organization, and deadlock among interests that it fosters. Aside from decentralization within the executive itself, the principle of checks and balances strongly militates against change. Whereas our socioeconomic system provides few leverages for effective change, it has numerous "choke points" at which change can be stopped or slowed down by antagonistic interests, bureaucratic lethargy, or strategically situated anachronistic values. At the heart of this situation is the perennial issue of political science—that of freedom versus control.

The paradox of this issue is that the absence of control in society does not provide freedom, since an individual's freedom would then be abridged by other individuals exercising theirs. Thus, to maximize freedom, society has to introduce a measure of control or, to put it differently, to strike a balance between the two. This balance does not necessarily have an eternal validity; as circumstances change, a new balance between freedom and control may be necessary.

The American system of government is based on the separation of powers and concomitant pluralism, whose purpose is to safeguard freedom. The value of pluralism built into our system has been proved time and again, Watergate being only one important example. However, on top of the pluralistic characteristics built into our governmental system, the impact of technology has overpluralized Washington and the entire nation. Excessive pluralism in the U.S. government presents a serious problem inasmuch as it impedes the government's ability to control the impact of technology and thus create conditions necessary for the enjoyment of freedom. This was precisely why, in chapter 11, I proposed greater centralization of the U.S. science and technology policy. But the issue is broader than control of technology alone. A new balance between centralization and pluralism in the U.S. government is needed if we are to become an enlightened, virtuous society and a high civilization. This new balance need not, and should not, reduce the pluralistic nature of our system to the point that its ability to safeguard freedom is jeopardized. But it would be expected to eliminate those excesses of pluralism, largely induced by technological impact, that seriously impede the ability of the government to function effectively and to introduce desirable changes.

As was the case with the goal of controlling technology and its impact, there will be no need for a radical change in the U.S. government to make it possible to achieve the objective of an enlightened, virtuous society and a memorable civilization. A reorganization of the Federal Executive, to enable the President to exercise a more effective control over the departments and agencies and special interests inside and outside the government, would be a necessary and important step toward this end.[29] However, reorganization alone will not be enough. Institutions are mere instrumentalities that can be given life and substance only through effective policy. Here we are coming to the crux of the matter—leadership.

To reorient the nation toward the goal of an enlightened, virtuous society and a memorable civilization would take a president who is an inspiring and strong leader of high personal standards and attributes whom the nation—and the world—could admire and follow. Studies of U.S. public opinion that show a high degree of cynicism and disaffection with principal institutions indicate that this disaffection is predominantly directed at the leadership of our institutions, and not the institutions themselves. The overwhelming majority of the American people believe that the government can be made to work efficiently, given proper leadership. Indeed there seems to be a longing in Americans for a strong and candid leadership and a hope that such leadership will come.[30]

There are thus reasons to believe that a strong and inspiring president could gain enough support to make the goal of the higher national purpose a realizable one. On the other hand, in view of the magnitude of the task, it is doubtful that merely an average president could move the nation to the new, more noble frontiers of America's national purpose. In fact he could do serious and perhaps irreparable damage to this goal and probably would do better not even to try.[31]

On the assumption that a leader of appropriate qualities is elected, and reorganizes the federal government to make it more efficient and responsive, what further steps toward the implementation of the new national purpose should follow? It is important to recognize the utility of institutions as modifiers of values and to use them accordingly. The President, acting through the top leaders of the executive agencies, could set higher standards in their respective organizations, whereby objectivity, high ethical standards, and loyalty to society and its purpose—at the expense of narrow loyalty to a constituency or parochial interests of the organization—would be encouraged and enforced. The federal government could use its leverages (such as grants) for modifying values through the nation's educational institutions. In particular it would be necessary to tone down the traditional stress on the legitimacy of self-interest and success in material terms, while responsibility to society and service to society would require encouragement.

We have seen that standards of a higher civilization find grass-roots support and, here and there, break through the entrapments created by materialistic values. This suggests that sprouts of leadership (perhaps based on enlightened self-interest) can also come from society itself and need not rely on the federal government alone. However, both the Presidency and enlightened leadership coming from society must support each other to be effective in moving the nation to higher goals.

The business world, because of its highly decentralized nature and because certain values are deeply entrenched there, is likely to remain the stronghold of such values as competition for material gain and of material values in general. However, there are hopeful signs for change in the business world, too, as manifested by the relatively few, yet influential, corporations that have for years demonstrated a genuine sense of public responsibility. Public responsibility in business can be further promoted by the governmental support of the public-service-oriented corporations. The case of Japanese business, where profits occupy a relatively low priority as compared to what is thought to be the interest of Japan, provides a hopeful indication that material values need not necessarily reign supreme in the business community.

We have seen in chapter 14 that there are two principal instrumentalities of power in reasonably well-governed postindustrial societies—an effective social organization (primarily the government) and intellectual tools relevant to the running of society and the promotion of its well-being. Rationality and intellectual pursuits would require an upgrading in the nation's hierarchy of values if our society is to solve its problems and reach higher levels as a civilization. Societal purposes and goals themselves are not constants, and as we look into the future they would have to be modified continuously as society achieves some and finds others irrelevant because of new circumstances. Intellectual resources of the highest caliber will be needed for this task, and a greater regard for rationality in society will be necessary to facilitate the general acceptance of societal purposes and goals.

The intellectual task of examining the national purpose and societal goals and the means for their implementation is not a challenge that belongs only to the future. I have barely touched upon the many questions which need to be answered now. How ready is the United States to pursue the national purpose of an enlightened, virtuous society and a memorable civilization? What are the best paths to pursue these goals and in what time sequence? What are the costs associated with them? What are their implications for our social institutions, especially the judicial system? What are the international implications of these societal goals?

An area that requires special attention is culture. Being the most advanced phase of human evolution, culture is an instrumentality for man's improvement and perfectibility. And yet, Daniel Bell and other writers point out that not everything is well with our culture. In recent years the avant-garde culture has taken over to become our popular culture, and there is a definite disjunction between it and our social structure. For how long can this disjunction exist, short of culminating in a crisis possibly toppling our institutions devoid of their ideal content? Just as important, modern trends in our culture—reflected in music, art, and literature—do not necessarily elevate man. On the contrary they tend to be debasing.[32]

Perhaps a little perspective is in order. We have recognized some time ago that war is too important to be left to the generals. It has taken us a little longer, but we have begun to recognize that science and technology are too important to be left to the scientists and the engineers. The case may well be that our culture is much too important to be left to the artist, the composer, and the writer, and to the commercial interests that frequently motivate or stand behind them.

The questions and subjects such as those pointed out above cannot be adequately examined and answered by a single discipline; it is the task of the intellectual community at large. But perhaps political science, a discipline whose traditional goal has been to develop a body of knowledge necessary for steering society, has a special responsibility in this regard. Rather than pursue the two divergent directions of "science" and "relevance," it could render important service to society by combining healthy empiricism with the frankly normative approach of the earlier political philosophers like Plato and Aristotle in an effort to answer the question: "Where are we going and where should we go?" When we look into the future, which increasingly assumes global dimensions, perhaps an even greater ultimate responsibility lies with the field of international relations, which is geared to these dimensions and has the potential of integrating and building upon the concepts and findings of other social sciences.

What is the outlook for the intellectual community's rising to the occasion and addressing itself to the question of building an enlightened and virtuous society and a memorable civilization? If one would wait for the initiative to come from the intellectual community itself, the outlook is not good at all. But there is a glimmer of hope. For some time now, there has been a feeling in the university community that its views are not sufficiently listened to in Washington and that its talents are not adequately used by the government. If an inspiring president respected by the academic community calls upon it to help to explore new horizons for the nation, he is likely to get a constructive response. This development may not necessarily lead to the fulfillment of Daniel Bell's prediction that the universities will become the ganglion of the postindustrial society, but it would certainly strengthen the contribution of the universities to society and enhance their stature.

Lastly, a question must be raised: Considering that we have lately seen other efforts in print to move society into certain directions and toward certain goals, in what way is my study different enough to carry a serious promise that it is not another exercise in futility? What is the outlook for the implementation of the principal proposals outlined here? These objectives are realistic and achievable, and there is substantial evidence for their desirability and support, but they do depend for their implementation on an uncommon, strong, farsighted leadership, coming both from our top elected officials and from various sectors of society and capable of inspiring the nation and crystallizing societal will. Therefore, the answer to our question must remain open-ended. Useful assistance may, however, come from technology. If, as a result of the failure to provide a timely solution to major problems created by technological advance, the nation is thrown into a grave crisis, such leadership may appear. Truly major crises tend to facilitate the emergence of men of great stature and to put them at the top to straighten things out, as the cases of Franklin D. Roosevelt and Winston Churchill, among others, suggest. If this happens, technology will perhaps have performed its single greatest service to our society.

The National Purpose in a Global Context

It might be appropriate at this point to raise the question of the future of America as a civilization. Is America as a world civilization, as some contend, a mere flash in the pan of history, now rapidly approaching its extinction? What is the relevance of the new national purpose discussed in this chapter to her role as a world civilization? To shed some light on this subject, it must be discussed in the context of the following three considerations: (1) the outlook for the rise of competitive civilizations of global significance in the future, (2) major trends in the development of world society, and (3) the factors that brought about America's rise as a world civilization.

A Shift in World Civilizations?

Western Europe is one possible candidate for resurgence as a world civilization. It is conceivable that Western Europe my be unified in the next 20 to 30 years, and it may demonstrate a substantial economic vigor. But even if these optimistic assumptions materialize, it is not likely to produce the degree of cohesiveness and spiritual drive necessary for a world civilization. Western Europe consists of a number of nationalities with strong historical roots and particularistic loyalties, and it would probably take a cataclysmic event to mold them into a spirit of collectivity that would generate dynamism of its own. Otherwise, the balancing process among the various nationalities and the remnants of class divisions, assisted by the pluralizing effect of technological advance, will continue to consume a good part of Western Europe's energies.

A shift to a semistationary society in the next century would favor Western Europe in the sense that the shift would, to a large extent, free it from dependence on outside sources of raw materials. However, it is not clear how successful Western Europeans will be in taking the necessary steps to make the transition to a semistationary society a smooth one, as distinguished from a succession of crises. It would take considerable strength and will at the top, which, even if unified, the conglomerate Western Europe may not be able to produce in a timely fashion.

In some respects Japan presents a greater promise than Western Europe as a major civilization. A highly cohesive, almost collectivistic society, with a great deal of dynamism, she is certainly more in tune with the future than Western Europe is. However, Japan is seriously handicapped in rising as a world civilization by at least four considerations:

1. The national goal of Japan of economic aggrandizement carries strong connotations of a zero-sum approach to world affairs in an era in which the non-zero-sum game is gaining in the international political process. Japan is likely to suffer significant setbacks unless she reorients her goals in a way so that they do not run against prevalent international trends. She appears to be rethinking and reorienting her goals, but their exact direction is not yet clear.

2. The characteristics of the Japanese national character—in particular, Japan's inward, self-centered orientation—do not as yet make Japan ready to assume the responsibilities of a world civilization.

3. The Japanese society and Japan's value system are in the process of change. Japan appears to be losing some of her strongest assets, such as her collectivistic spirit. Perhaps more important, it is not quite clear what Japan's value system will evolve into in coming decades and whether the transition in the Japanese society can take place without major instability.

4. Japan is constrained by her small territorial base and her high degree of dependence on the outside world. By themselves these characteristics are not necessarily prohibitive to Japan's becoming a world civilization, but when viewed along with the limitations enumerated above, they are seriously restrictive.

Having a limited territorial base and a high industrial and population density and being heavily dependent on foreign raw materials, Japan will probably be an early candidate for converting her economy to a closed-materials type. She is likely to maintain large investments outside her borders and continue drawing on them to sustain her economic well-being. This may reduce the pressure on Japan for an early reorientation of her values away from material pursuits, but sooner or later a degree of such reorientation will become necessary. The consensual decision-making process and inward orientation are likely to persist in Japan for some time, so that there is some question whether she will be able to adapt to new requirements in a timely fashion and produce a responsible and inspiring image that would be necessary for her to attain a stature commensurate with a world civilization. Cooperation with China, if it develops, could partially compensate Japan for her lack of a large territorial base. But it is doubtful that the degree of cooperation that could eventually be achieved would be close enough to create a Sino-Japanese civilization of world significance.

It is conceivable, however, that China, initially assisted by cooperation with Japan and leaning on her ancient culture and on the more modern purposiveness of her stoic ideology, may eventually raise a persuasive claim to a world civilization in her own right. But this development, if it takes place, would bring us somewhere near the end of the time frame covered by this book.

The prospects of the Soviet Union's becoming a leading civilization are beclouded by the uncertainty of its societal development. A country consisting of a number of distinct nationalities in which the leading nationality—the Russians—is likely to become a minority, with a population waking up to the attraction of consumer goods and pressuring for more, with the different components of the vast bureaucratic apparatus showing signs of developing vested interests of their own, and with the government expanding contacts with the "capitalist" world and attempting to assimilate advanced nonmilitary technology, the Soviet Union is distinctly vulnerable to the pluralizing impact of technology and its manifold corollaries. At the same time the government is tightening its control over society and expressing confidence in its ability to stay in control and achieve its objectives short of allowing pluralization of society to take hold.

Suppose we give the Soviet leadership the benefit of the doubt and assume it will successfully proceed toward the achievement of a model of society it appears to be striving for: an ideologically constrained, technology-oriented society. Given this assumption, it is remarkable how much the future appears to favor the USSR. The energy crisis will probably strengthen the Soviet Union. The emergence of semistationary societies, at least initially, is also likely to benefit the Soviet Union. It will open up an opportunity for the USSR to strengthen its economic growth and to enhance Soviet influence in the less advanced countries (LACs) and LDCs (see chapter 14). Moreover, when the USSR itself eventually becomes a semistationary society, it will not need to reorient the values of its population to ensure an effective functioning of its system of rewards. Although a highly materialistic society, the USSR has traditionally rewarded its people by prizes and privileges, and only sparingly with material goods.

But as a rising civilization, what would the Soviet Union have to offer the world? Certainly not an inspiring ideology: Marxism has lost its appeal to the masses of people in the USSR and outside. What remains then is an effective organization for power. For a while the sheer effectiveness of the organization can probably overcome the diffusion of power fostered by technological advance, but can it be maintained for a long time? Fundamentally, an organization for power operating in the international arena pursues a zero-sum game, although self-interest may dictate that some of its activities may have distinct non-zero-sum elements. Can the USSR pursue a zero-sum strategy for a lengthy period of time short of being "socialized" into a different mold in an international political process in which non-zero-sum elements are gaining in relative significance?

In sum, of all actual or potential great energy systems in the world, the USSR is likely to remain America's leading competitor for a place in the sun as a world civilization. There are a number of factors that favor the rise of the USSR in the future, but, even under "optimistic" (for the Soviets) assumptions, the model of the USSR envisioned by its leadership will face a number of challenges that will make it difficult to sustain its regime in the long run without some systemic modifications. With or without such modifications the USSR is not likely to be able to generate strong appeal in the outside world, except insofar as sheer manifestation of power might be appealing to some.

World Society: A Source of New Vitality?

As we have seen earlier (chapter 10), in recent decades the activities within the framework of our rudimentary world society began to develop a certain purposiveness. They have been mainly focusing around three goals: (1) maintenance of international peace, security, and stability; (2) satisfaction of material wants; and, most recently, (3) maintenance of the planet's ecological system. Is there reason to believe that, in the light of future developments, these or new activities might acquire vitality within the framework of world society, inspire peoples and nations, and energize them into a determined collective pursuit of these goals? If so, then perhaps America's national purposes and goals ought to be adjusted to these international developments so as to render them wholehearted and unqualified support. But the outlook for such a vitality is not particularly promising.

The United Nations will continue to play a role in international peace and security, but it will be marginal as in recent years. Wars will decline in relative significance in coming decades, but this will come largely from the constraints that the impact of technology generates rather than from the political concept of collective security as it evolved in the theory and practice of international organization. These constraints find their roots in the destructiveness of nuclear weapons, the tendency toward stalemate in the sub-nuclear military sector, the growing interdependence among nations, and, later, the difficulty that closed-materials economies are likely to encounter in continuing to serve as the war potential of nations.

Just as in the case of international peace, much of international activity aimed at the satisfaction of material wants will bypass international organizations. As before, this activity is likely to continue to flow through multiple channels: direct aid to the LDCs, independent and directed activities of private multinational corporations (MNCs), bilateral and multilateral regional programs of scientific and technological cooperation, and so forth. As some LDCs will be less successful than others in balancing population growth with adequate supplies of food, famines will occur; they will be handled, with varied success, through unilateral assistance and that of international organizations. This activity will, however, be sporadic; it is not likely that the international community will undertake a general crusade to improve material needs of its less fortunate members. As more and more LDCs become industrialized and material conditions generally improve, the satisfaction of material wants as a focus of international activity is likely to decrease in importance.

Unlike the other two activities the maintenance of the planet's ecological system will gain in relative significance. As pointed out in chapter 10, it is not likely—at least, not in the next 50 years—to develop into a strong international authority. Aided by the self-interest of nations to provide safeguards against pollution within their own borders, international activity in this area is likely to focus on monitoring developments and providing coordination of and, perhaps, guidelines to, national and regional efforts. It is possible, however, that if, for extraneous reasons like political instability or economic emergencies, certain nations are incapable of maintaining requisite ecological standards, and serious crises occur, the international community may be sufficiently energized to provide both remedies to solve the crises and "teeth" to international authority to prevent their recurrence. Otherwise, international activity in the ecological area may well become routinized until, perhaps, much later years, when such phenomena as excess waste heat from industrial activities might necessitate a broad-based and fundamental action on a global scale.

This outlook for the future of major trends in world society, although far from being apocalyptic, suggests some fundamentally disturbing questions. Where does the hope for a brighter future of world society lie? What happened to the expectations of the theorists of international organization to the effect that the growing international society aided by human rationality will continue building a system of regulatory procedures curtailing war and eventually culminating in a world government? As we look into the future, we do seem to discern a decline in the role of war, but almost by a freak, a succession of technological impacts not always logically interrelated. Moreover, we have no clear indication that this decline, if it takes place, will be "capped" by an appropriate organizational arrangement and thus be legitimized and given greater certainty. Nor are we left with a great deal of faith in functionalism, at least not as it operates through intergovernmental international agencies. Functional organizations will not free the world of crises; at best—especially in global ecology—they may save us from grave dangers.

In short, world society does not seem to open up new and bright horizons in the future toward which man can confidently stride. A quick look at the United Nations since World War II may provide us some clues for understanding the future and may help answer the questions raised in the preceding paragraph.

When, in the immediate post-World War II years, both the political and functional activites of the United Nations and its family of specialized agencies were making very limited, if any, progress, the reasons appeared to be readily available. The culprit was the Cold War, and the contributory factor was the lack of universality of the United Nations, which embraced at the time only a part of the world community. Then, more than 30 years after establishment of the United Nations, the Cold War was gone, the membership in the U.N. increased from the original 51 members to 146 and became truly universal, and yet the United Nations is probably at the lowest level of its effectiveness.

The political agenda of the United Nations is filled with items that the major powers consider to be of small or no importance, but the majority of the U.N. membership, consisting of small, largely less developed nations, is deeply concerned with them. To the extent that the more important questions are considered, they frequently get stalled between the newer members who have the votes and the major powers who have the veto. The problem cannot, however, be reduced simply to a division between the North and the South; there are significant differences among the LDCs as well as among the major powers. By and large, the mainstream of international relations bypasses the United Nations.

The best, idealistic international civil servants of earlier days have left the U.N., and the organization's bureaucracy is beset by low competence, cliques, and uncertain loyalty. The U.N. can claim better results in the social and economic areas than in the political, but this is not to say that everything is well in those areas and that functionalism is on the rise. The specialized agencies have relatively small budgets, are jealous of each other's power, and view with a touch of dismay the emergence of a new, powerful, yet somewhat unruly presence in the functional field—the MNC.[33]

One can provide a number of proximate explanations for the sad state of affairs in international organization—the organizational imbalance whereby those nations that have most votes lack resources and power, the tendency of some nations to use the international civil service as a training or dumping ground for their personnel, and so on. The more fundamental reasons for the weakness of international organization, however, will have to be sought elsewhere. No single factor can fully explain the situation, but if one tried to reduce the answer to a single most important factor, it could perhaps be found in the direct and indirect impact of science and technology.

Scientific and technological advance makes societies much more complex, more pluralistic, and much more difficult to govern. This is true enough for national societies; this truth becomes magnified in the case of world society, where there was enough complexity and pluralism to start with and where no effective authority existed in the first place. In this regard the impact of technology on the nation-state and world society is differential. Perhaps even more than in the case of national societies it brings into sharp relief the paradox: While the need for governing world society has considerably increased, the difficulty of governing it has increased much more.[34]

Pessimistic as this conclusion is, the case may well be that the outlook for a reasonably well-governed world society has been postponed considerably further into the future. This suggests that, if man is to seek the benefits that a well-governed society can bring, he may have to focus his efforts on a scale smaller than that of world society. However, the interdependence of the world being what it is, he cannot afford to be oblivious to it, although his endeavors in this area can bring but very imperfect results.

America as a World Civilization

America's remarkable rise as a world civilization was a product of primarily five factors:

1. The United States, more than any other nation, benefited from the application of science and technology to a continent-size territorial base rich in natural resources.

2. America's dynamic population, unlike that of European nations, was not splintered into classes or groups deadlocked in a struggle among themselves.

3. Similarly, unlike European nations, which had enemies on their borders, had to maintain large military establishments, and mutually neutralized a good part of each other's energies, the United States had no strong ene-

mies nearby to guard against. She could thus apply all of her energies to internal development.

4. The United States was born out of a revolution, a fight for freedom, and the promotion of freedom was firmly embodied in the American mythology. The ideal, the legend of America as a land of freedom, was an effective symbol that attracted millions of immigrants to the United States, served as a powerful weapon on the side of the American cause in many wars, and helped to sustain the confidence and dynamism of Americans in peacetime.[35]

5. World War II destroyed or greatly weakened old power centers of Europe and Asia and thus passed on to America the mantle of supreme world power.

It is indeed remarkable how much of these assets the United States managed to lose before even 30 years had elapsed. Some of this loss was inevitable, and some was not.

Because of worldwide scientific and technological advance, the United States lost the old advantage of not keeping a large military establishment in peacetime. To be sure, no major enemy emerged on America's borders, but the rise of the Soviet Union required a redefinition of where the physical frontiers of America's security were. Not only did the United States end up in the old predicament of European nations of maintaining the burden of military arms in peacetime, but also her burden was actually disproportionate to that of some of her allies, and this allowed them to capitalize on nonmilitary technology and gain in economic power.

It was inevitable that America's relative power position would decline from its peak right after World War II— the nations devastated in the war were bound to recover sooner or later. This decline was not necessarily detrimental to America's role as a world civilization; in some respects it was fully consistent with it. This was exemplified by U.S. economic assistance to the nations that suffered destruction in the war and, later, by economic aid to the LDCs, both of which contributed to a change in the world distribution of power. However, the full extent of America's decline in relative power was not inevitable. This, in particular, applies to the lack of a timely and rational delineation of priorities in science and technology and a general rationalization of science and technology policy, a weakness that tends to persist.

It was inevitable that, sooner or later, America's rich natural resources base would begin to be depleted. By itself, this development was not very critical—perhaps not even very important. However, when viewed in the context of technological trends and, in particular, America's technological potential, any detriment the nation suffers as a result of shortages of raw materials must be placed at the door of deficiencies in policy and timely decision making.

Post-World War II decades also witnessed two divergent phenomena in Western Europe and the United States that began to curtail some of America's earlier advantages. Western Europe, through economic integration and a general acceptance of the premise that a war among Western European nations is no longer tenable, began to release the energies previously locked in the intranational and the international balance of power. The United States, on the other hand, began to demonstrate a growing tendency of societal forces and interests to stalemate one another and thus tax the society's energies and curtail its capabilty to act. This phenomenon, reinforced by some extraneous factors such as Vietnam, was "inevitable" only insofar as society resigned itself to certain "deterministic" influences of technological advance and refused to counter it by intellectual insight, imagination, and will.

In the post-World War II decades America lost a great deal of her ideological appeal and inspiration, both to the outside world and to her own people. In part this stemmed from a conflict between considerations of national security and the nation's ideological preferences. National security was viewed as more important, and the United States found herself allied with and supporting regimes whose devotion to freedom was often questionable and in some cases completely nonexistent. To be sure, insofar as this was not running in direct conflict with the perceived interests of American national security, the United States remained basically loyal to her ideals, which found expression in the support of the concept of the open society. Not very precisely defined, it was applied in somewhat different forms to different areas. It was applied to the world at large through American policy in the United Nations; it included America's support of human rights, freedom of human travel and contacts, free trade, and economic development. With regard to the LDCs, it was backed by economic aid and technical assistance. In a different and a somewhat subdued form it was eventually applied to Eastern Europe and the USSR, where freer trade, the spread of consumer goods, and closer contacts were expected to produce a more pluralized, less power-oriented society.

Basically the concept of the open society was leaning on the autonomous forces of technology, which were raising the standard of living of peoples; stimulating freer exchange of goods, human contacts, and ideas; and promoting a more pluralistic world. When carried to its full development, the open society was to be a reflection of American society. But then, as the United

States began to face internal difficulties as the impact of technology was both creating problems and deadlocking societal will to solve them, the appeal of the open society began to tarnish rapidly.[36]

The loss of America's ideological appeal was a loss of a great asset that the United States could ill afford, but it did not attract so much attention as the crumbling of American material assets. Perhaps it was because, in our post-World War II national strategy, we placed our reliance overwhelmingly on the material means: the strength of our military power and the invincibility of our technologic-economic arm. The military policy of containment was by and large successful, but our resources were overextended, and a withdrawal followed. As the time was approaching for U.S. technologic-economic power to become a decisive element in U.S. national strategy by first contributing to the pluralization of the Soviet society and later outcompeting it, America's confidence in her power was given a jolt by the competitive vigor of streams of sophisticated consumer products from our allies, sold in the United States and the world market. At the same time there were problems at home. The image of America as the bulwark of freedom was much too weak now to bolster the U.S. world position. It was then that articles began to appear referring to the United States as a possible flash in the pan of world history.

As we look back at the developments in post-World War II decades, it appears that a number of assets we lost in those years are recoverable and that those not recoverable are not important.

The United States can regain her appeal and leadership. She can do so by the pursuit and eventual implementation of the two concepts extensively discussed earlier: control of technology and its impact and an enlightened national purpose. In combination these two concepts can provide greater social cohesiveness, help resolve internal deadlocks, release energies and resources for constructive pursuits, and contribute to a happier, more agreeable society. In this way the United States would provide a much more attractive model of society, capable of providing inspiration to other nations.

The enlightened national purpose is likely to be appealing to many LDCs as well as highly advanced nations. A number of LDCs are based on older civilizations with major cultural achievements of their own. Although poor and striving to develop, they have combined both an envy and a scorn in their response to an American policy that seemed to them as brandishing military power and material wealth but showing very little in the way of cultural achievement and spiritual values. By shifting our emphasis to the latter, these nations can be made participants and contributors to common goals in the development of humanity's higher values and not just recipients of what-

The National Purpose   325

ever aid we might be able to render them. The element of contributing, sharing, and cooperating in a mutual goal provides a much stronger foundation for acceptability of leadership than a mere giving of material assistance—of which the United States, in recent years, has had less to offer.

Some of the same considerations would apply to advanced nations as well, but here America's national purpose and goals will have another important dimension. The United States is technologically the most advanced society in the world. It is thus a trailblazer in charting its course for an uncertain future permeated by the impact of science and technology. America's role thus involves both responsibility and danger. We can be successful and show others the way, or we can fail and let those who follow us fail along with us or profit from our mistakes. Our adaptability and ability to select and implement the type of goals enabling us to meet the future effectively is thus critical not only to America, but to a good part of the rest of the world. No other major region in the world appears to offer a constructive example of how to meet the future. No useful guidance is likely to come from international organization; in fact, if anything, it needs an injection of constructive and inspiring leadership. There is thus a void to be filled, a role to play, and a responsibility to be carried out.

The United States cannot hope to recover its vast military superiority of earlier years, but such superiority is not needed. Military power will remain important in coming years, but it will be needed and important only insofar as it effectively serves as an instrument of international stability. When appropriately adapted to that task, it can serve adequately in sustaining America as a world civilization.

America's natural resources base can never be as plentiful as before, but improvements in our national decision-making process and increased vitality of our science and technology can go a long way in compensating us for whatever depletion in natural resources we may experience. And it is well within our means to improve the vitality of our technologic-economic sector. The material instrumentalities of our society cannot, however, be effectively improved and, even if improved, cannot effectively serve the nation unless we restore and enhance the ideal component of our national substance. Recent decades indicate that a single-minded faith in our technologic-economic potential and military power can be misleading.

There is thus no reason to believe that America as a world civilization is rapidly reaching its end. But it is not enough to express faith in America and proceed as usual on the assumption that somehow the nation will "muddle through." Faith in America may be fully justified, but those who express it will not sound convincing unless they show the ways and means to stimulate society's vitality and creativity, so essential for a world civilization.

Lastly, a question should be raised: Why would the United States want to be a world civilization? Is it because it is a worthwhile goal in its own right? The answer will have to be "No." At best it can be worthwhile only insofar as it is related to the fundamental validity of the nation's values and the requirement for the type of a world order in which those values can be preserved and enhanced. A look at world society suggests that, by itself, it cannot provide an international system that can serve that purpose. America's role as a world civilization thus provides a means of ensuring a suitable climate—at home and in the world, since these can no longer be separated—for the preservation and further development of her values.

If the United States pursues the purpose of an enlightened and virtuous society and a high civilization, the universality of the values underlying this purpose would make them suitable for possible acceptability on a global scale. Eventually, as America makes progress in the pursuit of this purpose, its underlying values—the improvement of man and the development of what is best in him—might be introduced on the international forum as an objective of international organization. After proving the workability of this objective in her own society, it might be a logical step for America to take.

If the United States were to adopt a memorable civilization as her overriding national purpose, I believe that she could, perhaps for the first time in human history, demonstrate that there is no deterministic inevitability in the cyclic rise and fall of civilizations and that human intellect and will are indeed capable of molding the future of societies. If the United States could successfully introduce this purpose on a global scale and it effectively proceeds to its fulfillment, such an accomplishment would strongly suggest that human fate need not be a product of deterministic forces or chance but that human advance can be sustained by a conscious and deliberate will of mankind itself. To prove these two propositions might be a befitting goal for a young nation like America—which has gone so incredibly far in her short history—to undertake upon the entry into her third century.

Conclusions: America and a Changing International System

We live in an era of a changing international system. Some of that change is within our capability to control, but most of it is not. But the difference between our capitalizing on what is controllable or neglecting to do so will spell the difference between, on the one hand, progressive evolution of mankind, a sense of direction, and greater amount of human happiness and, on the other, conflict and instability within and between societies, a senseless drift on the part of most nations and humanity itself, and alienation of the individual.

I have discussed earlier some important changes in the individual elements of the international system in the post-World War II decades. They included changes in the nature of international power, in the relative importance of its components, and in its world distribution; and emerging purposiveness of the international system—the search for peace and stability, satisfaction of human wants for material goods, and maintenance of the planet's ecological system; and, lastly, proliferation of actors in the international stage and the emergence of a transnational political process not confined to the interplay among nation–states but reaching both inside nations and extending itself on a global scale. The last phenomenon in particular has tended to change what has been generally referred to as "world order" or "international system"—to use the terms synonymously—into something akin to a "world system," although it is far from adequately integrated and not evolving around a central institutional framework. In short, the aggregate of changes in the elements of the international system in the past two decades or so has profoundly changed the system itself, and not just made changes within it.[37]

It appears that an even more fundamental transformation of the international system is likely to take place with the advent of the semistationary society. Being largely a product of internal societal change, the semistationary society would not affect the world as a whole at once but would have impact on individual nations one by one, depending on their stage of technological advance. Hence international system characterized by the prevalence of semistationary societies would evolve gradually. During its evolution several stages of societal development might exist side by side: preindustrial, industrial, postindustrial, and semistationary. From the point of view of American policy it is likely to be a difficult period, with a different set of constraints applying to nations within each period of technological advance. The constraints applicable to semistationary societies are likely to be greater than those pertaining to nations at a lower stage of technological advance and thus create potential perils for the United States.

The difference between an international system based on semistationary societies and that based on its predecessor, the postindustrial society, would be of a sufficient magnitude to create a dialectical situation in which the postindustrial society would have to adequately mature and synthesize some of its elements with those of the emerging newcomer before the latter's complete ascendancy and before certain features of the postindustrial society become obsolete and fall off into decay. Otherwise, collisions and instability would result.

For the nation leading in technological advance—as the United States is—this would require a reconciliation of certain elements of policy that are not entirely compatible. On the one hand the policy would have to capitalize on certain attributes of the postindustrial (and industrial, too, since in this case they would overlap) society, attributes like those involving acquisition of material wealth through technology to meet internal needs of society now and at a later time when material horizons would be narrowing and when material means would decline in the total system of societal rewards. A strong material base would also be needed to support the nation's external viability and vitality in an international system that would not as yet be able to safeguard the nations' security through its own mechanisms. On the other hand the United States would have to impose restrictions on certain forms of technological advance so as to safeguard its population as well as regional and world ecological systems and otherwise provide for a smooth transition to a higher technological stage.

This policy was referred to earlier as "control of technology and its impact." Quite apart from the difficulty of reconciling the various elements of such a policy within itself, it would not be easy to implement, because of the propensity of technological advance to create a growing number of competing interests. These interests are not readily amenable to resolution so as to make it possible to steer society in a desirable direction. American policy is likely to become even more difficult to carry out as the nation approaches the semistationary stage of its development which would increasingly necessitate the toning down of materialistic values as an element in society's system of rewards and their partial substitution by other values.

Suppose we assume that the existing international system (perhaps, by that time, a true world system) is overwhelmingly based on semistationary societies and that some nations are already approaching the stationary-state society. What would be its goals and a sense of direction, if any?

Satisfaction of material human wants would largely fade away as one of the goals. It may still be of concern to the small number of the remaining LACs, but it would be left to them to cope with. Peace would also diminish as a concern, since the characteristics of semistationary societies would discourage resort to warfare, although armed forces would probably be maintained. Stability, both internal and international, would remain high on the agenda of the new international system. Ecological concerns would likely rise in significance, especially in the area of thermal pollution and potential atmospheric and stratospheric imbalances. But the necessity of coping with them on the national and regional levels would largely alleviate the burdens that these considerations could impose on the global system.

As the older goals and concerns recede into the background and as some disappear, individual semistationary societies and the world system in general are likely to turn their attention increasingly to *homo sapiens* and to the question of evolution. This development is likely to take place because of the growing realization that ideas of a largely self-propelling evolution of mankind are no longer tenable. *Homo sapiens* has succeeded in releasing forces capable of not only impeding evolutionary advance, but bringing it to a halt, or, indeed, reversing it. Moreover, culture has become the principal vehicle of the evolutionary process, but culture does not always carry individual societies to new heights; in fact it may drag some of them down. Last but not least, attention to the development of man—including a higher value system as a part of his total existence—is likely to be stimulated by the pragmatic necessity for diverting societal growth away from material growth. Given these considerations, evolution of man and his perfectibility would have to become a product of collective rationality and will of nations as an integral component of the task of their governing themselves.

Since the task of producing progressive human evolution would largely necessitate advance in culture,[38] the national goal of America as a high civilization would therefore be an appropriate one for the technologically most advanced nation to initiate. Quite apart from its short-range benefit to the United States and its basic accord with longer range world trends, this goal would facilitate the voyage through the rock-studded rapids of the not-so-distant major change in the international system.

Notes

1
Introduction

1. William T. R. Fox pointed out that the term *international relations* does not accurately reflect the nature of the field; *world politics* would be a more appropriate term. But since there is no generally agreed-upon paradigm of world politics, the emphasis on "relations" in world politics provides a convenient criterion for selecting what is important for analysis. See his "The Study of 'Relations' in International Relations," in *The Search for World Order*, Albert Lepawsky, Edward H. Buehrig, and Harold Lasswell, eds. (New York: Appleton-Century-Crofts, 1971), pp. 386–87. In this study "international relations" and "world politics" are used synonymously in referring to the field; however, in substantive analysis, the stress is on "world politics" as an evolving process.

2. See, for example, James E. Dougherty and Robert L. Pfaltzgraff, Jr., *Contending Theories of International Relations* (Philadelphia: J. B. Lippincott Co., 1971), pp. 385–398; and E. Raymond Platig, *International Relations Research; Problems of Evaluation and Advancement* (New York: Carnegie Endowment for International Peace, 1966), chaps. 1 and 5.

3. For a recent analysis of the state of international relations theory, see Arend Lijphart, "The Structure of the Theoretical Revolution in International Relations," *International Studies Quarterly* 18 (March 1974): 41–74.

4. Harry Eckstein, in an effort to define the scope of political science, suggested that it be confined to the study of *authority patterns* or sets of "asymmetric relations among hierarchically ordered members of a social unit that involve the direction of the unit." But then Professor Eckstein ran into a problem: how to accommodate international relations into political science, since authority patterns in international relations are very limited and most relations are symmetrical. He proposed to solve that problem by including in the scope of political science those symmetric relations that "are manifestly concerned with the direction of the units (in short, intrinsic to authority)." But if the scope of symmetric relations *within* nations is broadening, then perhaps "concern with the direction of units" in symmetric relations within nation-states should receive greater attention as a legitimate subject of inquiry of political science, with a correspondingly lesser emphasis on "patterns of authority." What this suggests is that technological impact appears to eradicate distinction between domestic and international politics in conceptual terms and in terms of relevant intellectual tools of analysis.

For Eckstein's discussion, see his "Authority Patterns; A Structural Basis for Political Inquiry," *The American Political Science Review* 67 (Dec. 1973), especially pp. 1153–1160.

5. There is a definition of the camel as "a racing horse created by a committee." The camel, obviously, must have been created in the days when the impact of technology was relatively limited. He was thus lucky. If the period of his creation were more modern, he probably would never have seen the light of day or, if he had, he would have been a much more incongruous racing horse.

6. See Henryk Skolimowski, "Technology: The Myth Behind the Reality," *Architectural Association Quarterly* (London), July 1970, p. 30. Skolimowski presents a concentric definition, in

332   1. Introduction

which the narrowest definition ("totality of all man-made tools") is in the center, while each subsequent definition presented in successive outer circles includes the former, narrower one. Thus the spectrum of definitions is as follows: (a) totality of all man-made tools; (b) their use and function; (c) the material results of their application (technological products); (d) social impact of these products; (e) the influence of technological change on the life of particular individuals. See also Library of Congress, Congressional Research Service, Science Policy Research Division, *Science Policy; a Working Glossary* (Washington, D.C.: U.S. Government Printing Office [hereafter referred to as GPO], 1973), pp. 70–71.

2
The Levels of Technological Impact
1. The concept of strategic limited war is deliberately not considered here inasmuch as it is not directly relevant to the present analysis and its introduction would only result in unnecessary complexity through qualifications.
2. For a good summary of the rise of the nuclear umbrella, see the Washington Center of Foreign Policy Research, The Johns Hopkins University, "Developments in Military Technology and Their Impact on United States Strategy and Foreign Policy," in U.S. Senate, Committee on Foreign Relations, *United States Foreign Policy* (Washington, D.C.: GPO, 1961), pp. 718–730. The acquisition of a first-strike capability by a superpower willing to resort to first strike would be one possible way to change significantly the distribution of world power through nuclear weapons. This is, however, a very unlikely development.
3. For a more extensive discussion of the implications of the spread of nuclear weapons, see chapter 9.
4. See Warner R. Schilling, "Science, Technology and Foreign Policy," *Journal of International Affairs*, 13, no. 1 (1959): 8.
5. Henry A. Kissinger, *Nuclear Weapons and Foreign Policy* (New York: Council on Foreign Relations, 1957); and Robert E. Osgood, *Limited War; The Challenge to American Strategy* (Chicago: University of Chicago Press, 1957) were the two works that provided an early comprehensive analysis of limited war in its nuclear context and were perhaps the most influential publications in changing the shift of strategy toward the subnuclear-umbrella level.
6. See Anthony Verrier, "Strategic Bombing—the Lessons of World War II and the American Experience in Vietnam," *The Royal United Service Institution Journal*, 112 (May 1967): 157–158. The author states that "[t]hese defenses have been described by American crews at symposia conducted by U.S. Staff Colleges as worse than World War II and Korea. . . . We . . . find the United States over North Vietnam in a considerable fix, paralleling the David and Goliath situation on the ground."
7. On projections of Soviet military power, see Thomas W. Wolfe, "Soviet Military Strategy and Policy" in *The Soviet Impact on World Politics*, Kurt London, ed. (New York: Hawthorn Books, 1974), pp. 237–270.
8. The major advantage of the laser in military applications consists of its characteristics as concentrated light: an absolutely straight trajectory and an extremely high speed (186,000 miles per second). The obstruction of the atmosphere limits the effectiveness of laser weapons to a few kilometers, but their use with a complementary medium (e.g., television, which would provide long-range guidance while lasers would enable extreme precision in the vicinity of the target) is likely to contribute to the conduct of warfare with lasers over large distances (50–100 miles or more) and to an eventual demise of the highly maneuverable fighter and bomber aircraft. For application of lasers in future warfare, see "Visions of the Next War," *Newsweek* (April 22, 1974): 52–55. For projections and impact of electro-optical weapons I am principally indebted to Wal-

3. Trends and Prospects 333

ter R. Sooy, Science Applications, Inc., Arlington, Va. (March 1976); also, to Peter O. Clark, Advance Research Project Agency, Department of Defense (June 1976).

9. To illustrate: The initial industrial lead of Great Britain was undermined by the invention of the Bessemer process of steel production (1856), which, by making mass production of steel possible, had made the United States the leading steel producer in the world by the end of the 19th century. The Thomas–Gilchrist process (1879) enabled Germany to use the phosphoric iron ores of Alsace–Lorraine and thus make Germany the greatest European steel producer by 1893. Railroads were a major factor in the rise of the United States and Russia as great powers. Exploitation of hydroelectric power, chemical resources, telecommunications, and cheap water transportation were important in Japan's rise to power before World War II.

10. See Gordon J. F. McDonald, "How to Wreck the Environment," in *Unless Peace Comes*, Nigel Calder,ed. (New York: Viking, 1968), pp. 187–191. However, in September 1976 the Geneva Disarmament Conference agreed to forward to the U.N. General Assembly a draft Convention on the Prohibition of Military or any other Hostile Use of Environmental Modification Techniques. If adopted, this Convention would limit the extent to which climate and weather modification could change distribution of power.

3
Technological Trends and Prospects

1. "Steel Is Where You Make It—in Japan," *Resources* (January 1967): 18.
2. See "Japan Forges a Colossus in Steel," *Business Week* (April 5, 1969): 92–93.
3. As used here the term *technological superstructure* encompasses the aggregate of technological instruments of a society, such as railroads and other means of transportation, factories, and utilities. Technological superstructure does not include technologies in the stage of research and development.
4. Nominally the decrease was by 36 percent, from $5.70 to $3.60 for the first three minutes. Source: Office of Telecommunications, Department of Commerce. Also consultation with Comsat personnel about future trends.
5. For details on a satellite community TV network for India, see General Electric Company, Space Systems Organization, "Television Broadcast Satellite (TVBS) Study; TVBS Technical Report" (prepared for the National Aeronautics and Space Administration, Contract No. NAS 3-9708, Nov. 15, 1969), Vol. 3, pp. 2–19, 7–81, 7–106, 7–108. An agreement had been concluded between the United States and India in 1969 to establish a smaller experimental network for one year. On August 1, 1975, NASA's Application Technology Satellite 6 (ATS-6) was made available to the Indian Government to broadcast directly into more than 2400 receivers appropriately augmented and located in Indian villages. In addition local ground stations were retransmitting the programs into a number of conventional receivers. The programs, controlled and produced by the Indians, covered subjects such as agricultural technology, family planning, basic literature and health. See "India/United States Experimental Satellite Project," UN Document A/AC.105.72, Dec. 11, 1969, and "Earth-Pacing Satellite Moves for Good Views of India," NASA News, Release No. 75-159 (June 2, 1975). The original U.S. participation in the program was terminated on July 31, 1976, but India intends to continue the program under a different arrangement.
6. The cost of augmentation of TV sets for direct reception from satellites assumes mass production at 1975 prices. The lack of uniformity in TV systems complicates direct broadcasting through satellites. (Source of technical information: NASA, August 1976.) Moreover, direct satellite broadcasting would be subject to regulation of the International Telecommunications Union and, possibly, other legal restrictions which may be developed by the United Nations. For U.N. ac-

tivity in this area, see U.N. Working Group on Direct Broadcast Satellites, "The Various Implications of Space Communications: Report of the Working Group on Direct Broadcast Satellites" (A/AC.105/127, April 2, 1974), and UN Committee on the Peaceful Uses of Outer Space, "Report of the Legal Subcommittee on the Work of Its Fifteenth Session (3–28 May 1976)" (A/AC. 105/171, May 28, 1976), Annex 2.

7. See National Research Council, Committee on Atmospheric Sciences, *Weather and Climate Modification; Problems and Progress* (Washington, D.C.: National Academy of Sciences, 1973), pp. 3–8, 26–28.

8. "Experiment in Abu Dhabi Desert Greening—Japan Desert Development Institute to Build a Pilot Farm," *Information Bulletin* (Tokyo), 22 (Aug. 15, 1975): 15–16.

9. For details on Landsats and their potential, see *Proceedings of NASA Earth Resources Survey Symposium*, Houston, Texas, June 1975 (Houston: Lyndon B. Johnson Space Center, 1975), especially Vols. 1-A through 1-D and 3. The benefits of Landsats and their follow-on systems may amount to billions of dollars. Their functions include measuring and monitoring forestry, agriculture, water resources, marine resources, geology, and environment.

10. Source: U.S. Department of Commerce, Bureau of Mines, April 1976.

11. Source: Division of Conservation, U.S. Geological Survey, June 1976. As of that time, technological capability for exploitation of oil at a depth of more than 1000 feet (305 meters) existed.

12. See "Ocean Mining Comes of Age, Part 1," *Oceanology*, 6 (Nov. 1971): 34–41; and "Ocean Mining Comes of Age, Part 2," *Oceanology*, 6 (Dec. 1971): 33–38.

13. Source: U.S. Navy and the Embassy of France, Washington, respectively, April 2, 1976.

14. For projections for the future, I am indebted to a number of individuals consulted in August 1976, especially J. Morgan Wells, Diving Coordinator, National Oceanic and Atmospheric Administration (NOAA); James A. Lawrie, Vice President (Safety and Training), Ocean Systems, Inc., and Walter R. Bergman, Special Assistant for Diving, Naval Sea Systems Command, Dept. of the Navy.

15. See "Seaborg Foresees World Grid for Electric Power Distribution," *New York Times*, August 11, 1970, p. 21. Glenn T. Seaborg was at that time Chairman of the U.S. Atomic Energy Commission.

16. Conventional wheel track systems are not viable at speeds in excess of about 190 miles (300 km) per hour; even at speeds of 150 mph (240 km/h) the problem of track maintenance is serious. Therefore, if higher speeds are to be attained—such as 300 mph—more advanced systems must be used. Magnetic levitation systems (using magnetic forces for suspension and guidance systems and a linear induction motor to supply the drive) and air-cushion systems are studied for operational capability in the 1980s in Western Europe and Japan. However, the gravity–vacuum transit concept, which is similar to the operation of a change carrier shot through a pneumatic tube, has the advantage of even higher speeds than the expected 300 mph capability of magnetic levitation trains. A gravity–vacuum transit system could link downtown Washington with midtown Manhattan in 73 minutes, with a number of intermediate stops. The system's principal problems are the costs of tunneling and technical complexities of maintaining vacuum. However, the application of *thermal* nuclear energy to tunneling may drastically cut the cost (perhaps 50 percent by 1985, if this approach is financially supported), and thus brighten the outlook for the gravity-vacuum system. Source: U.S. Department of Transportation, Washington, D.C., August 1976. On magnetic levitation trains, see R. A. Hein, "Superconductivity: Large-Scale Applications," *Science* 185 (July 19, 1974): 219–221; on the gravity-vacuum concept, see "Urban Train Propelled by Gravity-Vacuum," *New York Times*, April 19, 1969, p. 43.

17. The millimeter waveguide is currently in the experimental stage of development. It consists of a pipe of 1–2 inches in diameter made of oxygen-free, high-conductivity copper and is capable

3. Trends and Prospects 335

of transmitting radio frequency energy of 1-millimeter wave length. In this stage of technology the millimeter waveguide closely competes with optical fibers as a system to be introduced in the early 1980s. Optical fibers—extremely thin threads of glass capable of replacing much bulkier copper wire—must be combined with light-emitting diode (LED) technology or lasers for communication purposes. Source: A T & T, Jan. 1976. On optical fibers, see "The Light Wave of the Future," *Business Week*, (Sept. 1, 1975): 48–50.

18. Sources: For computers, estimates by the International Data Corporation; for commercial network services, estimates obtained from the Institute for Computer Sciences, National Bureau of Standards.

19. A bit is the smallest amount information can be subdivided into, and, in computer usage, constitutes one-eighth of a character or a digit. Thus, a word of 5 characters is equivalent to 40 bits.

20. A breakthrough in bubble memories was made by IBM early in 1973. See "Why IBM Got a Jump in Bubble Memories," *Business Week* (Feb. 17, 1973): 38, 40.

21. For progress in computer technology, see Willis H. Ware, "Computers and Society—The Technological Setting," in *Proceedings of the AFIPS/Stanford Conference on Computers, Society, and Law: The Role of Legal Education*, June 25–27, 1973, Joseph E. Leininger and Bruce Gilchrist, eds. (Montvale, N.J.: AFIPS Press, 1973), pp. 14–19, and *passim*; David A. Hodges, "Trends in Computer Hardware Technology," *Computer Design* 15 (Feb. 1976): 77–85. I am indebted to Paul Meissner of the Institute for Computer Sciences, National Bureau of Standards, for additional information (September 1976).

22. The establishment of the unified commands after World War II—where the forces of the three services are unified for a specific function under the command of a general officer of any one of the three services—is an illustration of the integrative trend.

23. Ernst Haas points out that the physically integrative effect of communications and other forms of technology, when applied to the less developed peoples, tends to divide. Uprooted from the village style of life, people become "socially mobilized," but at the same time, they usually become dissatisfied and frustrated. Profiting from the new communications media, this mass of the newly mobilized people tends to organize around symbols and values stressing dignity, equality, and autonomy, typically resulting not in one, but in a series of competing nationalistic ideologies. Ernst E. Haas, *Beyond the Nation-State* (Stanford: Stanford University Press, 1964), p. 468.

24. This subject is more extensively discussed in chapter 10.

25. American investment abroad is not a new phenomenon—it goes back to the interwar period and earlier. What is new is that the post-World War II investment has been primarily in manufacturing industries (and not extractive, as before), and it has been on a very large scale. See chapter 10.

26. Soviet economic foreign aid to non-Communist less-developed countries between 1954 and 1974 amounted to about $9.6 billion, of which only a half had been delivered by the end of 1974. The delivery rate indicates the slow overall progress of Soviet programs. In terms of economic aid actually delivered, the Soviet aid is less than one-tenth of U.S. economic aid. Sources: U.S. Department of State, Office of Economic Research and Analysis, and U.S. Agency for International Development, Statistics and Reports Division.

27. The comparative figures were $4,469 million vs. $3,439 million. OECD, Development Assistance Committee, *Development Cooperation; 1975 Review* (Paris, Nov. 1975), p. 130. (Hereafter cited DAC, *1975 Review*.)

28. The ocean-going merchant fleets of the EC countries amounted to 69,883,000 gross tons in 1974, as compared with 12,503,000 gross tons under U.S. registry (which does not include many

3. Trends and Prospects

additional American-owned ships under foreign registry) and 13,533,000 gross tons flying the Soviet flag. Source: U.S. Maritime Administration, *Merchant Fleets of the World* (Washington, D.C., Sept. 1975), pp. 6–7.

29. Projections of scheduled revenue passenger–miles (charter and supplemental flights are excluded) by the Federal Aviation Administration for U.S. domestic and international air carriers anticipate a 92 percent increase between FY 1976 and FY 1986 (from 169.5 to 325.6 billion). U.S. Department of Transportation, Federal Aviation Administration, *Aviation Forecasts, Fiscal Years 1977–1988* (Washington, D.C., September 1976), p. 48. These figures are admittedly conservative.

30. Personal communication to the author from Thomas F. Malone, Director, Holcomb Research Institute, Butler University, dated June 10, 1976. Dr. Malone confirmed his earlier forecast to the effect that technology pertaining to large-scale modification of the climate could go through explosive developments in the 1975–1995 time frame. Among capabilities likely to be achieved by the year 2000, he mentioned the ability to identify conditions for controlling precipitation, significant interference in the hail process, interference in hurricanes, progress in the identification and manipulation of tornadoes, and, perhaps, progress in altering the large-scale circulations through the influence in the earth's albedo. See also National Research Council, U.S. Committee for the Global Atmospheric Research Program, *Understanding Climatic Change; A Program for Action* (Washington, D.C.: National Academy of Sciences, 1975).

31. "Statement of Dr. Robert L. Hirsch, Acting Assistant Administrator for Solar, Geothermal, and Advanced Energy Systems, ERDA," in U.S. Congress, Joint Committee on Atomic Energy, Subcommittee on Legislation, *ERDA Authorizing Legislation, Fiscal Year 1977; Hearings* (Washington, D.C.: GPO, 1976), part 3, pp. 2768–76. Time estimates are reasonable program goals at the present level of funding. On a crash-program basis involving higher costs, a demonstration fusion reactor could be built in the late 1980s. The estimates of the costs (in 1975 dollars) were provided by Dr. Hirsch in a personal communication to the author on October 22, 1975.

32. By comparison the largest thermal generating units at present have a capacity of about 1000 megawatts, and units of 500 megawatts are still considered to be very large. The estimate of the size of fusion reactors—between 2,000 and 10,000 MW(e)—was provided by Edwin E. Kintner, Director, Magnetic Fusion Energy Division, ERDA, May 25, 1976.

33. The critical problem in the utilization of thermonuclear energy is the ability to confine an intensely hot plasma long enough for a reaction to take place. Magnetic fusion relies on magnetic confinement, using the so-called magnetic bottles to confine the plasma for a period from 1–10 seconds to achieve the reaction; subsequently the plasma is continuously confined by a magnetic field produced by superconducting magnets. Laser fusion uses lasers to implode fuels to extreme densities; by reaching these high densities, a state of inertial confinement is achieved (i.e., the plasma, in effect, confines itself). Thus the basic difference between magnetic and laser fusion is "magnetic confinement" vs. "inertial confinement." On laster fusion, see John L. Emmett, John Nuckolls, and Lowell Wood, "Fusion Power by Laser Implosion," *Scientific American* 230 (June 1974): 24–37.

34. The advantage of a hybrid reactor is that fusion machines are powerful breeders, more efficient than conventional breeder reactors in producing fissile fuel from either uranium or thorium. See William D. Metz, "Fusion Research (III): New Interest in Fusion-Assisted Reactors," *Science* 193 (July 23, 1976): 307–308. For dual purpose reactors, see Fusion Systems Corporation, "Enhanced Energy Utilization from a Controlled Thermonuclear Fusion Reactor" (Rockville, Md., June 1976); Submitted to Electric Power Research Institute, Palo Alto, Calif., under ERPI Project #RP 471–1. If pure fusion encounters serious difficulties in development, one or both of the above approaches could provide a fall-back position.

35. In recent years, air-cushion vehicles (ACVs), using nonnuclear propulsion, have been rapidly growing in both military and civilian application. The United States has two test surface-effect

3. Trends and Prospects 337

ships, 100 tons each. One of them, SES–100B, set up a world speed record on June 30, 1976, moving at 89.7 knots (103.5 mph). At such speeds, an SES fleet could cross the Atlantic in one day. The United States is proceeding to building a prototype 3000-ton warship. (Source: U.S. Navy, Sept. 1976.) Whereas these ships are nonamphibious, the USSR has an amphibious test ACV as large as 220 tons, capable of unloading tanks and having a maximum speed of about 70 knots. On ACVs, see *Jane's Surface Skimmers, Hovercraft and Hydrofoils, 1975–1976* (New York: Franklin Watts, Inc., 1975), pp. 17, 96, 99–101, 127–128.

36. There are cases pointing in the other direction, that is, toward a smaller scale. Gas centrifuge technology promises to decrease the scale (if not the cost per unit of weight) required for the enrichment of uranium from more than $1 billion for a single plant to perhaps $200 million per plant. Intersecting storage rings (ISR) atom smashers, if they prove workable, may substantially decrease the cost of particle accelerators. But such cases are the exception rather than the rule.

37. Council on Environmental Quality, *Environmental Quality; the 5th Annual Report* (Washington, D.C.: GPO, December 1974), p. 221.

38. The reasons for the slowing down appear to be political, economic, and bureaucratic (e.g., lesser allocation of resources to R & D by the U.S. government; lack of available capital for implementation of technological innovations; the delays involved in approving the marketing of chemical products by appropriate authorities), and not those inherent in technology itself. *Cf.* "Technology Pace Is Found Declining," *New York Times*, April 20, 1976, p. 11. The slowing down of technological change may not necessarily apply to countries other than the United States, now or in coming years.

39. Source: National Center for Health Statistics, HEW. This is a provisional figure.

40. Leonard Hayflick, "Perspectives in Human Longevity," a paper presented at the Conference on Social Policy, Social Ethics, and an Aging Society, May 30–June 1, 1975, The University of Chicago, Chicago, Ill., p. 2. There is no firm consensus among gerontologists on the span of the biological clock for man; some give a figure as low as 80.

41. In FY 1976 appropriations for cardiovascular diseases amounted to $214.5 million, and for cancer, $761.7 million. Source: Budget Office, HEW.

42. Resolution of all cardiovascular diseases would increase life expectancy at birth by 11.8 years; cancer, by 2.5 years. Jacob S. Siegal, "Demographic Aspects of Aging and the Older Population in the United States," U.S. Bureau of the Census, *Current Population Reports*, Series P-23, No. 59 (May 1976), p. 35. Absolutely reliable figures of the increase in life expectancy due to elimination of a particular disease are not possible, since an individual who would no longer die from a heart attack may die from pneumonia a couple of years later. Progress in gerontology would, however, make the elderly less vulnerable to disease in general.

43. Not only funds, but a large percentage of able-bodied manpower would have to be diverted from productive pursuits and assigned to take care of the elderly and thus potentially seriously weaken the economy.

44. Alexander Comfort, formerly Director, Medical Research Council Group on Aging, University of London, believes that it is likely that by 1990 a means will be found to prolong vigorous life (and correspondingly youthful appearance) by 20 percent (about 15 years). See his "Changing the Life Span," in *The End of Life*, John D. Roslansky, ed. (Amsterdam: North-Holland Publishing Co., 1973), pp. 60–61. Johan Bjorksten, a major contributor to the crosslinkage theory of aging who has been doing empirical research on the theory for a number of years, believes that, by the year 2000, the average life expectancy could be increased by about 75 years. His projection is not based on the assumption of vast increases in funding but on a better focusing of existing research. Bjorksten and his associates have recently developed an enzyme of low molecular weight, not yet tested *in vivo*, but possibly capable of breaking up the crosslinked aggregates of molecules accumulated with age. Source: a letter to the author, June 16, 1976. See also his "The

Crosslinkage Theory of Aging: Clinical Implications," *Comprehensive Therapy*, 2 (Feb. 1976): 65–74. Other gerontologists, basing their projections on different theories of aging, consider it possible that, in the twenty-first century, the life span could be increased up to 200 and even 400 years. See Gene Bylinsky, "Science Is on the Trail of the Fountain of Youth," *Fortune*, 94 (July 1976): 140 and *passim*. The great majority of gerontologists are more pessimistic than these projections suggest, but they usually do not anticipate extensive funding of this field.

45. "One Adult Cell Used to Produce Frog," *New York Times*, Oct. 17, 1970, p. 30.

46. On cloning and its social implications, see Willard Gaylin, "The Frankenstein Myth Becomes a Reality—We Have the Awful Knowledge to Make Exact Copies of Human Beings," *New York Times Magazine*, March 5, 1972, pp. 12–13, 41, 43–44, 48–49; on progress in fundamental research in this area, see "Clues to Identity of Genetic 'Master Switch' Grow," *New York Times*, March 26, 1976, p. 1.

47. See M. S. Baldwin and C. C. Sterrett, "The Potential Impact of Superconducting Turbine Generators on the Electric Utility Industry," a paper presented at American Power Conference, 36th Annual Meeting, April 29–May 1, 1974, Chicago, Ill., pp. 7–8. The projected cost benefits assumed superconducting generators of 2000 MVA (about 1800 Mwe) and included incremental benefits from operating costs. For details on benefits from superconducting generators, see C. J. Mole and C. C. Sterrett, "A Superconducting Machine for Central Station Power Generation," a paper presented at American Power Conference, 35th Annual Meeting, May 8–10, 1973, Chicago, Ill., pp. 7–9.

48. See Berndt T. Matthias, "The Search for High-Temperature Superconductors," *Physics Today* 24 (Aug. 1971): 23–28; letter to the author from Berndt T. Matthias, dated June 29, 1972; and T. H. Gebelle, "New Superconductors," *Scientific American* 225 (Nov. 1971): 22–33. Some researchers suggested that room-temperature superconductivity might be eventually achieved. Matthias believes, however, that it is "pure science fiction," although he views superconductivity at temperatures of 25–30° Kelvin "a realistic possibility" (p. 23).

49. See Donald P. Snowden, "Superconductors for Power Transmission," *Scientific American* 226 (April 1972): 84–91. For superconductivity in its many potential applications, see Hein, "Superconductivity," 211–222.

50. In an MHD power station, direct conversion of thermal energy into electric energy is achieved by passing superheated (about 4000° F) gases through a magnetic field. The exhaust gases are further used to produce steam for conventional generation of electric power. The increase in efficiency results from using electric power from both of these sources, direct conversion and conventional generation. See Allen L. Hammond, "Magnetohydrodynamic Power; More Efficient Use of Coal," *Science* 178 (Oct. 27, 1972): 386–387.

51. The two surveys were conducted in October 1968 and March–April 1969. They interviewed 2058 individuals and were reported in Daniel Seligman, "A Special Kind of Rebellion," *Fortune* 79 (Jan. 1969): 67 ff.; and Jeremy Main, "Dissidence among College Students Is Still Growing, and It Is Spreading beyond the Campus," *ibid.*, June 1969, pp. 73–74. This and the following paragraphs are based on the data supplied by these two articles.

52. Kenneth E. Boulding, "The Emerging Superculture," in *Values and the Future*, Kurt Baier and Nicholas Rescher, eds. (New York: The Free Press, 1969), p. 347.

53. Seligman, "A Special Kind of Rebellion," p. 175.

54. The postindustrial society is discussed in chapter 14.

55. See Seligman, "A Special Kind of Rebellion," pp. 174–175.

56. This and the preceeding paragraph are based largely on Edmund Faltermayer, "Youth After the Revolution," *Fortune* 87 (March 1973): 145 ff. This is a follow-on article to those by Seligman and Main in *Fortune*, January and June 1969 respectively, cited in note 51, above. It

reports the results of extensive interviewing conducted at fourteen high schools and colleges across the country early in 1973.

57. As early as the mid-1970's, when the number of senior citizens were relatively small, pressure was building up to eliminate compulsory retirement at the age of 65. In February 1975, Representative Paul Findley introduced a bill in the U.S. Congress, co-sponsored by 46 other members of the House, to outlaw forced retirement at 65. See, e.g., his remarks in the House of Representatives, "Time to Apply 14th Amendment to Senior Citizens," *Congressional Record*, October 6, 1975, p. E5244.

58. See Karl Mannheim, *Essays on the Sociology of Knowledge* (New York: Oxford University Press, 1952), pp. 302–315. To Mannheim common experiences and participation in "a common destiny" were more important than being in the same generation. Thus he pointed out that several differentiated and antagonistic units can exist within a given generation.

59. Ronald Inglehart, "The Silent Revolution in Europe: Intergenerational Change in Post-Industrial Societies," *The American Political Science Review* 65 (December 1971): 991–1017. A more extensive study conducted in Japan strengthens Inglehart's findings. See chapter 7, note 32.

60. Seymour Martin Lipset and Everett Carll Ladd, Jr., "College Generations—from the 1930s to the 1960s," *The Public Interest*, No. 24 (Fall 1971): 110.

61. Peter F. Drucker, "The Surprising Seventies," *Harper's Magazine* (July 1971): 35–39.

4
The United States

1. For a discussion of the development and functioning of the contract system, see Don K. Price, *Government and Science* (New York: New York University Press, 1954), pp. 65–94.

2. U.S.-based manufacturers built 89 percent of the estimated $61 billion worth of computers installed in non-Communist countries by the end of 1974. Computed from "Estimated Worldwide Installations of General-Purpose and Dedicated Application Computers, Yearend 1974," *EDP Industry Report* 11 (Oct. 24, 1975): 2; and "Worldwide Computer Market (U.S.-Based Manufacturers)," *ibid.* 11 (April 30, 1976): 7.

3. The chemical industry has either kept generating new products (e.g., polyester fibers) or concentrated on the improvement of the already available products, such as plastics, drugs, and insecticides. However, for all its strength, the American chemical industry is about equal, and not superior, to that of Western Europe; Japan is third. For an overview of the American chemical industry, its growth, and a glimpse into the future, see a compilation of articles in the 50th anniversary issue of *Chemical and Engineering News* 51 (Jan. 15, 1973): 21–100.

4. In 1975 total direct U.S. industrial investment abroad amounted to $133.2 billion; of this figure, manufacturing accounted for $56.0 billion. U.S. Department of Commerce, Bureau of Economic Analysis, *News Release*, Aug. 25, 1976.

5. Tom Alexander, "Shipbuilding's Big Lift from Aerospace," *Fortune* 78 (Sept. 1, 1968): 78–83; and Rush Loving, Jr., "How Subsidies Launched a New Fleet of Superships," *Fortune* 84 (April 1973): 58 ff. A major change in shipbuilding was brought about by the 1970 Merchant Marine Act, which provided for large subsidies combined with incentives for standardized designs and improvements in productivity and thus stimulated rejuvenation in shipbuilding.

6. See Seymour Melman, *Our Depleted Society* (New York: Holt, Rinehart and Winston, 1965), pp. 7–9, 74–81. Melman contends that the stress on military production resulted in the drain of skilled personnel, research and development, and capital away from these industries. Whereas competition from defense industries has been a factor, the decline in these industries appears to

be a multifactor phenomenon. A part of the picture is the "aging" of the U.S. economy. See Charles P. Kindleberger, "An American Economic Climacteric?" *Challenge* 16 (Jan.–Feb. 1974): 35–44.

7. See "Reorganization Plan No. 1 of 1973," the White House press release of January 26, 1973, and Daniel S. Greenberg, "Science and Richard Nixon," *New York Times Magazine*, June 17, 1973, pp. 12 ff.

8. The information on the NSC organization for science and technology comes from my examination of the U.S. government organization for science and technology related to foreign policy for the Commission on the Organization of the Government for the Conduct of Foreign Policy in 1974–1975 and published in an abridged form as "U.S. Government Organization for Science, Technology, and Foreign Policy" in Commission on the Organization of the Government for the Conduct of Foreign Policy, *Appendices* (Washington, D.C.: U.S. Government Printing Office, 1975), Vol. 1, pp. 192–205. (Hereafter referred to as "Basiuk in Commission on the Organization of the Government for the Conduct of Foreign Policy.")

9. See OECD, *Reviews of National Science Policy: The United States*, (Paris, 1968), pp. 357–361, 451–461; "The Scientific Mafia," *Economist*, Jan. 13, 1968, pp. 55–56.

10. The figures were obtained or computed from *Toward a Science Policy for the United States; Report of the Subcommittee on Science and Astronautics*, U.S. House of Representatives (Washington, D.C.: GPO, 1970), p. 115; and the Federal R & D expenditures reported in *Science* 175 (Jan. 28, 1972): 391, and 183 (Feb. 15, 1974): 635–636.

11. Source of the figures: Bureau of Economic Analysis, U.S. Department of Commerce.

12. See "Statement by Peter G. Peterson, Secretary of Commerce, before the Subcommittee on Science, Research, and Development, House Committee on Science and Astronautics," *Second Series of Hearings on Science, Technology, and the Economy*, April 11, 1972 (U.S. Department of Commerce, Washington, D.C.), *passim*; "Statement by Hon. Maurice H. Stans, Secretary of Commerce of the United States," in U.S. House of Representatives, Subcommittee on Science, Research, and Development of the Committee on Science and Astronautics, *Science, Technology and the Economy; Hearings*, 92nd Congress, 1st session, July 27, 28, 29, 1971 (Washington: GPO, 1971), pp. 3–32 (hereafter cited as U.S. House of Representatives, *Science, Technology, and the Economy*); and Michael Boretsky, "Trends in U.S. Technology: A Political Economist's View," *American Scientist* 63 (Jan.–Feb. 1975): 70–82. Technology-intensive products (also referred to as "high-technology products"), as defined by Boretsky, include chemicals, nonelectrical machinery, electrical machinery and apparatus, including electronics, all types of transportation equipment, including aircraft and automobiles, and scientific and professional instruments and controls. Boretsky's research was principally responsible for governmental attention to this area.

13. See Claude E. Barfield, "Science Report: White House Views Intense Technology Hunt as Useful Exercise, though Few Projects Emerge," *National Journal Reports* 4 (May 6, 1972): 756–765.

14. See "Importance of our Investment in Science and Technology—Message from the President of the United States" (House Doc. No. 92–193), *Congressional Record* 118 (March 16, 1972): H2144–H2148.

15. See "Why the White House Shelved 'A Strong New Effort'," *Business Week*, March 3, 1973, p. 36. Some of the programs that eventually got off the ground included the Experimental Technology Incentives Program at the National Bureau of Standards, the Experimental R & D Incentives Program, and the Research Applied to National Needs (RANN) Program at the National Science Foundation. Their funding is small and effectiveness varies. The designation of the Department of Commerce as "the focal point" of the executive government for policy development in industrial R & D has not produced meaningful results.

4. The United States   341

16. See Federal Council on Science and Technology, *Report on the Federal R & D Program, FY 1976* (Washington, D.C.: GPO, 1975), p. 4 and *passim*. Civilian R & D increased from about $3 billion in FY 1966 to about $7 billion in FY 1976, defense R & D grew from about $7.5 billion to $11.5 billion, while outer space R & D declined from about $5 billion to under $3 billion. See also Robert Gilpin, *Technology, Economic Growth, and International Competitiveness* (Washington, D.C. GPO, 1975), pp. 14–25 and *passim*. Gilpin argues for stronger federal support of industrial R & D, although he does not quantify the extent of desirable support.

17. A 1972 Chase Manhattan Bank study estimated that, if then existing trends continued, the outflow of dollars to import sources of energy would be "in excess of $30 billion per year" by 1985. The Chase Manhattan Bank, *Outlook for Energy in the United States to 1985* (New York: June 1972), p. 51. The study, of course, did not foresee the quadrupling of the price of oil within two years, engineered by OPEC nations. In 1975, petroleum and products imported by the United States amounted to $24.8 billion, as compared with $4.3 billion in 1972. Source: Bureau of Economic Analysis, Department of Commerce.

18. See "Energy Self-Sufficiency: An Economic Evaluation," *Technology Review* 76 (May 1974): 22–55; and Robert Gillette, "Energy 'Blueprint' Sees Little R & D Impact before 1985," *Science* 186 (Nov. 22, 1974): 718.

19. See Energy Research and Development Administration, *A National Plan for Energy Research, Development and Demonstration: Creating Energy Choices for the Future; Summary* (Washington, D.C., June 1975).

20. See "Experts Skeptical of Ford's Energy Plan," *New York Times*, March 1, 1975, p. 1; "Energy: A Bill and a Prayer," *Newsweek*, January 5, 1976, p. 59. The Energy Conservation and Production Act of 1976, enacted on August 14, was an important step in the conservation of energy. Applicable only to buildings, it could result in energy savings amounting to the equivalent of 6 million barrels of oil a day by 1990 for new buildings. See Luther J. Carter, "Energy Conservation: Congress Acts on Building Standards," *Science* 193 (Aug. 27, 1976): 748–749.

21. See Energy Coordinating Group, Ad Hoc Group, "International Cooperation on Energy Research and Development; Report" (U.S. Department of State Document, ECG/ERD/36 final, June 6, 1974). This report became the basis of IEA cooperative programs in energy R & D. See also "New Agency Plans Strategy on Oil," *New York Times*, Feb. 6, 1975, p. 5.

22. "Can Agriculture Save the Dollar?" *Forbes* 3 (March 15, 1973): 32–44. Source of the projections: U. S. Department of Agriculture, Nov. 1975.

23. Joseph A. Yager and Eleanor B. Steinberg, *Energy and U.S. Foreign Policy* (Cambridge, Mass.: Ballinger, 1974), p. 303.

24. This, reportedly, is an estimate of the World Bank, as of July 1976. At that time the U.S. Treasury Department used its own estimate of $200 billion in 1974 prices, which would amount to about $300 billion in current prices. Source: U.S. Department of the Treasury.

25. For details on activities in the field of the environment, see the sixth and seventh annual reports of the Council on Environmental Quality, entitled *Environmental Quality* (Washington, D.C.: GPO, 1975 and 1976, respectively), *passim*.

26. See U.S. Senate, Committee on Rules and Administration, Subcommittee on Computer Services, *Technology Assessment for the Congress: Staff Study* (Washington, D.C.: GPO, 1972); and Office of Technology Assessment, *Annual Report to the Congress* (Washington, D.C.: GPO, 1976).

27. U.S. House of Representatives, Select Committee on Committees, *Committee Reform Amendments of 1974; Report to Accompany H. Res. 988* (Washington, D.C.: GPO, 1974), pp. 35–37, 43–45, and *passim*.

28. See the statement of William D. McElroy, then Director, National Science Foundation: "I don't think the scientific community has ever really worried about science policy. If the money is

flowing in to support good scientists and to get work done, I do not think they care what the policy is." Quoted in Glenn T. Seaborg, "New Frontiers of the Mind," *Science Policy Review* 4, No. 1 (1971): 30.

29. Some aspects of this reciprocal relationship are discussed in greater detail in chapter 12.

30. See *Science, Growth and Society: A New Perspective*, Report of the Secretary-General's Ad Hoc Group on New Concepts of Science Policy (Paris: OECD, 1971), pp. 65–66. (Hereafter cited as OECD, *Science, Growth and Society*.)

31. See chapter 11 for a discussion of present governmental institutions with reference to both an appropriate balance between centralization and pluralism and a better organization of societal will and purposiveness on the international forum.

32. "Uncle Sam, the Banker?" *Forbes* 108 (Aug. 1, 1971): 15–17. This article cites cases of changing views in favor of the government's stepping into the private credit picture and concludes: "Is this good or is it bad? The answer is that it's probably inevitable. Given the demands for credit and the life-or-death importance of who gets it and who doesn't, there is probably no way for the Government to stay out of the situation" (p. 17).

33. For a more extensive discussion of the government-business relationship, see chapter 12.

34. Harold and Margaret Sprout, "The Dilemma of Rising Demands and Insufficient Resources," *World Politics* 20 (July 1968): 660–693.

35. See OECD, *Science, Growth and Society*, p. 19.

36. See the testimony of Milton Katz and Emmanuel Mesthene in U.S. House of Representatives, Committee on Science and Astronautics, Subcommittee on Science, Research, and Development, *Technology Assessment; Hearings* (Washington, D.C.: GPO, 1970), pp. 173–175, 240–244.

37. See "Auto's Future: Cutback in Size and Frills," *New York Times*, March 30, 1975, sec. 1, p. 1.

38. See, for example, "Welfare: Billions to Pay, and a Spreading Revolt," *Time* 106 (Sept. 1, 1975): 7–8. The article describes the U.S. welfare system as "the world's worst welfare mess. . . . It is a monster [which] costs some $45 billion a year at all levels of government, delivers benefits to 25 million people and requires a quarter of a million government employees to administer it" (p. 7). See also Alfred M. Skolnik and Sophie R. Dales, "Social Welfare Expenditures, Fiscal Year 1974." *Social Security Bulletin* 38 (Jan. 1975): 3–19.

39. See Robert Hamrin, "Are Environmental Regulations Hurting the Economy?" *Challenge* 18 (May/June 1975): 29–38.

5
The Soviet Union

1. The shift to higher quality output was significant for the Soviet Union, since this was an aspect of technology badly neglected in the prewar Five-Year Plans, when the emphasis was on quantitative output.

2. See P. I. Lyashchenko, *Istoriya narodnogo khozyaistva SSSR* [History of the National Economy of the USSR] (Moscow, 1956), Vol. 3, p. 517; also, John R. Deane, *The Strange Alliance* (New York: Viking, 1947), p. 100.

3. Total lend-lease shipments to the Soviet Union amounted to $11.3 billion. Of the total approximately one-third were industrial materials and products for the expansion and relocation of Soviet industry, but many items listed as "munitions" had potential economic significance (e.g., the Soviets received 375,800 trucks). See U.S. President, *Twenty-First Report to Congress on Lend-Lease Operations* (Washington, D.C.: GPO, 1946), p. 25; and U.S. President, *Twenty-Third Report to Congress on Lend-Lease Operations* (Washington, D.C.: GPO, 1946), p. 25.

4. See William H. Gregory, "Soviet Union Seeks Balance in Technology Growth," *Aviation Week and Space Technology* 88 (March 18, 1968): 82–83.

5. R. Amman, M. J. Berry, and R. W. Davies, "Science and Industry in the USSR," in OECD, *Science Policy in the USSR* (Paris, 1969), p. 394. (Hereafter cited as OECD, *USSR*.)

6. See Dr. F. Wetter, "The Natural Gas Potential and Its Interlocking Pipeline System in the East Bloc," *Chemistry and Industry*, Nov. 30, 1968, p. 1680 and *passim;* and J. Richard Lee, "The Soviet Petroleum Industry: Promise and Problems," in U.S. Congress, Joint Economic Committee, *Soviet Economic Prospects for the Seventies; A Compendium of Papers* (Washington, D.C.: GPO 1973), p. 284. (Hereafter cited as JEC, *Soviet Economic Prospects.*)

7. For details on the Soviet petroleum policy and output in the 1950s and the 1960s, see Theodore Shabad, *Basic Industrial Resources of the USSR* (New York: Columbia, 1969), pp. 12–19. See also Lee in JEC, *Soviet Economic Prospects*, p. 284 and *passim*. The change in the Soviet energy base between 1960 and 1972 is illustrated by the following share of fuels in national output (in percent):

	1960	1966	1972
Coal	53.9	40.7	33.8
Crude oil	30.5	36.7	42.6
Natural gas	7.9	16.5	19.6

8. See Shabad, *Basic Industrial Resources*, pp. 27–34; and J. J. O'Connor, "Siberia Takes Lead in USSR Energy-Systems Growth," *Power* 112 (Nov. 1968): 203–210. Advance in technology, which made possible the construction of large generating units (single generators of 500,000 kw were built), was an important factor in this development. The availability of cheap oil and gas distributed via trunk pipelines to the newly built central stations were other factors that made the new policy of power generation possible.

9. See Holland Hunter, *Soviet Transport Experience* (Washington, D.C.: Brookings, 1968), pp. 61–64. This increase in traffic was achieved "with only a modest increase in the length of the network, very little rise in the operating labor force, and no increase at all in the number of locomotives" (p. 61).

10. G. W. Hemy, "The Soviet Chemical Industry," *Chemistry and Industry*, Feb. 17, 1968, p. 208. In the same period, U.S. output increased by 73 percent, that of the United Kingdom by 62 percent.

11. For a background discussion of investment as a factor in Soviet economic growth, see Raymond P. Powell, "Economic Growth in the USSR," *Scientific American* 219 (Dec. 1968): 22–23.

12. See Rush V. Greensdale, "The Soviet Economic System in Transition," in U.S. Congress, Joint Economic Committee, *New Directions in the Soviet Economy* (Washington, D.C.: GPO, 1966), part I, pp. 4–9 (hereafter cited as JEC, *New Directions*); and Abram Bergson, "Soviet Economic Perspectives: Toward a New Growth Model," *Problems of Communism* 22 (March–April 1973): 2–4.

13. See Loren R. Graham, "Cybernetics," in *Science and Ideology in the Soviet Union*, George Fisher, ed. (New York: Atherton Press, 1967), pp. 83–106; see also John P. Hardt, et al. (eds.), *Mathematics and Computers in the Soviet Planning* (New Haven, Conn.: Yale University Press, 1967), *passim*.

14. For details, see John P. Hardt, Dimitri M. Gallik, and Vladimir G. Treml, "Institutional Stagnation and Changing Economic Strategy in the Soviet Union," in JEC, *New Directions*, part 1, pp. 50–52.

15. For a more extensive analysis of the reforms of the Soviet Academy of Sciences, see Loren R. Graham, "Reorganization of the USSR Academy of Sciences," in *Soviet Policy-Making*, Peter

5. The Soviet Union

H. Juviler and Henry W. Morton, eds. (New York: Praeger, 1967), pp. 137–158; and Helgard Wienert, "The Organization and Planning of Research in the Academy System," in OECD, *USSR*, pp. 201–205. The status of science in Soviet ideology has also undergone change. In the 1930s, science was viewed as a part of the "superstructure"; that is, it was a product of the "base" consisting of technology, manpower, natural resources, etc., with their corresponding social relationship (e.g., capitalist, socialist). Now science began to be viewed as an independent force and a part of the base.

16. This is the present name of the Committee, which it has held since 1965. The Committee's evolution, under various names, can be traced back to 1947, but it did not emerge in a form similar to the present until 1961. It was then known as the State Committee for Coordination of Scientific Research, and its authority was more limited than that of the present Committee. See Eugène Zaleski, "Central Planning of Research and Development in the Soviet Union," in OECD, *USSR*, pp. 52–57.

17. *Ibid.*, pp. 56–63, and interviews (April–May 1976) with U.S. scientists and government officials who had contacts with State Committee officials and staff members. American observers were generally impressed by the high caliber of the State Committee's personnel.

18. Premier Kosygin, speaking before the 23rd Party Congress, in March 1966, stated: "The course of the economic competition between the two world systems depends on the rate of development of our science, and on the scale on which we use the results in production." Quoted on John Ledes, "Soviet R & D," *Science and Technology*, March 1969, p. 9. A similar theme, stressing both economic and military power and invoking the authority of L. I. Brezhnev, was raised by G. Volkov, docent of C. P. Central Committee's Academy of Social Sciences, in "Nauka—proizvoditel'naya sila obshchestva" [Science Is a Productive Force of Society], *Pravda*, Feb. 4, 1969, pp. 2–3. The theme of science as a productive force of society has been continuously emphasized in Soviet publications in recent years.

19. See G. M. Dobrov, *Nauka o nauke* [Science of Science] (Kiev, 1966), pp. 15–16, 18, 250–251, 254–255; I. Manyushis, "Effektivnost' proizvodstva: problemy i perspektivy" [Effectiveness of Production: Problems and Prospects], *Kommunist* (Moscow), No. 5 (May) 1969, pp. 63–74. See also "Report on Science of Science Activity," *SSF Newsletter* 3 (Aug. 1968): 3–4. In more recent years, technology, along with science, began to receive emphasis, especially with regard to productivity. See "Proizvoditl'nost' truda—glavnyi faktor razvitiya sovetskoy ekonomiki" [Productivity of Labor—Principal Factor in the Development of the Soviet Economy], *Planovoe Khozyaystvo*, No. 7 (July) 1974, pp. 3–6.

20. See Manyusnis, "Effectiveness of Production," p. 63; Academician I. Gerasimov, "Nuzhen general'nyi plan preobrazovaniya prirody nashei strany" [A General Plan for the Transformation of our Country's Natural Environment Is Needed], *Kommunist* (Moscow), no. 1 (January) 1969, p. 70. I. Gerasimov is the Director of the Institute of Geography of the Academy of Sciences of the USSR.

21. See, for example, John P. Hardt and George D. Holliday, *U.S.–Soviet Commercial Relations: The Interplay of Economics, Technology Transfer, and Diplomacy*; prepared for the Subcommittee on National Security Policy and Scientific Developments of the Committee on Foreign Affairs (Washington, D.C.: GPO, June 10, 1973), pp. 31–33 and *passim*.

22. For example, the agreement on a chemical barter between the Soviet government and a consortium headed by the Occidental Petroleum Corporation, activated in June 1974, involves a total of $20 billion in 20 years. "Occidental Signs Deal with Soviet," *New York Times*, June 29, 1974, p. 35.

23. See Jerry F. Hough, "The Soviet System: Petrification or Pluralism?" *Problems of Communism*, 21 (March–April 1972), p. 30. "CPSU Central Committee Draft Directives for 1976–1980

Five-Year Plan," *Foreign Broadcast Information Service, Daily Report; Soviet Union* (Supplement), 3 (Jan. 2, 1976): 7.

24. See Academician T. Khachaturov, "Povyshenie effektivnosti obshchestvennogo proizvodstva v novoy pyatiletke" [Increase of Effectiveness of Public Production in the New Five-Year Plan], *Voprosy Ekonomiki,* No. 4 (April) 1971, pp. 22–23; Academician A. Rumyantsev, "O sovershenstvovanii ekonomicheskogo mekhanizma upravleniya proizvodstvom" ["On Improvement of the Economic Mechanism for the Management of Industry"], *Voprosy Ekonomiki,* No. 11 (Nov.) 1975, p. 7; V. Lebedev, "Planirovanie i nauchno-tekhnicheskiy progress" [Planning and Scientific-Technological Progress], *Planovoe Khozyaystvo,* No. 11 (Nov.) 1975, pp. 10–17.

25. Peter G. Peterson, *U.S.–Soviet Commercial Relations in a New Era* (Washington, D.C.: U.S. Department of Commerce, Aug. 1972), p. 33.

26. See Powell, "Economic Growth in the USSR," p. 23; Khachaturov, "Increase of Effectiveness," p. 30. Industrial auxiliary workers in particular comprise a major potential for improving labor productivity. It has been estimated that there are 85 auxiliary workers for every 100 basic workers in Soviet industry (a total of about 8 million auxiliary workers), while in the United States there are 38 auxiliary per 100 basic workers (amounting to about 4 million auxiliary workers). Whereas the level of labor productivity of basic workers is 70–75 percent of the level of their counterparts in the United States, Soviet auxiliary workers are only 20–25 percent as productive as U.S. auxiliary workers. See Murray Feshbach and Stephen Rapawy, "Labor Constraints in the Five-Year Plan," in JEC, *Soviet Economic Prospects,* p. 488.

27. See "France and Soviet Sign a 10-Year Accord," *New York Times,* Oct. 28, 1971, p. 1; Jean Ross-Skinner, "The Russians Are Coming," *Dun's* 100 (Aug. 1972): 35–36, 88; and Marshall I. Goldman, *Détente and Dollars: Doing Business with the Russians* (New York: Basic Books, 1975), pp. 133–42, 298–300.

28. See D. Chevignard, *Computers and the Soviet Economy,* JPRS Report No. 64479 (Arlington, Va., April 3, 1975), *passim.*

29. David W. Bronson, "Scientific and Engineering Manpower in the USSR and Employment in R & D," in JEC, *Soviet Economic Prospects,* pp. 580–581.

30. "Soviet Is Reconciling Its Quest for Modernization with Communist Ideology," *New York Times,* Dec. 14, 1973, p. 12.

31. Only 12 percent of Soviet R & D workers are employed in industrial enterprises; moreover, in terms of quality, industrial R & D personnel is inferior to that of research institutes. Largely because of the separation between research institutes and industry, the average time from research to production in Soviet industry is 5 to 10 years, while more than 90 percent of the research projects in U.S. industry are brought to production stage in less than 5 years. Bronson, "Scientific and Engineering Manpower," pp. 580–581.

32. "Soviet Industrial Reform Borrows from U.S. Setup," *New York Times,* April 4, 1973, p. 57; "Soviet Puts New Stress on Research Based Units," *New York Times,* Nov. 12, 1973, p. 53.

33. See Leon Smolinski, "Towards a Socialist Corporation: Soviet Industrial Reorganization of 1973," *Survey* 20 (Winter 1974): 24–35. In appraising the introduction of production associations, Smolinski goes as far as to say: "The current reorganization may be viewed as potentially the most important of such measures in four decades" (p. 24).

34. See "Vysokaya effektivnost'—osnova resheniya problem desyatogo pyatiletnego i dolgosrochnogo perspektivnogo planov" [High Efficiency Is the Foundation of the Resolution of Problems of the Tenth Five-Year and the Long-Range Plans], *Planovoe Khozyaystvo,* No. 7 (July) 1974, pp. 3–6; and T. S. Khachaturov, "Long-Term Planning and Forecasting in the USSR," *American Economic Review* 62 (May 1972): 444–455.

35. This also applies to experimentation with new technologies. For example, under the Environmental Policy Act of 1969, an environmental impact statement is required for conducting experiments in atmospheric modification in the United States, which frequently involves public hearings and otherwise complicates procedures for experiments.

36. Information on Soviet technological capability was obtained from multiple sources, including interviews with scientific and technical personnel. For some published sources on this subject, see "Technicians Hail Soviet (MHD) Generator," *New York Times*, January 20, 1972; "Civilian Atom Blasts in the Soviet Found to Surpass the U.S. Effort," *New York Times*, March 5, 1970, p. 1; "Soviet Operating Breeder Reactor," *New York Times*, July 17, 1973, p. 9; John D. Holmfeld, "Resource Allocation in the Soviet Space Program," in U.S. Senate, Committee on Aeronautical and Space Sciences, *Soviet Space Program, 1966–1970; Staff Report* (Washington, D.C.: GPO, 1971), pp. 107–114; Charles S. Sheldon, II, *United States and Soviet Progress in Space: Summary Data through 1973 and a Forward Look* (Washington, D.C.: Congressional Research Service, Jan. 8, 1974); and "Electron Beam Fusion: Soviets Claim Advance," *Science News* 109 (April 3, 1976): 212–213.

37. See Gerasimov, "A General Plan," p. 70.

38. "Rivers Will Learn to Run Backwards," *Business Week*, June 13, 1970, pp. 43 and 46. Soviet planners and engineers have been talking about this project for a number of years, but there is no evidence of a firm commitment to move ahead. However, there are signs that the Soviets are making some progress with the project on an experimental and piecemeal basis. See, e.g., "Atom Device Used for Soviet Canal," *New York Times*, February 16, 1975, Sect. 1, p. 8.

39. "Computers: Great Bleep Forward," *Time* 102 (July 16, 1973): 52–53. For details on the Ryad Series, see Wade B. Holland, "Unwrapping the ES Computers," *Soviet Cybernetics Review* 3 (Sept. 1973): 7–22.

40. This increase in state employees followed a period of stability in numbers for a decade. See Gertrude E. Schroeder, "Soviet Economic Reforms at an Impasse," *Problems of Communism* 20 (July–Aug. 1971): 36–46.

41. In 1973 planned output growth of consumer goods was reduced by 44 percent, while growth of producers goods was reduced by 17 percent. Paul K. Cook, "The Political Setting," in JEC, *Soviet Economic Prospects*, p. 4.

42. A. H. S. Candlin, "Events in the Middle East and Asia," *The Army Quarterly and Defense Journal* 104 (Jan. 1974): 152.

43. Theodore Shabad, "The Resources of a Nation," in *The Soviet Union: The Fifty Years*, Harrison E. Salisbury, ed. (New York: Harcourt, 1967), p. 228.

44. See John P. Hardt, "West Siberia: The Quest for Energy," *Problems of Communism* 22 (May–June 1973): 26–30; and J. Richard Lee, "The Soviet Petroleum Industry: Promise and Problems," in JEC, *Soviet Economic Prospects*, pp. 283–290.

45. A CIA study estimated that in 1974 the USSR had a trade surplus of between $500 million and $1 billion. The study's projections indicate that, at least for 1975–1980, the USSR should be able to pay for imports of Western technology short of relying on foreign credits. Central Intelligence Agency, "USSR: Long-Range Prospects for Hard Currency Trade" (Washington, D.C.; Jan. 1975), pp. 4–5.

46. "Red North Star? Plans to Import Soviet Natural Gas Quietly Hatch," *Barron's* 56 (May 31, 1976): 9 ff.

47. Hardt, "West Siberia," pp. 30–36.

48. "Soviet Is Proudly Pushing Rail Line Across Untapped Region in Siberia," *New York Times*, April 4, 1975, p. 2. Besides economic, BAM obviously has potential strategic significance.

49. See Gerasimov, "A General Plan," p. 71. Only 25 percent of the total arable land currently under cultivation is found in regions naturally favorable for agricultural production. On the other extreme, 10 percent of the land now under cultivation is found in climatically highly unfavorable desert zones.

50. "Statement of D. Gale Johnson, Department of Economics, University of Chicago," in U.S. Congress, Joint Economic Committee, *Soviet Economic Outlook; Hearings* (93rd Cong., 1st session, July 17, 18, and 19, 1973), p. 70.

51. "Brezhnev Unveils a Vast Effort to Farm Steppes of Northern Russia," *New York Times*, March 16, 1974, p. 6.

52. See Keith Bush, "Environmental Problems in the USSR," *Problems of Communism* 21 (July–Aug. 1972): 21–31; and David E. Powell, "The Social Costs of Modernization; Ecological Problems in the USSR," *World Politics* 23 (July 1971): 618–634.

53. M. Loyter, "Effektivnost' kapital'nykh vlozheniy v okhranu prirodnoy sredy" [Effectiveness of Capital Investments in Safeguarding the Environment], *Voprosy Ekonomiki*, No. 1 (January), 1976, p. 34.

54. *Ibid.*, p. 38.

55. "Kosygin Says Soviet Tops Western Economic Rivals," *New York Times*, Feb. 2, 1976, p. 1.

56. The comparative figures on Soviet roads were computed from data supplied in Imogene U. Edwards, "Automotive Trends in the USSR," JEC, *Soviet Economic Prospects*, p. 312. Unsurfaced dirt roads comprise 60 percent of the total road mileage in the USSR but only 23 percent of total U.S. road mileage. According to Soviet projections there will be a total of about 10 to 11 million cars in the USSR by 1980, of which about 8 million will be private. By the year 2000, the total will grow to 40 million (*ibid.*, p. 313.)

57. See Feshbach and Rapawy, "Labor Constraints," pp. 491–493.

58. *Ibid.*, p. 489.

59. B. Bolotin, "Glavnaya zadacha desyatoy pyatiletki" [The Main Task of the Tenth Five-Year Plan], *Mirovaya ekonomika i mezhdunarodnye otnosheniya*, No. 3 (March), 1976, pp. 19–20. During the Ninth Five-Year Plan, Soviet industry grew 5.9 percent, and GNP 3.9 percent, annually.

60. *Ibid.*, p. 20.

61. See "Soviet Defense Cost Higher, CIA Says," *New York Times*, May 19, 1976, p. 4; "U.S. Agencies Rift on Size of Soviet Arms Budget," *ibid.*, May 8, 1976, p. 4; "U.S. Statistics on Soviet Question Extent of Threat," *ibid.*, April 24, 1976, p. 3. CIA estimates (11–13 percent of the Soviet GNP allocated to defense) are lower than those of the DIA, which apparently uses 15 percent as a minimal figure. Differences in the estimates arise principally from differences in the methodologies of computing.

62. Stanley H. Cohn, "The Economic Burden of Soviet Defense Outlays," in JEC, *Economic Performance and the Military Burden*, pp. 178–179.

63. See John P. Hardt, "Brezhnev's Economic Choice: More Weapons and Control or Economic Modernization," *Parameters* 1 (Fall 1971): 46–47; Andrew Sheren, "Structure and Organization of Defense-Related Industries," in JEC, *Economic Performance and the Military Burden*, pp. 123–132.

64. See Thomas W. Wolfe, "Soviet Interests in SALT: Political, Economic, Bureaucratic and Strategic Contributions and Impediments to Arms Control" (The RAND Corporation, Santa Monica, Calif., Sept. 1971, p-4702), pp. 22–24; Malcolm Mackintosh, "The Soviet Military: Influence on Foreign Policy," *Problems of Communism* 22 (Sept.–Oct. 1973): 3–4. The previous name of the Defense Council was the Higher Military Council. In the Soviet hierarchy it stands

above the Council of Ministers and apparently handles some of the Politburo's work on military affairs.

65. Hardt, "Brezhnev's Economic Choice," p. 48. The more traditional and the single most powerful group within the Soviet elite is, of course, the Party bureaucracy. For a more extensive analysis of the emergence of various interest groups in Soviet society that has eroded its once totalitarian nature, see H. Gordon Skilling and Franklyn Griffiths, eds., *Interest Groups in Soviet Society* (Princeton, N.J.: Princeton University Press, 1971), *passim.*

66. In the debate on the course of Soviet economic development appearing in print in recent years, the wisdom of allocating priority to heavy industry has been challenged by some writers, but the military sector proper was spared from such suggestions. This does not necessarily indicate its complete immunity—after all, Khrushchev did successfully divert some resources from the military sector. However, since Stalin's death, Khrushchev was an exception among Soviet leaders in his propensity to challenge the basic interests of a number of power centers. Perhaps the present leaders have learned some caution from his experience.

67. A possible impact of technology on internal stability of the USSR is briefly considered in chapter 8.

68. The possibility of a military takeover of power in the USSR, although conceivable, will not be discussed here.

69. For a discussion of the New Economic Mechanism (NEM), introduced in Hungary in January 1968 and tightened up after the Soviet invasion of Czechoslovakia in August 1968, see Barnabas Racz, "Political Changes in Hungary After the Soviet Invasion of Czechoslovakia," *Slavic Review* 29 (Dec. 1970): 633–650; Laszlo Jotischky, "Hungary, Fifteen Years After," *The Political Quarterly* 42 (April–June 1971): 158–164; and Tamas Nagy, "The Hungarian Economic Reform, Past and Future," *The American Economic Review; Papers and Proceedings* 61 (May 1971): 430–435.

70. See Merle Fainsod, "Roads to the Future," *Problems of Communism* 16 (July–Aug. 1967): 23.

71. For a description of the Economic System of Socialism (ESS) and its predecessor, the New Economic System (NES), see Thomas A. Baylis, "Economic Reform as Ideology: East Germany's New Economic System," *Comparative Politics* 3 (Jan. 1971): 211–230; and Melvin Croan, "After Ulbricht: The End of an Era?" *Survey* 17 (Spring 1971): 74–92.

72. See Hough, "The Soviet System," pp. 42–44. Hough suggests that, according to the theory of "the circular flow of power," which is intended to explain the dominance of the General Secretary over other members of the Politburo, it takes the General Secretary several years to consolidate his power. However, once he consolidates his power, it can be compatible with a degree of pluralism in a complex society.

73. See Zev Katz, "Insights from Emigrés and Sociological Studies on the Soviet Economy," JEC, *Soviet Economic Prospects*, pp. 89–93 and *passim*. This study is based on interviews of Soviet emigrés. It also suggests that planning in the USSR may not be as rational or central a process as appears from the outside. In many cases there is much pushing and in-fighting of various interests around the Plan. What matters for each interest is to get the kind of a plan it wants, and ingenious economic theories, calculations, and lobbying are resorted to in order to get a desirable plan. The interests involved may be economic (to get a plan one can comfortably fulfill and overfulfill), but in many cases they are also power interests or those of local nationalism. For example, the Latvians are reported to oppose strongly new large-scale industries in the republic, because usually they bring an influx of Russians and other foreigners. The Ukrainians want large-scale industry because this way they hope to arrest the outflow of their labor force to work in other republics. (*Ibid.*, pp. 93–94.)

74. "Moscow Backs New Economists," *New York Times*, June 5, 1973, p. 1; "Soviet Economists Split on Flexibility in Planning," *New York Times*, Oct. 9, 1973, p. 65. A group of mathematical economists headed by R. P. Fedorenko proposed a "system of optimal functioning of the economy" (SOFE), encompassing a set of flexible long-term development goals, to be worked out by a new planning agency. Officials of the State Planning Commission (Gosplan)—which, in effect, would be partially supplanted by the new planning agency—did not think it was a good idea.

75. See Wolfe, "Soviet Interests in SALT," pp. 40–41.

6
Western Europe

1. For a more nearly comprehensive discussion of the impact of the Marshall Plan on Western Europe, see *Sixth Report of the Organization for European Economic Cooperation: From Recovery Towards Economic Strength* (Paris, 1955), passim; for the organizational efforts to implement the Plan, see Harry B. Price, *The Marshall Plan and Its Meaning* (Ithaca, N.Y.: Cornell University Press, 1955), pp. 71–86.

2. See Angus Maddison, *Economic Growth in the West* (New York: Twentieth Century Fund, 1964), pp. 58–59.

3. For a discussion of "the transport revolution" in Western Europe, see J. Frederic Dewhurst et al., *Europe's Needs and Resources* (New York: Twentieth Century Fund, 1961), pp. 279–311.

4. See Andrew Shonfield, *Modern Capitalism* (New York: Oxford University Press, 1965), especially pp. 62–67, 121 ff. The extent of governmental management of the economy varies. Germany is perhaps the most "capitalistic" (in the traditional sense of the word) in her adherence to some of the basic tenets of market ideologies. However, even Germany is devoted to market ideologies much less than the United States is.

5. See Robert Gilpin, *France in the Age of the Scientific State* (Princeton, N.J.: Princeton University Press, 1968), pp. 151–238, 326–331.

6. For some representative discussion of the technology gap, see Gilpin, *ibid.*, pp. 17–38, 377–460; Jean-Jacques Servan-Schreiber, *The American Challenge*. Translated by Ronald Steel (New York: Atheneum, 1968); John Diebold, "Is the Gap Technological?" *Foreign Affairs* 46 (Jan. 1968): 278–291; Christopher Layton, *European Advanced Technology* (London: Allan and Unwin, 1969), passim.

7. Harvey Brooks, "What Is Happening to the U.S. Lead in Technology?" *Harvard Business Review* 50 (May–June 1972): 112–115; and OECD, *Research and Development in OECD Member Countries: Trends and Objectives* (Paris, October 1971), p. 72. The United States allocated only 9 percent of its R & D budget to economic objectives in 1969. France, the next lowest nation after the United States, allocated 17 percent of her government-funded R & D to economic development in 1970.

8. U.S. Department of Commerce, Office of the Assistant Secretary for Science and Technology, *Technology Enhancement Programs in Five Foreign Countries* (Washington, D.C., Dec. 1972), p. 1. (Hereafter referred to as U.S. Department of Commerce, *Technology Enhancement Programs*.)

9. With the output per man hour for manufacturing equal to 100 for 1967, the index for the United States increased from 107.4 in 1969 to 128.3 in 1973. For France, however, it rose from 112.7 to 144.4; for West Germany, from 113.8 to 139.0; for Italy, from 112.6 to 149.4. The United Kingdom, however, with an increase from 108.3 to 127.0, lagged behind the United States. Source: Bureau of Labor Statistics, U.S. Department of Labor.

10. See Raymond Vernon, "Rogue Elephant in the Forest: An Appraisal of Transatlantic Relations," *Foreign Affairs* 51 (April 1973): 578–580.

11. For a more extensive discussion of France's science and technology policy under De Gaulle, see Gilpin, *France in the Age*, pp. 188–217, 340–341, 378–391, and *passim*.

12. The EC Commission provided the following guidelines for annual growth in real GDP (Gross Domestic Product) for the period 1976–1980: Germany, 4–5 percent; France, 5.5 percent; Italy, 4.5–5 percent; U.K., 4–4.5 percent; EC as a whole, 4.5–5 percent. Commission of the European Communities, "Fourth Medium-Term Economic Policy Programme; Draft" (Brussels, Oct. 8, 1976; COM[76] 530 final), p. 37.

13. See U.S. Department of Commerce, *Technology Enhancement Programs*, pp. 81–83; "France Is Taking Aim at German Economy," *New York Times*, Jan. 13, 1973, p. 37; personal interview (March 1976) with Edgar L. Piret, former U.S. Science Attaché in Paris.

14. See U.S. Department of Commerce, *Technology Enhancement Programs*, pp. 127–128; D. S. Greenberg, "Germany: Booming Research Effort Turning to Space and Computers," *Science* 164 (April 18, 1969): 281–283; and "Dr. Leussink's New Ministry," *Nature* (London) 226 (June 13, 1970): 1019–1020. For 1976, appropriations of the Federal Ministry for Research and Technology declined by 3.9 percent. U.S. Department of State Airgram A-571, Dec. 9, 1975; Subj.: FRG Science/Technology Policy, p. 1. For details on programs, see *ibid.*, pp. 3–14.

15. Italy has a Minister (without portfolio) for the Coordination of Scientific and Technological Research, whose function is to obtain funding and to coordinate and promote research. However, nuclear and health research is not within his jurisdiction. The Minister has a staff of about 100 people. Besides being rather amorphous, the Italian science and technology scene is highly politicized. There is, however, a growing realization in Italy of the need for strengthening its organization in this field. (Source: Embassy of Italy, Washington, June 1976.)

16. For details on Britain's science and technology policy in post-World War II years, see Layton, *European Advanced Technology*, pp. 61–63; U.S. Department of Commerce, *Technology Enhancement Programs*, pp. 31–34, 293–297; and John Walsh, "British Science Policy: Assuming a Lower Profile," *Science* 193 (July 9, 1976): 132–134. For additional information, I am also indebted to Holand Smith, Counsellor for Science and Technology, British Embassy, Washington, D.C. (March 1976).

17. On July 1, 1967, the European Coal and Steel Community (ECSC), the European Economic Community ("the Common Market"), and the European Atomic Energy Community (Euratom) were merged into the European Communities under a single Commission and a single Council of Ministers. To avoid ambiguity, the term "European Community" (EC) is used here in singular, rather than the technically more correct plural.

18. The Spokesman of the ECSC's High Authority and European Community Information Service, *Europe and Energy* (Luxembourg, 1967), p. 14. These figures are for the then six EC nations.

19. Commission of the European Communities, *Energy Policy: Problems and Resources, 1975–1985* (Brussels, Oct. 4, 1972; Document # COM(72) 1201 Final), pp. 6, 12, and 22. By contrast, in the United States, about 60 percent of petroleum is used for automobile fuel production.

20. Eugene Bacot, "Checkmate for Energy Master Plan," *Vision* (Geneva), Dec. 1975, pp. 35–38; see also William D. Metz, "Science in the European Community: Deadlock on Fusion," *Science* 192 (Apr. 2, 1976): 37.

21. Daniel Yergin, "European Energy: A Policy Evolves?" *European Community*, No. 192 (Jan.–Feb. 1976): 32–36. See Henri Simonet, "Energy and the Future of Europe," *Foreign Affairs* 53 (April 1975): 458–460.

22. See "Pollution in Europe; The Community Acts," *European Community*, No. 169 (Oct. 1973): 19–22.

23. For EC activity in this area, see *Ninth General Report of the Activities of the European Communities* (Brussels, Feb. 1976), pp. 136–139. (Hereafter cited as EC, *Ninth General Report*.)

24. See Caryl P. Haskins, *The Scientific Revolution and World Politics* (New York: Harper and Row, 1964), p. 71.

25. See René Foch, "Europe and Technology," *The Atlantic Papers* (No. 2, 1970), pp. 23–29.

26. "Aviation: A Troubled Bird," *Newsweek* (June 2, 1976): 73 and 75. Quite apart from the objectionable noise and resultant difficulties in obtaining landing rights in some countries, the Concorde consumes a great deal of fuel and is expensive (about $70 million per aircraft of a 100-passenger capacity as compared with the subsonic Boeing 747, carrying 330 passengers and a price tag of about $35 million).

27. See Warren H. Donnelly, *Commercial Nuclear Power in Europe: The Interaction of American Diplomacy with a New Technology* (Washington, D.C.: GPO, 1972), pp. 133–134; and John Walsh, "Uranium Enrichment: Both the Americans and Europeans Must Decide Where to Get the Nuclear Fuel for the 1980's," *Science* 184 (June 14, 1974): 1160–1161.

28. See Robert H. Hayes, "Europe's Computer Industry: Closer to the Brink," *The Columbia Journal of World Business* 9 (Summer 1974): 113–120; Werner Van de Walle, "Reply to the American Challenge," *European Community*, No. 179 (Aug.–Sept. 1974): 14.

29. *Ibid.*, p. 15; "European Computer Makers Gang up on IBM," *New York Times*, Feb. 24, 1973, pp. 40 and 42; Hayes, "Europe's Computer Industry," pp. 121–122.

30. Cf. Theodore Geiger, "A New U.S. Hegemony in Western Europe?", *Looking Ahead* 22 (Feb. 1974): pp. 2–3; "Europeans Look Inward, Political Unity Idea Fades," *New York Times*, March 3, 1975, sec. 1, p. 1.
The phenomenon of a narrow nationalistic stance in response to the demands of domestic interests for protection or support has to be distinguished from the more traditional nationalism where compromise was resisted because of considerations of sovereignty, national pride, or power. Under the traditional nationalism, a particular interest was backed by the government to enhance national power or economic strength. In today's quasi-nationalism in Western Europe a particular interest is supported because it is a constituency that may effect the government's survival, directly or indirectly. Except for France, the traditional nationalism is probably declining in Western Europe. It must be recognized, however, that in actual practice it is difficult to distinguish between the two types of nationalism. They frequently combine and thus reinforce each other and make agreement much more difficult.

31. France has at the top the most purposeful and dynamic science and technology policy, but its arrangements lack the intimate association between universities and industry found in Germany, an association that bridges the gap between science and its application and creates a strong foundation for technological capability. On the other hand the exalted status of the professor in both France and Germany introduces an element of rigidity into scientific research and inhibits the teamwork so important for advance in modern science and so successfully used by the British. France could learn something from both Britain and Germany about the advantages of greater decentralization in science and technology—in particular the greater degree of freedom of university scientists from a somewhat stifling dependence on the government. The German experience on the university–industry level and the French setup on the policy formation level could help the British to bridge the dichotomy between science and technology that still persists in the United Kingdom.

32. "Is it Goodbye to ESRO and ELDO?" *Nature* 226 (June 13, 1970): 1011.

6. Western Europe

33. See European Space Agency, *ESA Annual Report 1975* (Neuilly-sur-Seine, 1976), *passim*. The ESA Convention appears on pp. 183–197.

34. See John Walsh, "In a Hard Year in Brussels, Things Look Up for Science," *Science* 184 (May 31, 1974): 962–967; and EC, *Ninth General Report*, pp. 180–184. A potentially important development took place in 1975, when CREST completed plans for a Joint European Torus (JET) project, intended to develop commercial fusion and emphasizing magnetic confinement.

35. Source: European Community Information Service, Washington, D.C., August 1976. For the text of the EC convention on patents, see "Legislation," *Official Journal of the European Communities*, L 17, Vol. 19 (January 26, 1976).

36. Source: European Community Information Service, Washington, D.C., August 1976. See also "Draft Convention on the International Merger of Sociétés Anonymes," *Bulletin of the European Communities*, Supplement 13/73 (European Communities Commission, 1973), pp. 9–28, and Edmund Fawcet, "European Companies," *European Community*, No. 188 (July–August 1975): 3–6.

37. Richard Howe, "Uncle Sam Outfaced by Europe's Global Giants," *Vision* (May 1976): 57–60.

38. "North Sea Oil and Gas," in United Kingdom, Central Office of Information, *The Survey of Current Affairs; Economic and Scientific Affairs* (July 1974), p. 219. Further discoveries being allowed for, it was estimated that production from the British continental shelf in 1980 could be in the range of 100–140 million tons.

39. See *ibid*. and Royal Norwegian Ministry of Finance, *Parliamentary Report No. 25* (1973–1974); *Petroleum Industry in Norwegian Society* (Oslo, 1974), pp. 5–6. In 1978 production of oil is expected to reach 35 million tons, and by 1981–1982, about 50 million tons. The latter figure is considerably below the projected production of Britain of 100–140 millions by 1980.

40. Jonathan Radice, "Going European Isn't Enough," *Vision* (Feb. 1973): 43–45.

41. Christopher Redman, "The European Aircraft Industry; Competition and/or Cooperation Across the Atlantic," *European Community*, No. 182 (Dec. 1974): 9.

42. "Servan-Schreiber Updates His 'Challenge,' " *Business Week* (Oct. 14, 1972): 63, 66, and 68. His position is motivated by a perceived need to control jointly multinational corporations.

43. See "Soviet Union May Enter World Uranium Market," *New York Times*, May 3, 1973, p. 63. The Soviet magazine *Vneshnyaya Torgovlya* [Foreign Trade] implied that in about 1977–1978, when the world demand for uranium will exceed supply, the time would be ripe for the entry of the Soviet Union into the world uranium market as a major supplier.

44. See Christopher Layton, "Hope of Flying," *European Community*, No. 192 (Jan.–Feb. 1976): 28–30. The situation with the aircraft industry was viewed as "critical" in the mid-1970s. In 1975, the EC Commission proposed an action program to enable it to compete.

45. See "The European Challenge in Nuclear Fuel," *Business Week* (April 28, 1973): 101 and 104; and Walsh, "In a Hard Year," pp. 1160–1161.

46. Both West Germany and Japan have, of course, central political institutions, but for the past 15 years they have been timid in using their considerable technological and economic capability for political purposes. The picture appears to be changing now.

47. See Ferdinando Riccardi, "State of the Community; EC Commission President Defines New Goals for the Old World," *European Community*, No. 184 (March 1975): 3–4.

48. See Harold and Margaret Sprout, "The Dilemma of Rising Demands and Insufficient Resources," *World Politics* 20 (July 1968): 660–693. The Sprouts' discussion focuses on internal social needs versus external requirements for maintaining states' positions in the world and does not address itself to the trends in and demands of nonmilitary technology in particular. However, the requirements of nonmilitary technology strengthen the Sprouts' analysis.

49. There is a precedent for a potential U.S.–European rivalry for the resources of the Atlantic Ocean in terms of claims, if not an actual exploitation of resources. See Laurance Reed, *Ocean-Space—Europe's New Frontier* (London: The Bow Group, 1969). This monograph was written under the auspices of the Bow Group, the influential political research society of younger members of the British Conservative Party. Among other things, the monograph calls for appropriation by Western Europe of the Atlantic seabed up to the 4,000-meter (13,500-foot) isobath, to be owned and developed by a truly supranational European body, the Oceanic Development Commission. This acquisition would have given Western Europe an additional 5 million square miles of territory.

7
Japan

1. "Real GNP in 1975 Showed 2% Gain; Fourth Quarter Growth Stayed Flat," *The Japan Economic Journal* (Tokyo) 14 (March 9, 1976): 2. The Japan Economic Research Center projected Japan's real growth for 1975–1985 at 7 percent; other sources give somewhat lower projections. Hisao Kanamori, "Industry to Become 'Knowledge-Intensive' Type in 7% Growth," *ibid.*, p. 10.

2. Kazuo Noda, "Postwar Japanese Executives," in *Postwar Economic Growth in Japan*, Ryutaro Komiya, ed. (Berkeley: University of California Press, 1966), pp. 233–234.

3. Howard F. Van Zandt, "The Japanese Culture and the Business Boom," *Foreign Affairs* 48 (Jan. 1970): 347–348.

4. See Daniel L. Spencer, "The New Technology in Japan," *World Affairs* 132 (June 1969): 14–15.

5. Warren S. Hunsberger, *Japan; New Industrial Giant* (New York: American-Asian Educational Exchange, 1972), pp. 18–19.

6. The figures are for 1965–1974. U.S. Arms Control and Disarmament Agency, *World Military Expenditures and Arms Transfers* (Washington, D.C.: GPO, 1976), pp. 15, 34, and 50.

7. James C. Abegglen, "The Economic Growth of Japan," *Scientific American* 222 (March 1970): 33–35; Ryutaro Komiya, "Economic Planning in Japan," *Challenge* 18 (May–June 1975): 9–20.

8. See Abegglen, "Economic Growth of Japan," pp. 33–35; Van Zandt, "Japanese Culture," pp. 347–349; "Japan's Drive to Outstrip U.S.," *U.S. News and World Report*, April 6, 1970, pp. 26–27.

9. Chie Nakane, *Human Relations in Japan* (Tokyo: Ministry of Foreign Affairs, 1972), pp. 15–20. This is a translated and abridged version of Professor Nakane's *Tateshakai No Ningen Kankei* ("Personal Relations in a Vertical Society") (Tokyo 1967). In Japan loyalty and human relations revolve around what Chie Nakane calls "frame" (family as a household, including servants; locality; company; nation), not "attributes" (sibling relationship; occupation; profession).

10. Van Zandt, "Japanese Culture," pp. 350–351; Abegglen, "Economic Growth of Japan," p. 33.

11. See Nomura Research Institute of Technology and Economics, "The Key Factors in Japanese Economic Growth" (Sept. 1969), pp. 12–13.

12. Kotaro Horisaka, "Trading Companies: Business Worsens from Slow Trend of World Trade; Many Concentrate on Trimming Fat," *Industrial Review of Japan, 1976* (Tokyo: *The Japan Economic Journal*, 1976), pp. 129–130; "Volume, not Profits," *Forbes* 109 (May 1, 1972): 26–31.

13. See Japan, Economic Planning Agency, *Economic Survey of Japan (1966–1967)* (Tokyo: *The Japan Times*, 1967). pp. 64–67; Spencer, "New Technology," pp. 17–19, 22.

14. See T. Dixon Long, "Policy and Politics in Japanese Science; The Persistence of a Tradition," *Minerva* 7 (Spring 1969): 446.

15. *Ibid.*, pp. 438–444; Tsuneo Asai, "Science and Technology: Private Sources Meet 70% of R & D Cost; Gov't Stress Is on Atomic Energy, Space," *Industrial Review of Japan, 1976*, pp. 42–43; and Japan, Prime Minister's Office, Science and Technology Agency, *Science and Technology Agency: An Outline* (Tokyo, 1975), *passim*. In 1975 STA had a total of 455 employees (in addition to 1600 in its laboratories) and a budget of $566 million.

16. Masami Shimizu, "Shipbuilding: Global Excess of Bottoms Inevitably Forces Shipbuilders to Reduce Rate of Operations," *Industrial Review of Japan, 1976*, p. 70.

17. See "Japanese Chemicals Take a Global View," *Chemical Week*, June 3, 1970, pp. 42–46.

18. "Computer Firms Are Moving Out to Revamp Their Setups for Exports," *The Japan Economic Journal* 13 (Dec. 9, 1975): 8; "NEC, Toshiba Will Unveil Ultra Large Type Computers," *ibid.* 14 (April 6, 1976): 8. In 1976, the computer industry has been completely liberalized by the government, thus setting the Japanese firms to compete on their own.

19. For an overview of Japan's allocation of resources to science and technology, see Tsuneo Asai, "Science and Technology: Private Sources," pp. 42–43, 142. Source of budget statistics for FY 1976: The Embassy of Japan, Washington, D.C.

20. NASA will launch these satellites on reimbursable basis. "Engineering Test Satellite 'Kiku' Orbited; Japan's 7th Artificial Satellite," *Information Bulletin* (Tokyo) 13 (Oct. 15, 1975): 11–12. Additional information was obtained from the Office of International Affairs, NASA, April 1976.

21. Tsuneo Asai, "Science and Technology: New Energy Development Looms Large in R & D Investment," *Industrial Review of Japan, 1975*, (Tokyo: *The Japan Economic Journal*, 1975), p. 39; and Asai, "Science and Technology: Private Sources," 42. Figures here, as elsewhere, are approximate because of flotation of the dollar.

22. See "Giant 'Think Tank' to Be Formed by Government and Private Quarters," *Information Bulletin* (Tokyo) 20 (Aug. 31, 1973): 217–220.

23. "Recession Reveals Fragile Nature of 'Think Tanks' Firms in Japan," *The Japan Economic Journal* 13 (March 4, 1975): 12.

24. The deposits were described as possibly "one of the ten largest (underwater oilfields) in the world." See "Japanese Oil Find Poses Title Problem," *New York Times*, Aug. 28, 1969, p. 1. However, because of the dispute about the title to the deposits (Japan, China, and Taiwan are involved), the discovery has not been adequately explored, let alone exploited.

25. For a discussion of the economic potential of manganese nodules, see chapter 10.

26. "New Industrial Robots Replace Workers in Monotonous Tasks," *Japan Report* 21 (Sept. 16, 1975): 6. In 1975 about 10,000 robots, most of them equipped with memory devices, were used in Japan.

27. Japan, Office of the Prime Minister, Bureau of Statistics, *Statistical Handbook of Japan, 1975* (Tokyo, 1975), p. 69.

28. See Kakuei Tanaka, *Building a New Japan: A Plan for Remodeling the Japanese Archipelago* (Tokyo: The Simul Press, Inc., 1973); and Louis Kraar, "Japan Sets Out to Remodel Itself," *Fortune* 87 (March 1973): 98–104, 184, and 186.

29. Hisao Kanamori, "Growth Potential Is Still Ample," *Industrial Review of Japan, 1975*, p. 5.

30. Yoshio Matsumoto, "Extension of Aid to Middle-Eastern Nations Increases as Sequel to 'Petro Crisis,'" *Industrial Review of Japan, 1975*, pp. 48–49; Yoshio Matsumoto, "Recession Seriously Affected Flow of Money from Japan to Developing World," *Industrial Review of Japan, 1976*, pp. 40–41.

31. See Van Zandt, "Japanese Culture," pp. 344–357; *cf.* James C. Abegglen, *The Japanese Factory: Aspects of Its Social Organization* (Bombay: Asia Publishing House, 1958), pp. 95–96 and *passim*.

32. See Nobutaka Ike, "Economic Growth and Intergenerational Change in Japan," *The American Political Science Review* 67 (Dec. 1973): 1194–1203.

33. See George B. Ringwald, "Slowdown: The Painful Dilemma for Japan," *Industrial Review of Japan, 1976*, pp. 13–14; and Fumiaki Shiraishi, "Many Companies Fought Losing Battles; Bankruptcies Reach Postwar High," *ibid.*, pp. 50–51.

34. Charles F. Gallagher, "The Environment in Japan; From Awareness to Action," *Field Staff Reports*, East Asia Series 20 (March 1973): 4.

35. "Japan's Economy in Transition," *Business Week*, July 7, 1975, p. 46.

36. There have been growing indications to this effect in Japan in recent years. Observers of the Japanese scene noted that, in 1973, Japanese domestic politics were increasingly assuming the characteristics of confrontation politics, with a tendency toward polarization. See, for example, J. Rey Maeno, "Japan 1973: The End of an Era?" *Asian Survey* 14 (Jan. 1974): 63–64, and *passim*.

37. "Tanaka Urges Japanese to Learn from Criticism," *New York Times*, Jan. 22, 1974, p. 2. In January 1974 anti-Japanese riots greeted Premier Kakuei Tanaka in Indonesia, demonstrations in Thailand, and milder expressions of distrust in the Philippines, Singapore, and Malaysia.

38. Kazushige Hirasawa, "Japan's Emerging Foreign Policy," *Foreign Affairs* 54 (Oct. 1975): 155–172.

39. See *Industrial Review of Japan, 1975, passim*; Takafusa Nakamura, "The Turning Point of the Postwar Japanese Economy," *Japan Echo* 2, No. 2 (1975), pp. 48–62; and Kanamori, "Industry to Become 'Knowledge-Intensive,'" pp. 10, 19.

40. See "State of the Economy in Japan: 1975 White Paper," part 2, *Information Bulletin* (Tokyo) 22 (Sept. 15, 1975): 4–10; "Science, Technology White Paper," *The Japan Times* (Tokyo), Aug. 23, 1975, p. 2; and "Press Comment: Technology Development," *The Japan Times*, Aug. 24, 1975, p. 10. The last article noted that "even today the ratio between the technology in imports and exports is 100 to 12."

41. Nakane, *Human Relations in Japan*, pp. 82–83.

42. Japan (along with the Soviet Union) refused to agree to a 10-year moratorium on whaling and thus seriously endangered the survival of whales. Multiple appeals to the Japanese government were not producing any results. In the spring of 1974 girls from the United States, Canada, and Sweden paid a call on Premier Tanaka and presented him with 75,000 letters from children around the world urging Japan to agree to the moratorium. The situation led *The New York Times* to state in an editorial: "If the gentle pressure of letters from boys and girls does not evoke a responsive chord in Tokyo, adults may resort to other and harsher forms of pressure," "Japan Under Pressure," *New York Times*, June 6, 1974, p. 30.

43. See Zbigniew Brzezinski, *The Fragile Blossom; Crisis and Change in Japan* (New York: Harper and Row, 1972), pp. 64–68. Professor Nakane (*Human Relations in Japan*, pp. 84–86) points out that the Japanese prefer emotions to logic, and in their personal relationships (as distinguished from official and business relations) emotions and affection dominate. When a serious topic requiring rationality and logic is brought into a social situation in Japan, conversation dies. According to Nakane, this is an important reason why the Japanese "suffer from a lack of power of an international communication."

44. Kei Wakaizumi, "To Act or Not: Dilemma," *Foreign Policy*, No. 16 (Fall 1974): 35–47.

45. Herman Kahn, *The Emerging Japanese Superstate; Challenge and Response* (Englewood Cliffs, N.J.: Prentice-Hall, 1970), pp. 174–175.

46. See Taketsugu Tsuratani, "The Causes of Paralysis," *Foreign Policy*, No. 14 (Spring 1974): 126–141.

8

Technology and Distribution of Power

1. For a discussion of certain characteristics of governmental institutions and values that handicap effective use of technology in the United States, see chap. 11.
2. Computed from U.S. Department of Commerce statistics. In 1975, U.S. GNP was $1,516.3 billion.
3. There is, at present, a tendency among Western European countries to join the United States in bilateral programs in science and technology. As noted in chapter 6, multinational programs among Western European states are not especially successful, although there are exceptions.
4. OECD, Development Assistance Committee, *Development Cooperation: 1975 Review* (Paris, 1975), pp. 10–12. There are, however, major disparities in growth among LDCs.
5. See M. S. Marzouk, "The Brazilian Economy: Trends and Prospects," *Orbis*, 18 (Spring 1974): 277–91; Bruce Handler, "Flying High in Rio," *New York Times Magazine*, June 8, 1975, pp. 16 ff.; and *Conjunctura Economica* (Rio de Janeiro) 30 (March 1976): 89. It is noteworthy that Brazil's decrease of growth to 4 percent under the influence of a worldwide recession was still well above Japan's low of a *minus* 1.2 percent growth in 1974.
6. *Partners in Development;* Report of the Commission on International Development (New York: Praeger, 1969), pp. 32–36; see also Clifton R. Wharton, Jr., "The Green Revolution: Cornucopia or Pandora's Box?" *Foreign Affairs* 47 (April 1969): 464–476.
7. On potential contribution of science and technology to food production in the future, see Gene Bylinsky, "A New Scientific Effort to Boost Food Output," *Fortune* 91 (June 1975): 99 ff.; on population growth in the LDCs, see Lester R. Brown, "World Population Trends: Signs of Hope, Signs of Stress," *Worldwatch Paper*, No. 8 (Oct. 1976), p. 33 and *passim*.
On the basis of 1975 figures, Worldwatch Institute projected total world population for the year 2000 at 5.4 billion, as distinguished from 6.3 billion, based on 1970 data.
8. "World Conferences and Science in Developing Nations," U.S. Department of State, Bureau of Oceans and International Environmental and Scientific Affairs, *International Science Notes*, No. 35 (Jan. 1976): 14–16.
9. This development appears to be under way. In 1975 the OPEC nations increased loans and grants to LDCs (mostly Moslem countries but poor in oil) from $4 billion in 1974 to $6 billion. In addition the OPEC nations lent $4 billion to the International Monetary Fund and other international financial agencies. "OPEC's U.S. Investment Falls," *New York Times*, Feb. 24, 1976, p. 49.
10. "Giant Coal Deposit Found," *New York Times*, December 12, 1974, p. 74; "Campos Oil Strike Brightens Future," *Brazilian Bulletin* (New York) 27 (January 1975): 12. Rather conservative projections of the Brazilian government, based on deposits already discovered by the beginning of 1976, anticipate a peak of approximately 48 percent self-sufficiency in oil and natural gas in 1980, and a decline afterwards. Computed from Republica Federativdo Brasil, Ministerio Das Minas e Energia, *Balanço Energetico Nacional* (Brasilia, 1976), pp. 26 and 30. For geopolitical prospects of Brazil in the context of her foreign policy, see Norman A. Bailey and Ronald M. Schneider, "Brazil's Foreign Policy: A Case Study in Upward Mobility," *Inter-American Economic Affairs* 27 (Spring 1974): 3–22.
11. See, for example, Louis Kraar, "The Shah Drives to Build a New Persian Empire," *Fortune* 90 (Oct. 1974): 145 ff.
12. See "Ferment in the Land of the Nile," *Forbes* 114 (Aug. 1, 1974): 20–24. Egypt has a total population of about 36 million, of which 12 million are literate. Many Egyptians are highly skilled managers and technicians. In the summer of 1974 the Egyptian government passed an at-

tractive investment law for multinational companies that stimulated interest in American business circles. In addition Egypt is hopeful for an "Arab Marshall Plan," an idea that was boosted by pledges of Arab nations of $1 billion of aid to Egypt in early 1976. "Oil Lands Pledge Millions to Cairo," *New York Times*, March 1, 1976, p. 1.

13. See Arthur G. Ashbrook, Jr., "China: Economic Policy and Economic Results, 1949–71," in U.S. Congress, Joint Economic Committee, *People's Republic of China: An Economic Assessment*; A Compendium of Papers (Washington, D.C.: GPO, 1972), p. 11.

14. "China as an Oil-Exporting Nation," *New York Times*, Jan. 5, 1975, sec. 3, p. 3. The high estimate of 50 billion metric tons, placing China first in the world, includes offshore deposits.

15. "Peking's Oil Strategy," *ibid.*

16. See Dwight H. Perkins, "Looking inside China: An Economic Reappraisal," *Problems of Communism* 22 (May–June 1973): 4; and "China's Green Revolution," *Newsweek*, Oct. 14, 1974, p. 65. Actually China has been a substantial net exporter of food for several years now, but since she has been importing large quantities of fertilizers, she is not considered self-sufficient in food. In 1973 China concluded agreements for imports of 13 large chemical fertilizer plants, some of which are already operational.

17. See Carl Djerassi, "The Chinese Achievement in Fertility Control," *The Bulletin of the Atomic Scientists* 30 (June 1974): 17–24; and Brown, "World Population Trends," p. 34.

18. Perkins, "Looking inside China," pp. 3–5. China appears to be more successful than the Soviet Union in avoiding rigidity and bureaucratization in economic development. In particular a degree of local autonomy exists in both industry and agriculture that helps to provide incentives and flexibility in solving local problems. See A. Doak Barnett, *Uncertain Passage; China's Transition to the Post-Mao Era* (Washington, D.C.: Brookings, 1974), pp. 21–22, 25–26, and 139–143.

19. See Gene T. Hsiao, "Prospects for a New Sino–Japanese Relationship," *The China Quarterly*, No. 60 (Dec. 1974): 720–749.

20. According to a study by Haldi Associates, Inc., between 1970 and 2000 hours of work in the United States will decline by 19.6 percent for men and 15.9 percent for women. This may take the form of (1) reduced weekly hours of work; (2) reduced annual hours of work; or (3) earlier retirements. Haldi Associates, Inc., "Alternative Work Schedules; A Technology Assessment," Preliminary Final Draft, prepared under Grant No. NSF/GI-40456 (New York, October 1975), Tables 15-2 and 15-3; also, a personal letter to the author from John Haldi, dated April 9, 1976. Projections of leisure vary; some are considerably higher than the above. Leisure is defined here as the time free from gainful employment.

21. Totalitarian and autocratic states have been known to attempt to provide content for leisure in an effort to create an ideologically more cohesive and loyal society. The effort has been directed mainly at young people, but not exclusively so (e.g., the Soviet Union has "Red Corners" in the factories, flavored by propaganda items, to fill the leisure time for adults; movies and the theatre have ideological content, etc.). While totalitarian and autocratic societies thus seek to *increase* their power by giving content to leisure, democratic societies might face the necessity of providing content to leisure—without regimentation—in order to *maintain* their stability and viability. For a comprehensive discussion of leisure, see Max Kaplan, *Leisure: Theory and Policy* (New York: John Wiley and Sons, 1975).

22. Emmanuel G. Mesthene, "How Technology Will Shape the Future," *Science* 161 (July 12, 1968): 137.

23. See "Student Protest" (a chapter from "The Politics of Protest: Violent Aspects of Protest and Confrontation," a Staff Report to the National Commission on the Causes and Prevention of Violence, submitted in March 1969 by Jerome Skolnick), *AAUP Bulletin* 55 (Sept. 1969): 311. Addressing itself to the changing values of students, the Staff Report states: "Educated youth in

8. Distribution of Power

the advanced countries perceive the irrelevance of commercial, acquisitive, materialistic, and nationalistic values in a world which stresses human rights and social equality and requires collective planning."

24. Peter F. Drucker, "The Surprising Seventies," *Harper's Magazine*, (July 1971), pp. 35–39.

9
Technology and International Stability

1. For example, Morton Kaplan regards bipolar systems as inherently less stable than multipolar systems, while Kenneth Waltz, observing the post-World War II international scene, suggests that a bipolar system can provide greater world stability than a multipolar one. See Morton A. Kaplan, *System and Process in International Politics* (New York: Wiley, 1957), pp. 42–44; and Kenneth N. Waltz, "Stability of the Bipolar World," *Daedalus* 93 (Summer 1964): 881–909.

2. Michael Haas, "International Subsystems: Stability and Polarity," *American Political Science Review* 64 (March 1970): 98–123.

3. The point that domestic stability or instability is an important factor in international stability (or instability) has been emphasized in Richard N. Rosecrance, *Action and Reaction in World Politics: International Systems in Perspective* (Boston: Little, Brown and Co., 1963), pp. 304–306. In particular Rosecrance relates international instability to the domestic insecurity of elites.

4. See Rosecrance's (*Action and Reaction*, pp. 280 and 285) discussion of what he calls "resource patterns" and international stability. Rosecrance refers to the availability of favorable militarily relevant resources that provide temptations to action; in an era when international rivalry gravitates to other sectors (political, economic, and technological), a superior technologic and economic capability is likely to have a similar effect.

5. When equipped with MIRV, each missile has several individual warheads capable of separating at a certain point of the trajectory and hitting different targets and thus greatly expanding the striking effectiveness of a single missile.

6. For the various considerations involved in deploying the ABM and MIRV, see Robert Rothstein, "The ABM, Proliferation, and International Instability," *Foreign Affairs* 46 (April 1968): 487–502; William C. Foster, "Prospects for Arms Control," *Foreign Affairs*, 47 (April 1969): 422–432; Harold Brown, "Security through Limitations," *ibid.*, pp. 422–432; D. G. Brennan "The Case for Missile Defense" *ibid.*, pp. 433–448.

7. For the text of the 1972 agreements, see "Strategic Arms Limitation Agreements," *The Department of State Bulletin* 66 (June 26, 1972): 918–921.

8. At present the United States and the USSR are in a state of approximate parity in the strategic nuclear field. For details of strategic balance and programs, see U.S. Department of Defense, *Report of the Secretary of Defense Donald H. Rumsfeld to the Congress on the FY 1977 Budget and Its Implications for the FY 1978 Authorization Request and the FY 1977–1981 Defense Programs* (Washington, D.C.: January 27, 1976), pp. 41–49, 52–57, and *passim*.

9. See Fred Charles Iklé, "Can Nuclear Deterrence Last Out The Century?" *Foreign Affairs* 51 (Jan. 1973): 267–285.

10. See McGeorge Bundy, "To Cap the Volcano," *Foreign Affairs* 48 (Oct. 1969): 9–10. McGeorge Bundy points out that political leaders and strategic thinkers differ significantly in their understanding of what constitutes "acceptable damage." While "think-tank analysts" can calculate "acceptable damage" in terms of the tens of millions of lives, a much lower level of destruction would be viewed as a catastrophic blunder by political leaders in the United States or the USSR in real-life decision making, Bundy asserts.

9. International Stability 359

11. See Thomas A. Halsted, "Is the Dam About to Burst?" *The Bulletin of the Atomic Scientists* 31 (May 1975): 8–11.

12. The Conference merely issued a nonbinding declaration that, among other things, urged the parties to abide by the terms of the treaty. There was a strong division at the Conference between the nuclear powers and the nonnuclear states. See "Final Declaration of the Review Conference of the Parties to the Treaty on the Non-Proliferation of Nuclear Weapons," *The Department of State Bulletin* 72 (June 30, 1975): 924–929; and William Epstein, "Failure at the NPT Review Conference," *The Bulletin of the Atomic Scientists* 31 (Sept. 1975): 46–48.

13. See Joseph I. Coffey, "Nuclear Guarantees and Nonproliferation," *International Organization* 25 (Summer 1971): 836–837; Robert Gillette, "How Safe the Safeguards?" *Science* 184 (June 28, 1974): 1352; and Robert Gillette, "India and Argentina: Developing a Nuclear Affinity," *ibid.*, pp. 1351–1353.

14. In May 1974 India signed a 5-year cooperative agreement with Argentina providing for joint research projects in nuclear energy (see Gillette in *ibid.*, p. 1351). In June 1975 Brazil followed by signing an $8-billion, 10- to 15-year agreement with West Germany, involving transfer of nuclear technology (including a uranium enrichment facility) in exchange for a supply of uranium. See Robert Gillette, "Nuclear Proliferation: India, Germany May Accelerate the Process," *Science* 188 (May 30, 1975): 911–914.

15. "U.S. Joins in Pact on Atomic Curbs with Six Countries," *New York Times*, Feb. 24, 1976, p. 1; "Testimony of Henry A. Kissinger, Secretary of State," in U.S. Senate, Committee on Government Operations, *Export Reorganization Act of 1976; Hearings* (Washington, D.C.: GPO, 1976), pp. 768–769, 775, 792–793.

16. See Herman Kahn and Anthony J. Wiener, *The Year 2000* (New York: Macmillan, 1967), pp. 246–247. This is not, of course, to say that proliferation of nuclear weapons is desirable, because, in some respects, their effect may be stabilizing.

17. See Michael Howard, "The Relevance of Traditional Strategy," *Foreign Affairs* 51 (Jan. 1973): 265–266.

18. See Barbara Ward, *The Home of Man* (New York: W. W. Norton Co., 1976), pp. 192–212 and *passim*; "Habitat: U.N. Conference on Human Settlements, Vancouver, 31 May to 11 June 1976. Global Review of Human Settlements" (U. N. Document A/Conf. 70/A/1), pp. 43–44 and *passim*.

19. To a degree that it exists, coordination of development assistance is a pluralistic process to which several organizations—such as the Development Assistance Committee of the OECD, U.N. Industrial Development Organization, and the World Bank—contribute through publication of statistics and studies. But there is no coordination in the sense of concerted action or strategy for development, called for by the Pearson Report several years ago. See *Partners in Development; Report of the Commission on International Development* (New York: Praeger, 1969).

20. See "Long Process Seen in Revising Economic Order to Aid Poor Lands," *New York Times*, June 1, 1976, p. 3, and OECD, Development Assistance Committee, *Development Cooperation; 1976 Review* (Paris, November 1976), *passim*.

21. *Cf.* the concept of "elite ethos" in Rosecrance, *Action and Reaction*, pp. 279–281. Rosecrance notes that the most violent forms of international conflict have usually been associated with wide divergencies in elite ethos. Technological impact tends to bring about a degree of uniformity in elite ethos.

22. Obviously Nazi Germany and Imperial Japan did not fit this pattern. In both cases the prevalent ideology stressed the need for "living space" and conquest as a prerequisite to material well-being. The experience of the present-day Germany and Japan indicates that the real solution lay in effective utilization of technology, high productivity, and economic organization.

23. See, e.g., W. W. Rostow, *The Stages of Economic Growth* (Cambridge, England: Cambridge University Press, 1960), p. 133, and Zbigniew Brzezinski and Samuel P. Huntington, *Political Power: USA/USSR* (New York: Vintage, 1964), pp. 10–13, 424–428.

24. See Rosecrance, *Action and Reaction*, p. 281. According to Rosecrance, one of the major determinants of stability and instability is the influence of what he calls "control patterns," that is, the security of tenure of the national elite. When there are fundamental changes in such security of tenure, major international instability arises.

25. See Andrei Amalrik, *Will the Soviet Union Survive until 1984?* (New York: Harper and Row, 1970), *passim*.

26. See chapter 10 and footnote 35 on the resource potential of manganese nodules.

27. For a more extensive discussion of technology as a dependent and an independent variable in molding a future regime of the oceans, see Victor Basiuk, "Marine Resources Development, Foreign Policy, and the Spectrum of Choice," *Orbis* 12 (Spring 1968): 52–55, 58–65.

28. See William D. Coplin, "International Law and Assumptions about the State System," in *International Politics and Foreign Policy*, James N. Rosenau, ed. (New York: The Free Press, 1969), pp. 144, 152.

10
Technology and World Society

1. For an excellent analysis of the balance of power concept discussed, in part, in a historical context, see Ernst B. Haas, "The Balance of Power: Concept, Prescription or Propaganda," *World Politics* 5 (July 1953): 459–474.

2. For a conceptual discussion of bipolarity, see Kenneth N. Waltz, "The Stability of a Bipolar World," *Daedalus* 93 (Summer 1964): 881–909. Waltz's analysis of bipolarity as a system is based primarily on the distribution of physical power, in particular, the vast difference in power between the two superpowers and the secondary powers. This leads him to conclude (pp. 898–899): "Unless some states combine or others dissolve in chaos, the world will remain bipolar until the end of the present century."

3. See Oran R. Young, "Political Discontinuities in the International System," *World Politics*, 20 (April 1968): 369–392; Marshall D. Shulman, *Beyond the Cold War* (New Haven, Conn.: Yale University Press, 1966), pp. 80–85, 107–109, and *passim*.

4. See J. L. Brierly, *The Law of Nations*, 4th ed. (Oxford, England: The Clarendon Press, 1949), pp. 42–46 and *passim*.

5. See Frederick S. Dunn, *The Protection of Nationals* (Baltimore: The Johns Hopkins Press, 1932), pp. 21, 29, and *passim*.

6. See William D. Coplin, "International Law and Assumptions about the State System," in *International Politics and Foreign Policy*, James N. Rosenau, ed. (New York: The Free Press, 1969), pp. 143–144.

7. I am indebted to Thomas W. Wolfe of the RAND Corporation for suggesting this scenario.

8. There are growing difficulties along nationality lines in the USSR even now. See Teresa Rakowska-Harmstone, "The Dialectics of Nationalism in the USSR," *Problems of Communism* 23 (May–June 1974): 1–22.

9. The Soviets claim their own breed of international law, "socialist international law." While not differing in form from general international law, it is considered by Soviet jurists as a "higher" type of law that will eventually succeed that of present international law. Its major difference from Western international law appears to be in the aim it seeks to achieve: namely, the promotion of the interests of the USSR and the fostering of a Soviet-type society. For a more ex-

10. Technology and World Society 361

tensive discussion of the Soviet view of international law, see F. I. Kozhevnikov, ed., *Kurs mezhdunarodnogo prava* ["A Course of International Law"] (Moscow, 1966), particularly pp. 106–118; Oliver J. Lissitzyn, "Western and Soviet Perspectives on International Law—A Comparison," *Proceedings of the American Society of International Law* 53 (1959): 21–30; John N. Hazard, *Communists and Their Law* (Chicago: University of Chicago Press, 1969), pp. 6–9, 526–527, and *passim*.

10. See "East-West Trade: Red Dividend," *Newsweek*, Dec. 22, 1969, p. 76; and "Soviet Opens Big Natural Gas Field Relatively Near Markets in Europe," *New York Times*, Feb. 20, 1974, p. 51. At the 24th Soviet Communist Party Congress (April 1971), Premier Alexei N. Kosygin proposed the construction of electric power grid to encompass Europe as a whole. *New York Times*, April 7, 1971, p. 1.

11. See his *A Working Peace System* (Chicago: Quadrangle Books, 1966), especially pp. 13–99. Aside from the foreword, this volume consists of reprints of Mitrany's previous publications from 1946 on.

12. For a concise and excellent critique of functionalism, see Inis L. Claude, *Swords into Plowshares*, 3rd ed. (New York: Random House, 1964), pp. 351–367.

13. This can, perhaps, be explained by the early development of Mitrany's theory (between the early 1930s and 1940s) when technology was not as significant in international affairs as it is now. The subsequent writings of Mitrany and others apparently followed the earlier approach and also tended to overlook implications of technology. Ernest B. Haas, who subjected the theory of functionalism to a careful and penetrating scrutiny relatively recently, does not address himself to the impact of technology in particular. See Ernst B. Haas, *Beyond the Nation–State; Functionalism and International Organization* (Stanford, Calif.: Stanford University Press, 1964), *passim*.

14. Mitrany did envision the use of functional instruments for a limited political goal, but the value of such action for him lay in its having an effect as an independent variable, that is, transcending its original intent. This, in particular, applied to the realm of security. Thus, he stated that, "in a period of transition," such acts as Anglo–American naval cooperation or agreement on a mineral sanction by the countries controlling main sources of supply would serve the functional cause inasmuch as they cut across physical regions and provide experience in functional cooperation as well as a stepping stone toward eventual world security by organizing security on an interlocking regional basis. See Mitrany, *A Working Peace System*, pp. 71–72.

15. See, for example, Richard N. Gardner, *In Pursuit of World Order*, rev. ed. (New York: Praeger, 1966), *passim*.

16. See Claude, *Swords into Plowshares*, p. 366.

17. See Robert O. Keohane and Joseph S. Nye, eds., *Transnational Relations and World Politics* (Cambridge, Mass.: Harvard University Press, 1972). In their "Transgovernmental Relations and International Organization" (*World Politics* 27 [October 1974]: 39–62), Keohane and Nye made a distinction between "transnationalism," encompassing activities of nongovernmental organizations, and "transgovernmentalism," involving a weakening of the effective control by national governments over their functional agencies and their sub-units and the resultant emergence of multinational coalitions of bureaucrats. Transnationalism is used in this book as a political process encompassing both activities. In this broader sense transnationalism finds its conceptual origin in pre-World War II writings of Charles Merriam. See William T. R. Fox, "Pluralism, the Science of Politics, and the World System," *World Politics* 27 (July 1975): 597–599.

18. See Claude, *Swords into Plowshares*, p. 360.

19. For an example of a proposal to substitute a managerial basis for a legal one in international decision making (in this case, with regard to fisheries), see L. F. E. Goldie, "The Ocean's

10. Technology and World Society

Resources and International Law—Possible Developments in Regional Fisheries Management," *The Columbia Journal of Transnational Law* 8 (Spring 1969): 1–53. The author argues that a regime in fisheries directed to achieving managerial goals would have a number of advantages in ensuring the biological survival of the stock and the maintenance of its population level, both of which are important in the light of the needs of mankind for food and the highly efficient modern technology for catching fish.

20. These considerations apply to a lesser degree to international cooperation in science (as distinguished from technology or control of technological impact). Many areas of science have also the advantage of not being directly relevant to national power or commercial advantages. Accordingly science presents a more promising field for institutionalized cooperation on a worldwide basis.

21. An interesting example of a subsystem cooperation among advanced nations, bridging the East and the West, has been provided by the establishment of the International Institute of Applied Systems Analysis (IIASA) in Vienna in the Fall of 1972. The institute is run by scholarly organizations of 12 countries—the United States, the USSR, the United Kingdom, Canada, France, Italy, Japan, West Germany, the German Democratic Republic, Czechoslovakia, Poland, and Bulgaria. Its function is to use systems analysis and computer technology to work on problems common to highly developed nations. For the establishment of IIASA, see Constance Holden, "East–West Think Tank is Born," *Science* 178 (Oct. 13, 1972): 143.

22. For example, in the case of the Convention on the Prevention of Marine Pollution by Dumping of Wastes and Other Matter (more generally known as "Ocean Dumping Convention") agreed upon in London in November 1972, the Intergovernmental Maritime Consultative Organization (IMCO), and not UNEP, was designated in December 1975 as responsible for the Convention's secretariat duties. (For the text of the Convention, see *The Department of State Bulletin* 67 [Dec. 18, 1972]: 711–717.) An example of a subsystem activity, emerging independently from UNEP (but, possibly, later coordinated by it), is a convention signed in March 1974 in Helsinki by the seven countries bordering on the Baltic Sea, aimed at reducing its pollution. A separate commission and a secretariat were established to oversee the convention. "7 Nations on Baltic Sign Pact Designed to Cut Its Pollution," *New York Times*, March 23, 1974, p. 8.

23. On Earthwatch, see Clayton E. Jensen, Dail W. Brown, and John A. Mirabito, "Earthwatch," *Science* 190 (October 31, 1975): 432–438. On the establishment and activities of UNEP in general, see "A Special Report: What Happened at Stockholm," *Bulletin of the Atomic Scientists* 28 (Sept. 1972): 16–56, and "Environment Council Ends Fourth Session; Commitment to Cause Said Reaffirmed," *U.N. Chronicle* 13 (May 1976): 21–23. In a number of ways, UNEP's role as yet remains to be defined.

24. Land satellites can monitor great numbers of phenomena, including density and flow of population, water level of rivers, direction of currents in the ocean, the extent of pollution of fresh and salt waters, air, etc. They can become a principal tool of gathering information for the entire ecological system of the planet.

25. Stated in a letter to the author by Karl P. Sauvant, Associate Transnational Corporations Affairs Officer, Center on Transnational Corporations, The United Nations, dated March 30, 1976.

26. William K. Chung, "Sales by Majority-Owned Foreign Affiliates of U.S. Companies, 1973," *Survey of Current Business* 55 (Aug. 1975): 23.

27. See, for example, George Modelski, "The Corporation in World Society," *The Year Book of World Affairs* (1968), p. 78; see also Raymond Vernon, "Economic Sovereignty at Bay," *Foreign Affairs* 47 (Oct. 1968): 119–120.

28. Jack N. Berman pointed out that in a number of instances the U.S. government was unable to control the outflow of U.S. technology and exports by affiliates of U.S. companies abroad to Communist states. (See his "The Multinational Enterprise and Nation States: The Shifting Bal-

10. Technology and World Society   363

ance of Power," a paper presented before the Association for Education in International Business, Chicago, Ill., Dec. 28, 1968, pp. 9–12.) In actual practice, however, perhaps more important are those cases where awareness of the longer range implications of a particular technology transfer does not exist in the government, and MNCs become carriers of technology eventually backfiring at the originating nation.

29. A rather unusual case occurred in Venezuela, where the government expropriated the holdings of Owens–Illinois Glass for giving in to financial demands of the leftist kidnappers of the company's local manager. "Caracas Seizes U.S. Firm for Giving in to Kidnappers," *Washington Post*, April 7, 1976, p. A-23.

30. In 1975 the U.N. created a Commission on Transnational Corporations and an executive body for the Commission, the Center for Transnational Corporations. For their activities, see "Toward a Code of Conduct for Transnational Corporations," U.N. Document OPI/CESI Features ESA/164, Feb. 19, 1976. OECD has been involved with the issue of MNCs for quite some time. However, even this body, whose members are on a fairly equal level of economic development and do not include LDCs, has had serious difficulties in formulating a code of conduct for MNCs, the United States favoring minimal requirements. "U.S. Opposes Proposed Code for Multinationals," *Washington Post*, April 7, 1976, p. D-1.

31. The United States submitted a draft convention on the international seabed area in May 1970. See "Draft U.N. Convention on the International Seabed Area: U.S. Working Paper Submitted to U.N. Seabed Committee," *The Department of State Bulletin* 63 (Aug. 24, 1970): 209–218. It played an important role in leading to the 1975 ISNT. The concept of internationalization itself was introduced in the U.N. by Ambassador Arvid Pardo of Malta in 1967.

32. See United Nations, Third Conference on the Law of the Sea, "Revised Single Negotiating Text" (A/CONF.62/WP.8/Rev.1/Part I, May 6, 1976), pp. 1–2, 15, 18, 22 and *passim*; and U.S. Department of State, "The Third United Nations Conference on the Law of the Sea, New York, August 2–September 17, 1976; U.S. Delegation Report" (Washington, D.C., Sept. 1976).

33. For details on manganese nodules and their exploitation, see John L. Mero, *The Mineral Resources of the Sea* (Amsterdam: Elsevier Publishing Co., 1965), pp. 178–180, 277–279, and *passim*; David B. Brooks, "Ocean Mining: Political Opportunities and Economic Consequences," *Technology Review* 71 (July–Aug. 1969): 23–29; and "Ocean Mining: Bonanza or Bust?" *Forbes* 117 (May 1, 1976): 34 and 36.

34. J. L. Bischoff and F. T. Manheim, "Economic Potential of the Red Sea Heavy Metals Deposits," in *Hot Brines and Recent Heavy Metals Deposits in the Red Sea*, Egon T. Degens and David A. Ross, eds. (New York: Springer–Verlag, 1969), p. 535. This estimate, in 1967 prices, applies to Atlantis II Deep alone, by far the largest and the richest of the three deeps discovered. Metal-bearing sediments were also found on the floor of Discovery Deep, the second largest deep.

35. The deposits are located in the Red Sea between Sudan and Saudi Arabia, closer to the former than to the latter. In addition to these two states, Ethiopia also raised a claim to the deposits, based on a claim to certain islands in the Red Sea.

36. David A. Ross, "Red Sea Hot Brine Area: Revisited," *Science* 175 (March 31, 1972): 1455–1456.

37. "NOAA, University Scientists Discover First Metallic Mineral Field in Mid-Atlantic; Link Continental Drift, Oceanic Minerals," *U.S. Department of Commerce News*, April 10, 1974. The deposits cover some 40 square miles (100 square kilometers) and are officially referred to by NOAA scientists (who discovered them) as the Trans-Atlantic Geotraverse (TAG) Hydrothermal Field. Drilling is required to determine whether metallic deposits exist under the discovered manganese deposits.

364   10. Technology and World Society

38. Unknown, and perhaps very large, quantities of petroleum have been discovered at great depths (2.5–3 miles) off the African coast in the Atlantic, in the Gulf of Mexico, and the Mediterranean. These deposits, discovered as an undesirable by-product of scientific exploration, are too deeply situated to be exploitable for many years. See "Oil in Sea Floor is 2 Miles Down", *New York Times*, Sept. 1, 1968, p. 45; and "Oil-Rich Salt Domes Suspected in Sediments Beneath the Atlantic" *New York Times*, May 13, 1969, p. 29. However, the outlook for oil is considerably greater on the continental shelf and the continental margin than in the deep ocean areas. Oil is found primarily in sedimentary formations; sediments are thin in the deep ocean areas, and this fact significantly downgrades their petroleum potential. See National Petroleum Council, *Petroleum Resources under the Ocean Floor*, (Washington, D.C., 1969), pp. 22–23.

39. William T. R. Fox, "Science, Technology and International Politics," *International Studies Quarterly* 12 (March 1968): 14–15.

40. See Stephen H. Schneider, *The Genesis Strategy; Climate and Global Survival* (New York: Plenum Press, 1976), pp. 135–36, 180.

41. These include, among others, the so-called nongovernmental organizations (NGOs) whose international activity and influence have markedly increased in recent years. For an interesting analysis of the activity of nongovernmental organizations at the U.N. Conference on the Human Environment at Stockholm in 1972, see Ann Thompson Feraru, "Transnational Political Interests and the Global Environment," *International Organization* 28 (Winter 1974): 31–55. In the case of the environment, the organizations concerned included Friends of the Earth, Sierra Club, Commonwealth Human Ecology Council, International Council of Scientific Unions, Boy Scouts World Bureau, etc. Many of these organizations emerged as a result of a direct or indirect impact of technology.

11
Concepts, Institutions, and Values

1. At this point the distinction between technology as a dependent (controlled) and an independent (uncontrolled) variable becomes blurred. It can be argued that if the growth of a particular privately developed technology is closely followed by responsible national authorities and its actual and potential impact is fully appreciated, then the given technology is a dependent variable.

2. One example is the use of peaceful nuclear explosions for mining or excavation purposes that, in a given setting, might upset the ecological balance and/or have undesirable repercussions on foreign relations.

3. There is a considerable literature on decision-making process, and no particular originality is claimed here for the distinction between the "rational" and the "constituency" method. The "rational" method is close to what other writers called "the rationalistic approach" or "the rational-comprehensive model." The "constituency" method is similar to the "incrementalist" model or the "muddling through" approach. Although the constituency model and the incrementalist model share some characteristics in common, the constituency model is founded on a pattern of behavior created by interaction of groups or organizations ("constituencies") with vested interests of their own and is thus not limited to incrementalism, while the founders of the incrementalist model focus on incremental decision making as being a type of its own with virtues of its own. For a discussion of the incrementalist model, see David Braybrooke and Charles E. Lindblom, *A Strategy of Decision* (New York: Free Press, 1963), *passim;* for a criticism of incrementalism, see Yehezkel Dror, "Muddling Through—'Science' of Inertia?" *Public Administration Review* 24 (Sept. 1964): 153–157; see also Amitai Etzioni, "Mixed-Scanning: A 'Third' Approach to Decision-Making," *Public Administration Review* 27 (Dec. 1967): 385–392. For an excellent summary of various concepts and models of bureaucratic decision making, see Graham T. Allison, *Essence*

of Decision; Explaining the Cuban Missile Crisis (Boston: Little, Brown and Co., 1971), pp. 67–68, 144–162.

The constituency model (rather than the incrementalist or related models of bureaucratic decision making) is especially applicable in the context of the present analysis for two reasons: (1) This model is not limited to bureaucratic decision making in Washington but appears to reflect more accurately *national* decision making (culminating in Washington but finding its roots outside of it), where constituencies vie for resources and other preferential treatment. (2) It is particularly applicable to decisions involving science and technology where resources and resource allocation are important and thus stimulate a behavior and decision-making characteristic of the competitive interplay among constituencies (interest groups).

4. A historical example of a lack of an articulate constituency is provided by the case of America's interest in upholding the balance of power in Europe during World War I, which later led to accusations that the United States' entry into the war was a right decision for a wrong reason. See Edward H. Buehrig, *Woodrow Wilson and the Balance of Power* (Bloomington: Indiana University Press, 1955), pp. 106–169, 269–275. Until comparatively recently, the consumers in the United States did not comprise an articulate constituency; even now this constituency is in the process of formulation and development.

5. The idea of a spectrum was suggested to me by Harvey Brooks of Harvard University.

6. See Victor Basiuk, "U.S. Government Organization for Science, Technology, and Foreign Policy," in Commission on the Organization of the Government for the Conduct of Foreign Policy, *Appendices* (Washington, D.C.: GPO, 1975), Vol. 1, pp. 196–198.

7. *Ibid.* pp. 197–198.

8. In the Army the principal interest groups are infantry, artillery, armor, air mobile forces, air defense forces, and Army aviation. In the Navy they consist of aviators, submariners, and surface line officers. In the Air Force the principal division is between strategic and tactical air forces; the main interest groups within the strategic category are missile and bomber forces, while in tactical air it is general-purpose forces (ground forces support) and air defense forces (continental defense). Many of these groups compete for appropriations across Service lines more than within the Service itself. Thus there is a keen competition between Army tactical air and Air Force tactical air for close air support; Navy Polaris competes with Air Force ICBM, etc.

9. The above is based on personal experience in the Pentagon and on many discussions with Defense officials over a number of years.

10. Some analysts emphasize that the adherence to cost-effectiveness in the Pentagon has significantly strengthened rationality of choice in national defense. This is only partially true, since cost-effectiveness is a limited instrument inasmuch as it can help to make a choice between two or more weapon systems intended to perform a particular mission, but it is of practically no help in making a decision between weapon systems designed for different missions. Overall priorities must be determined by an evaluation of all principal factors relevant to national security. It is precisely here that constituency influence can be effectively applied, since the criteria involved are not nearly as exact as cost-effectiveness (which, at times, is also not a very exact criterion). See the discussion of priorities in chapter 12.

11. The above three paragraphs are principally based on my examination of the Department of State conducted for the Commission on the Organization of the Government for the Conduct of Foreign Policy in the fall of 1974 and first half of 1975. The results of this examination were published by the Commission in an abridged form in Victor Basiuk, "U.S. Government Organization," pp. 198–202, 205. For an historical background of SCI, see Franklin P. Huddle, *Science and Technology in the Department of State* (Washington, D.C.: GPO, 1975), pp. 28 ff.

12. One can also point out some other, less fundamental factors. The President's Science Advisory Committee was an organization consisting of part-time "outsiders" that met to consider indi-

vidual issues with which the President was faced; it was thus not suitable for formulating and planning a comprehensive policy in science and technology. As a full-time staff organization, the Office of Science and Technology was more suitable for this task, but it was largely immersed in day-to-day problems and issues and did not engage in overall planning and solid conceptual analysis.

13. In this connection the role of the Federal Council for Science and Technology (FCST), which consisted of the heads of the independent technical agencies (NASA, AEC, and NSF) and Assistant Secretaries for Science and Technology or their equivalent of the various Departments, should be mentioned. Chaired by the Science Adviser and staffed by OST, it was a part of the White House machinery for science and technology policy and continued to exist after Reorganization Act No. 1 of 1973, but outside the White House. Being an interagency committee, it suffered from the usual weaknesses of constituency decision making, although it was useful for exchange of information and coordination in secondary matters. Officially abolished in 1976, in effect it exists under a new name and on a statutory basis (see note 17, *infra*).

14. See Christopher Wright, "Scientists and the Establishment of Science Affairs," in *Scientists and National Policy-Making*, Robert Gilpin and Christopher Wright, eds. (New York: Columbia University Press, 1964), p. 278.

15. See chapter 4 for a discussion of the Magruder Exercise.

16. See Basiuk, "U.S. Government Organization," pp. 193–196.

17. For the text of the National Science and Technology Policy, Organization, and Priorities Act of 1976 (originally H.R. 10230), see Act of May 11, 1976, Pub. L. no. 94–282, 90 Stat. 459. The Act, in effect, preserved the old Federal Council for Science and Technology, but gave it a new name, the Federal Coordinating Council for Science, Engineering, and Technology (FCCSET).

18. For example, the President's Science Advisory Committee consisted almost exclusively of scientists and engineers. The position of the Science Adviser to the Secretary of State (Director of SCI) was unfilled for years because a scientist of suitable stature could not be found; eventually the Acting Director, Herman Pollack, a career State Department civil servant and not a scientist, was appointed to the post. One reason for insisting on the appointment of personnel from a particular group is, of course, the consideration of expertise; however, the question of acceptance of a particular appointee(s) by a given constituency plays an important role.

19. For a more extensive discussion of the values of the scientists, see Don K. Price, *The Scientific Estate* (Cambridge, Mass.: Harvard University Press, 1965), p. 83: Eugene B. Skolnikoff, *Science, Technology, and American Foreign Policy* (Cambridge, Mass.: The M.I.T. Press, 1967), p. 244; Warner R. Schilling, "Scientists, Foreign Policy, and Politics," in Gilpin and Wright, eds., *Scientists and National Policy-Making*, pp. 158–163.

20. Those scientists who do not hold important positions with the government or its affiliated institutions are explicitly reluctant to accept the necessity for science to be subordinated to social ends as determined by public authority. See Price, *The Scientific Estate*, p. 83.

21. See Paul Ritterband, "Economic Realia, Values and Migration" (Department of Sociology, Columbia University, 1969), pp. 16–21; and Everett Carll Ladd, Jr., and Seymour Martin Lipset, "Politics of Academic Natural Scientists and Engineers," *Science* 176 (June 9, 1972): 1094. Also, personal observations.

22. There is a remarkable paucity of literature on the values and attitudes of diplomats with respect to science and technology (which, in varied degrees, is true about other groups as well, with the exception of the scientists); thus, most conclusions have to be reached inferentially or from personal observations. *Cf.*, however, Schilling in Gilpin and Wright, eds. *Scientists and National Policy-Making*, p. 154; Andrew M. Scott, "The Department of State: Formal Organiza-

tion and Informal Culture," *International Studies Quarterly* 13 (March 1969): 2–4; William Attwood, *The Reds and the Blacks* (New York: Harper and Row, 1967), pp. 304–310.

23. This paragraph is mainly based on my broader research from multiple sources on the relationship between technology and the military. A general discussion of military values and attitudes, not specifically related to technology, can be found in Morris Janowitz, *The Professional Soldier* (Glencoe, Ill.: The Free Press, 1960), pp. 261–277.

24. For a general discussion of the more traditional American business ideology and values, see Francis X. Sutton *et al.*, *The American Business Creed* (Cambridge, Mass.: Harvard University Press, 1956), pp. 395–396 and *passim*. For more recent changes in attitudes among business executives, see David W. Ewing, "Who Wants Corporate Democracy?" *Harvard Business Review* 49 (Sept.–Oct. 1971): 12 ff.; and "For Many Corporations, Social Responsibility is Now a Major Concern," *Wall Street Journal*, Oct. 26, 1971, p. 1.

Traditional views are still strongly entrenched on key issues with regard to government policy. A poll by *Dun's* magazine revealed that corporation executives believe that the Congressional guarantee of bank loans to Lockheed set a dangerous precedent for American business. Loss of incentives and government control were feared. "Lockheed Aid Seen a Bad Precedent," *New York Times*, Sept. 11, 1971, p. 33.

25. See Peter F. Drucker, *The Age of Discontinuity* (New York: Harper and Row, 1969), pp. 171–172, 212–214. Referring to the pluralism and bigness of our institutions and the resultant diffusion of power, Drucker says: "To most of us today, the power of central government seems to be unchallenged—whether we applaud or deplore it. Tomorrow's historian may cause to call our era 'the twilight of central government.' Impotence rather than omnipotence may well appear to him the most remarkable feature of government in the closing decades of the twentieth century." (p. 171) A parallel argument has been raised by Don K. Price in "Purists and Politicians," *Science*, 163 (Jan. 3, 1969), p. 29. See also OECD, *Science, Growth, and Society*, p. 66. Focusing on the narrower issue of the two models of science policy—the pluralistic and the centralized model, the OECD study concludes: "Any viable system of science policy must involve some blend of the two approaches. Each tends, with time, to develop its own rigidities and limitations, and some shift in the mix from time to time may actually be desirable in order to open new lines of communication and linkage. Such periodic 'shake-ups' probably help the overall system to adapt to new social goals and thus to be more responsive to the changing needs of society." The study favored greater centralization in the present period.

26. See Zbigniew Brzezinski, *Between Two Ages; America's Role in the Technetronic Era* (New York: Viking, 1970), p. 259.

27. See, for example, National Academy of Sciences, *Ad Hoc* Committee on Science and Technology, *Science and Technology in Presidential Policy-making* (Washington, D.C., June 1974); "Organization for Science and Technology in the Executive Branch; AAAS White Paper," *Science* 187 (March 7, 1975): 810–814.

28. See Eugene B. Skolnikoff and Harvey Brooks, "Science Advice in the White House? Continuation of a Debate," *Science* 187 (Jan. 10, 1975): 38. This article points out that "basic researchers and academic scientists have a professional bias which assumes that if only the facts and understanding are made available, society will automatically appreciate their implications and act accordingly." It cites a number of cases where reports by PSAC failed to produce impact, which could also be traced to deficiencies in analysis and "deep intellectual gulf" between the scientific analysis and the policy pressures and options faced.

29. An incisive book that argues for the need for long-range planning in the U.S. government, with particular reference to foreign affairs, is that by Robert L. Rothstein, *Planning, Prediction and Policymaking in Foreign Affairs* (Boston: Little, Brown and Co., 1972). Although the analysis

of this and the following paragraphs on planning does not always parallel that of Professor Rothstein, his book and my personal discussions with him were most helpful in formulating my thinking on this subject, a debt I am happy to acknowledge.

30. The distinction between "substantive" and "instrumental" planning finds its parallel in Karl Mannheim's distinction between "substantial" and "functional" rationality. "Substantial" rationality is concerned with ends (goals) determined on the basis of intelligent insight into the interrelations of events in a particular situation. "Functional" rationality involves a logical series of steps leading to a previously defined end (goal) that in itself does not have to be rational. Mannheim's distinction is cited in Rothstein, *Planning, Prediction*, pp. 91–92.

31. In July 1969 a National Goals Research Staff was established at the White House. It operated under significant political and bureaucratic restrictions, and it produced a rather bland report in 1970, after the issuance of which the NGRS was abolished. A revival of the National Goals Research Staff or the establishment of a similar organization with a broader mandate could be an important instrument in strengthening the rational (as distinguished from constituency) type of decision making in the United States.

32. In later years, as science and technology increasingly pervade society and rise in importance, a stronger machinery may have to be designed. The Science and Technology Council could be patterned on the National Security Council and the Domestic Council. It would be headed by the President and include the Secretaries of State, Treasury, Defense, and Commerce; Chairmen of the CEA and the Council on Environmental Quality; and heads of NSF, ERDA, and NASA. Other organizations may be invited on an *ad hoc* basis. A single Science Adviser to the President would be the Council's Executive Director, while the staff organization would continue functioning as proposed above.

12
The Home Front

1. See Robert Gilpin, "Technological Strategies and National Purpose," *Science* 169 (July 31, 1970): 445–446.

2. It may be appropriate at this point to clarify the distinction between *technological strategy* and two other related terms, *control of technology and its impact* and *science and technology policy*. *Control of technology and its impact* is the most embracing term and normally includes the other two. It can be used from the point of view of an individual nation (in which case it is always "imperfect") or from that of global society (in which case it may be very far from perfection but, at least in theory, has the potential for coming close to perfection). *Science and technology policy* usually refers to the activity of a nation and encompasses both domestic and international aspects. Like *control of technology and its impact, science and technology policy*, to a degree, deals with the impact of technology also, but, depending on its characteristics and effectiveness, it may or may not result in control of technology. *Technological strategy* is a highly purposeful and structured activity; it may be absent in a given nation's science and technology policy if that policy is not well organized, but it is a component of an effective control of technology if such control is developed. *Technological strategy* encompasses decisions with regard to, and employment of, science and technology itself in conjunction with other instruments of policy (political, military, economic); it does not include activity with regard to the impact of technology, unless technological instruments themselves are used to deal with a particular impact of technology.

3. See chapter 2.

4. In application of military power, actual or potential, technological capability is strongly conditioned by geographic considerations, which in turn influence costs and resource allocation. For example, the question of United States presence in the Indian Ocean is not one of a certain

fixed allocation of resources but one of variable technologies, variable forces, variable costs, and a variable time scale. United States "presence" can be maintained by conventional ships patrolling the area, by limited ground bases supported by long-range airlift, by naval bases in Australia in combination with the development of high-speed (surface-effect and hydrofoil) ships that could deploy in the Indian Ocean on short notice, etc.

5. In fact U.S. national strategy in recent years has included a number of moves in the economic and technological spheres whose effect is to serve the interests of military deterrence. These include the agreement to sell wheat to the Soviet Union, the agreement between the U.S. government and the USSR on cooperation in nonmilitary technology, and the numerous agreements between U.S. private companies and the Soviet government on technological cooperation and investment of capital in the USSR. If the Soviets, say, attack the U.S. forces in the Middle East, the abrogation of all these activities would comprise part of the price the USSR would have to pay for the attack.

6. See Richard R. Nelson, "Statement," in U.S. House of Representatives, Committee on Science and Astronautics, Subcommittee on Science, Research, and Development, *Science, Technology, and the Economy; Hearings* (92nd Congress, 1st Session, July 27–29, 1971), pp. 98–99.

7. See Executive Office of the President, Office of Science and Technology, Energy Study Group, *Energy R & D and National Progress* (Washington, D.C.: 1964). The study projected developments to the year 2000.

8. This information is based on my interview of Dr. Ali B. Cambel, who was the Executive Director of the study, on Sept. 25, 1973.

9. In some cases existing knowledge is inadequately used. For example, it has been estimated that losses in the United States due to corrosion amount to $15 billion annually, of which $5 billion could be saved by application of existing knowledge in material effectiveness. National Commission on Materials Policy, *Materials Needs and Environment Today and Tomorrow* (Washington, D.C.: GPO, 1974), p. 10–10. There has been a growing awareness in the nation of the importance of materials and the need for a comprehensive materials policy. See *ibid.* and F. P. Huddle, "The Evolving National Policy for Materials," *Science* 191 (Feb. 20, 1976): 654–659.

10. See chapter 14 for a discussion of why recycling deserves a higher priority.

11. See National Research Council, Committee on Atmospheric Sciences, *Weather and Climate Modification: Problems and Progress* (Washington, D.C.: National Academy of Sciences, 1973), *passim*.

12. See John P. Craven, "Industry–Government Relations in Offshore Resource Development," a paper (No.OTC-1919) presented at the Fifth Annual Offshore Technology Conference, Houston, Texas, April 29–May 2, 1973. According to this study, which compared the United States with Canada, France, Germany, Japan, and Taiwan, the United States is the least effectively organized to develop ocean resources in the national interest.

13. John Diebold, "Business, Government, and Science: The Need for a Fresh Look," *Foreign Affairs* 51 (April 1973): 555–572.

14. A proposal to establish a National Institute of Technology, with a somewhat different mission, was made as far back as 1967. See Richard R. Nelson, Morton J. Peck, and Edward D. Kalachek, *Technology, Economic Growth, and Public Policy* (Washington, D.C.: Brookings, 1967), especially pp. 177–183. In the Department of Defense, Advanced Research Projects Agency (ARPA) looks into future military technologies and their promise. There is no corresponding body for nonmilitary technologies.

15. See chapter 11 and note 32.

16. The figure is for 1974. It covers private residential construction only, new housing units put

12. The Home Front

in place. If additions and alterations are included, then the total for private residential building put in place in 1974 amounted to $47 billion. Source: U.S. Department of Commerce, Bureau of the Census, *Construction Reports*, Feb. 1976, p. 6.

17. The value of nonresidential private buildings put in place in 1974 amounted to $29.6 billion; that of public buildings in the same year was $14.9 billion. *Ibid.*, pp. 6–7.

18. This estimate, provided to the author in May 1975, covers complete house packages. If manufacturers for prefabricated partial components are included, the figure may amount to 30 percent. Statistics on what comprises "the housing industry" are not very well defined, not precise, and not readily available.

19. For details on Operation Breakthrough, see *HUD Challenge*, Vol. 3 (June 1972). The entire issue is devoted to this project. Also "New Ways to Build Take Hold in City," *New York Times*, July 29, 1973, sec. 8, p. 1.

20. The need for a broad-based and continuous change in society is gaining recognition, although the general tendency is to seek change on an *ad hoc* and compartmentalized basis. A somewhat broader effort is exemplified by Common Cause, a citizens' lobby devoted to reform of the political system and solution of national problems. See Theodore Jacquency, "Washington Pressures—Common Cause Lobbyists Focus on the Structure and Process of Government," *National Journal Reports* 5 (Sept. 1, 1973): 1294–1304.

21. The Port Authority of New York and New Jersey (established in 1921 as the Port of New York Authority) provides an example of a successful product of cooperation between the states of New York and New Jersey, but it is still a quite unusual case of interstate cooperation. The Authority was incorporated as an interstate compact that was accepted by the legislatures of New Jersey and New York and approved by the Congress as required in the Constitution. The procedures involved illustrate the complexity of achieving interstate cooperation. See Elmer E. Smead, *Governmental Promotion and Regulation of Business* (New York: Appleton-Century-Crofts, 1969), pp. 474–478.

22. See Dale L. Hiestand, *Changing Careers After 35—New Horizons Through Professional and Graduate Study* (New York: Columbia University Press, 1971), especially chapters 1 and 7.

23. See Andrew Shonfield, *Modern Capitalism* (New York: Oxford University Press, 1965), pp. 298–300.

24. Thus, in addition to the Department of Transportation, three independent commissions are responsible for regulation and policies in the field of transportation: the Interstate Commerce Commission (railroads, trucking companies, bus lines, water carriers, and oil pipelines), the Civil Aeronautics Board (civil aviation), and the Federal Maritime Commission (merchant marine). Also, the Environmental Protection Agency is increasingly exercising its specialized authority over the domestic components of the transportation sector. For an excellent background analysis of the policies, interests involved, and problems pertaining to transportation, see Merle Fainsod, Lincoln Gordon, and Joseph C. Palamountain, Jr., *Government and American Economy*, 3rd ed. (New York: W. W. Norton and Co., Inc., 1959), pp. 289–316.

25. See, for example, "Overhaul Sought for U.S. Agencies," *New York Times*, Aug. 10, 1970, p. 1. The thrust of the Ford Administration's efforts with regard to the regulatory commissions has been not in changing their structures, but in modifying, principally through appropriate legislation, the substance of policies in which the commissions are involved. Richard E. Cohen, "Regulatory Report 1: Federal Policy Shifting in Its View of Economic Regulation," *National Journal Reports* 7 (Feb. 22, 1975): 267–278.

26. See Peter F. Drucker, *The Age of Discontinuity* (New York: Harper and Row, 1969), pp. 149–150. Nelson, Peck, and Kalachek, *Technology, Economic Growth, and Public Policy*, represents an early effort to fill some gaps in this field.

27. For example, the federal government could pursue with greater vigor the policy to ensure that the fruits of federally financed R & D are transferred to all corners of the economy. (H.R. 9379, introduced by Rep. J. E. Roush in June 1971, attempted to achieve this result by providing for an Office of Federal Technology Transfer, but the bill failed to pass Congress.) Military R & D and procurement policies, if appropriately modified, could play an important role in enhancing and modernizing the nonmilitary sector of the economy. In the spring of 1972 the Department of the Navy established a policy for a systematic transfer of appropriate Navy-developed technology to the civilian sector and for cooperation with the civilian sector in the development of technologies of mutual interest. If such a policy were adopted by the Department of Defense as a whole and energetically pursued, it would have a truly invigorating impact on the economy. The military buys vast quantities of products and services from the civilian economy; its procurement policies could be a powerful instrument of technological change. See Emile Benoit, "R & D Convertability with a Pentagon Transformer," *Columbia Journal of World Business* I (Fall 1965): 95–96; Edward E. Furash, "The Problem of Technology Transfer," in *The Study of Policy Formation*, Raymond A. Bauer and Kenneth J. Gergen, eds. (New York: The Free Press, 1968), pp. 282, 298–300.

28. See John H. McArthur and Bruce R. Scott, *Industrial Planning in France* (Cambridge, Mass.: Harvard University Press, 1969), pp. 547–553. Since, in effect, such a policy would involve assistance to some industries to help them to grow, while others would be allowed to decline, its development is not likely to be easy politically. Also, when instituted, safeguards against potential mistakes and abuses on the part of governmental authorities would have to be provided.

29. Actually, in one way or another, the policies and decisions of the federal government have been instrumental in modifying the technological composition of the economy for many years. A protective tariff or a subsidy granted to a particular industry affects technological composition of the economy. The federal response to the Soviet nuclear missile challenge and the decision to land man on the moon, in effect, created entirely new industries. However, these actions by the federal government did not involve systematic effort to orchestrate the technological composition of the economy in the sense in which it is used here, that is, requiring a purposeful policy to phase out the old and obsolescent technological superstructure and phase in new technology as a means of stimulating the nation's vitality and economic growth.

30. William D. Carey, "Government, Science and Technology," a paper presented at the annual meeting of the Industrial Research Institute, Inc., June 2, 1973, pp. 2–3.

31. See Emmanuel G. Mesthene, *Technological Change* (Cambridge, Mass.: Harvard University Press, 1970), p. 48.

32. Harold and Margaret Sprout, "The Dilemma of Rising Demands and Insufficient Resources," *World Politics* 20 (July 1968): 660–693.

33. Alvin Toffler, *Future Shock* (New York: Random House, 1970).

13
The Outside World

1. The reader who is not conceptually inclined may prefer to skip this introduction and focus his attention on the discussion of specific policies. The conceptually inclined reader may want to refresh his memory by referring to a related discussion in Chapter 1, where the distinction between technology as a dependent and an independent variable is treated at graater length.

2. The term *international system* is used here almost synonymously with *world order*, since both encompass a (primarily global) pattern of institutions and politically relevant conditions combined

13. The Outside World

with a behavior of international actors influenced by that pattern. However, the term *world order* has been frequently employed in a normative sense, implying internationl stability and the rule of law and accompanied by advocacy thereof (discussed later in this chapter). To avoid possible ambiguity, the more neutral term *international system* is therefore used.

3. Since American policy evolved over a period of years, no single monograph gives it a sufficiently comprehensive coverage. The two articles by George Kennan, "The Sources of Soviet Conduct," *Foreign Affairs* 24 (July 1947): 556–582, and "America and the Russian Future," *Foreign Affairs* 29 (April 1951): 351–371; and Harold D. Lasswell's " 'Inevitable' War: A Problem in the Control of Long-Range Expectations," *World Politics* 2 (Oct. 1949): 1–39, are probably the most significant early publications; some of their tenets remained valid for several decades. For later evolution of policy, see Walt W. Rostow, "American Strategy on the World Scene," *The Department of State Bulletin* 46 (April 16, 1962): 625–631; Zbigniew Brzezinski, "The Framework of East–West Reconciliation," *Foreign Affairs* 46 (Jan. 1968): 256–275.

4. For the texts of the Agreement on Cooperation in the Fields of Science and Technology of May 24, 1972, and other related agreements signed at the Moscow summit, see "Agreements signed at Moscow during President Nixon's Visit," *The Department of State Bulletin* 66 (June 26, 1972): 918–926. A series of other important agreements related to cooperation in science and technology was signed in Washington in June 1973. See "Agreements Signed during General Secretary Brezhnev's Visit to the United States," *The Department of State Bulletin* 66 (July 23, 1973): 159–169.

5. See Rogers Morton, Secretary of Commerce, *The United States Role in East–West Trade* (Washington, D.C.: U.S. Department of Commerce, Aug. 1975), *passim*.

6. The conclusions of this paragraph are based on Howard Raiffa, "General Research Strategy and Initial Programs," in The International Institute for Applied Systems Analysis, *IIASA; Background Information* (Schloss Laxenburg, Austria, 1974). Information on current activities of IIASA was supplied by Augustus Nasmith, U.S. National Academy of Sciences. See also chap. 10, footnote 21.

7. See Marshall D. Shulman, "Toward a Western Philosophy of Coexistence," *Foreign Affairs* 52 (Oct. 1973): 49–51.

8. See P. M. Borisov, "Can We Control the Arctic Climate?" *Bulletin of the Atomic Scientists* 25 (March 1969): 43–48; and Ye. Gushchenkov, "Can the Pacific Warm the Arctic?" *ibid.* (May 1969), especially p. 49, for elaboration of this concept.

9. A complete discussion of these models can be found in Chapter 6.

10. This is not to say that American support of a functionally oriented Europe might not eventually lead to political unification or that the United States should not support such a unification at all. To the extent that these two objectives overlap, an American policy aimed at European unification would be justifiable.

11. For the outcome of these negotiations, see Energy Coordinating Group, Ad Hoc Group, "International Cooperation on Energy Research and Development; Report" (U.S. Department of State Document, ECG/ERD/36 final, June 6, 1974), and "Statement by Robert C. Seamans, Administrator, Energy Research and Development Administration, June 30, 1976," *Hearings* before the Subcommittee on Domestic and International Scientific Planning and Analysis and the Subcommittee on Energy Research Development and Administration Concerning International Cooperation in Energy R & D (Washington, D.C.: ERDA, 1976), pp. 6–10 and *passim*. The largest multilateral agreement concluded within the framework of IEA was the construction of a fluidized combustion coal test facility, to be built in the United Kingdom at an estimated cost of $24 million.

12. See Victor Basiuk, "U.S. Government Organization for Science, Technology, and Foreign Policy" in Commission for the Organization of the Government for the Conduct of Foreign Pol-

icy, *Appendices* (Washington, D.C.: GPO, 1975), Vol. 1, pp. 201–202. This study analyzes AEC, but the situation in ERDA improved little, as compared with AEC.

13. Zbigniew Brzezinski suggested that a trilateral cooperation between the United States, Western Europe, and Japan "must now become the central priority of U.S. policy." (See his "U.S. Foreign Policy: The Search for Focus," *Foreign Affairs* 51 [July 1973]: 722–727.) Whether one accepts Brzezinski's thesis or not, cooperation between these three regions in advanced energy technology appears a logical activity.

14. For example, the United States (represented by the Woods Hole Oceanographic Institution) cooperates with France in the French–American Mid-Ocean Undersea Study (FAMOUS), which explores the depths of the Atlantic Ocean. In August 1974 Project FAMOUS made dives to the depth of 9000 ft. "9,000-Foot Dive to a Valley in Atlantic" *New York Times*, Aug. 12, 1974, p. 1.

15. For a conceptual discussion of a U.S.–European cooperation in marine R & D and exploration and exploitation of the Atlantic Ocean, see Victor Basiuk, "Marine Resources Development, Foreign Policy, and the Spectrum of Choice," *Orbis* 12 (Spring 1968): 67–69. Here, as in the case of energy, programs would have to be delineated with due regard to commercial sensitivities, but the obstacles likely to arise are not necessarily insurmountable.

16. See, for example, a proposal for a joint U.S.–Japan–Australia–New Zealand program (possibly including Indonesia) in marine resources development in *ibid.*, pp. 69–70.

17. To illustrate some of the concerns of these organizations and their compartmentalization: Bureau of International Organization Affairs (IO) in the Department of State deals only with certain types of international organizations (intergovernmental and universal but not regional, like OECD, or private, like MNCs). Arms Control and Disarmament Agency (ACDA) is concerned with international stability insofar as it might be affected by military technology and armaments. Agency for International Development (AID) is concerned with international stability insofar as it is affected by economic underdevelopment. Both ACDA and AID affect world distribution of power in the long run, whereas the Department of Defense is concerned with world distribution of power in the short as well as the long run.

18. It might be argued that the Assistant to the President for National Security Affairs should properly perform the function of advising on a future international system. No doubt, various incumbents of that office at various times gave advice to the President on different aspects of the system and some of them may have had a particular future international system firmly in mind throughout their activity. However, the function of advising the President and formulating policies on a future international system is much too big and much too important now to be left assigned merely as a collateral duty to an adviser preoccupied with other matters.

19. For example, Albert Lepawsky, Edward H. Buehrig, and Harold Lasswell, eds., *The Search for World Order* (New York: Appleton-Century-Crofts, 1971), and Robert W. Cox, "On Thinking about Future World Order," *World Politics* 28 (Jan. 1976): 175–196.

20. The foregoing three paragraphs benefited from a discussion with Oliver J. Lissitzyn of Columbia University.

21. See Victor Basiuk, "Statement," in U.S. Congress, House of Representatives, Committee on Science and Technology, Subcommittee on Domestic and International Scientific Planning and Analysis, *Technology Transfer to the Organization of Petroleum Exporting Countries; Hearings* (Washington, D.C.: GPO, 1976), pp. 315–328. In U.S. policy, MNCs are expected to provide industrial development to the LDCs, but their contribution in this regard is considerably short of expectations.

22. Allocation of resources for the purpose of shaping a future international system is not anything new. Foreign aid, U.S. contribution to the U.N. budget, and participation in certain international scientific and technological programs provide a few examples. One advantage of having an Adviser to the President for a Future International System would be his potential ability to

stimulate a better awareness among the members of the Congress and elsewhere of a systemic purpose of the various allocations of resources where multiple U.S. interests overlap and this purpose is obscured or completely overlooked.

14
The Changing Nature of Advanced Societies

1. For example, Zbigniew Brzezinski, in his *Between the Two Ages: America's Role in the Technetronic Era* (New York: Viking, 1970), uses the term *the technetronic society.*
2. The most recent and most nearly complete source on Daniel Bell's concept of the postindustrial society is his *The Coming of Post-Industrial Society; A Venture in Social Forecasting* (New York: Basic Books, 1973; rev. ed., 1976). Since the concept evolved over a period of years and has undergone modifications, only this latest work is used here as the source of Bell's views.
3. *Ibid.*, p. 344.
4. *Ibid.*, pp. 116–117.
5. *Ibid.*, pp. 380–386.
6. *Ibid.*, p. 426.
7. *Ibid.*, p. 119.
8. *Ibid.*, p. 345.
9. *Ibid.*, p. 366. See also pp. 297–298, 364–365, 469.
10. *Ibid.*, p. 477.
11. *Ibid.*, p. 482.
12. Although Bell says that it was "capitalism" that promoted a hedonistic way of life and thus created "the antinomian culture," the phenomenon appears to be much more fundamental to be narrowly ascribed to "capitalism" and Madison Avenue as its instrumentality. It can be observed in the demands for consumer goods and luxuries in the Soviet Union, in spite of the official resistance to this trend. In fact, Bell's "antinomian culture" finds its very close parallel in the advanced phase of Sorokin's "sensate" (empirical) culture, which, as Sorokin pointed out, is not confined to the United States. See chap. 15, note 32, for Sorokin's analysis of the rise and development of the sensate culture.
13. Bell, *The Coming of Post-Industrial Society* p. 483.
14. *Ibid.*, p. 480.
15. *Ibid.*, pp. 366–367.
16. Herman Kahn's publications usually appeared as a collective product with his Hudson Institute colleagues, with Kahn being the principal author. For the sake of stylistic simplicity, the Hudson Institute publications will be identified here with Kahn himself, and references to his colleagues will be omitted. The principal books dealing with the concept of a post-industrial society are Herman Kahn and Anthony J. Wiener, *The Year 2000* (New York: The Macmillan Co., 1967) (hereafter referred to as *The Year 2000*) and Herman Kahn and B. Bruce-Briggs, *Things to Come* (New York: The Macmillan Co., 1972) (hereafter referred to as *Things to Come*). Herman Kahn, William Brown, and Leon Martel, *The Next 200 Years* (New York: William Morrow and Co., 1976) is primarily a problem-oriented book, dealing with food, energy, population, etc. Insofar as it addresses itself to societal stages, it deviates from Daniel Bell and Kahn's earlier publications by introducing a "superindustrial economy" as the next stage of advanced societies, to be eventually followed by a "post-industrial economy." The borderline between the two is not clearly defined. The analysis in this book leans on the earlier two books as the more conceptual and using a more standard definition of the post-industrial society.

17. *The Year 2000*, p. 194.
18. *Ibid.*, pp. 198–202, 213–217.
19. *Things to Come*, p. 229.
20. *Ibid.*, p. 97.
21. *Ibid.*, p. 227–228.
22. *Ibid.*, p. 87. See also pp. 230–231 and *The Year 2000*, pp. 217–218.
23. *Things to Come*, p. 87.
24. *Ibid.*, p. 250.
25. In *The Year 2000*, an entire chapter (chap. 10) is devoted to "Policy Research and Social Change." But this chapter does not really deal with major changes in the direction of societal development; it largely deals with methodological approaches to studies and issues.
26. See Ken'ichi Tominaga, "Post Industrial Society and Cultural Diversity," *Survey* 16 (Winter 1971): 70. Bell agrees with this point.
27. When Daniel Bell emphasized "a process of direct and deliberate contrivance" as "perhaps the most important social change of our time," a process that includes the transformation of entire societies, the truly telling examples he gave were the industrialization of Japan, the USSR, and the colonial world—all related to the preindustrial and industrial stages and not the postindustrial. With regard to "the older Western societies" (which are closer to the postindustrial stage) Bell mentioned such innovations as the introduction of planning, economic growth and full employment. *The Coming of Post-Industrial Society*, pp. 345–346. The simple fact is that, for the reasons discussed above, a postindustrial society is much, much more difficult to transform and mold than a preindustrial one, tools of industrialization being available.
28. *Ibid.*, pp. 344–345.
29. This phenomenon has historical precedents. Marx's theory had limited effect until Lenin changed it and applied it to Russian conditions, at considerable violence to its original theoretical purity. Again, Trotsky's more orthodox Marxist concept of "permanent revolution" succumbed to Stalin's pragmatic "socialism in one country." In terms of power, Lenin and Stalin were certainly greater beneficiaries than Marx and Trotsky. In recent years Japan has provided a good example of how far a nation can go by capitalizing on adaptation of knowledge originated elsewhere.
30. See Robert L. Rothstein, *Planning, Prediction, and Policymaking in Foreign Affairs* (Boston: Little, Brown, 1972), pp. 13–18 and *passim*. Rothstein, however, suggests that, if complementarity between theorists and practitioners in the same field is capitalized upon, improvements are possible.
31. "Plans and Policy Elude Economists," *New York Times*, Dec. 29, 1973, p. 31.
32. See Karl W. Deutsch, "On Political Theory and Political Action," *The American Political Science Review* 65 (March 1971): 18–19.
33. Evron M. Kirkpatrick, " 'Toward a More Responsible Two-Party System': Political Science, Policy Science, or Pseudo-Science?" *The American Political Science Review* 65 (Dec. 1971): 989.
34. For a discussion of this relatively new movement in political science, see David Easton, "The New Revolution in Political Science," *The American Political Science Review* 63 (Dec. 1969): 1051–1061.
35. One might argue that, if objective factors in the postindustrial society favor the leadership of the universities and if we speak in terms of decades and not just a few years, the universities will eventually readjust themselves to ensure this leadership. This is not necessarily so. The rate of change in the future will be rapid, and leadership will require being abreast of the time. Unless the various disciplines within the university—especially in the social sciences—make a deliberate and continuous effort to provide intellectual leadership to society, it will not happen.

14. Advanced Societies

36. Bell, *The Coming of Post-Industrial Society*, p. 486.

37. *Things to Come*, p. 62.

38. The fact that this industry would be operating for profit might be an important factor in keeping it efficient and dynamic. One of the problems of governmental service bureaucracy is that it is difficult to measure its productivity and thus keep it efficient, since there are no products involved and no overall criteria of performance.

39. Donella H. Meadows, Dennis L. Meadows, Jorgen Randers, and William W. Behrens, III, *The Limits to Growth* (New York: Universe Books, 1972). Hereafter cited as *The Limits to Growth*.

40. Herman E. Daly, "Toward a Stationary-State Economy," in *Patient Earth*, John Harte and Robert H. Socolow, eds. (New York: Holt, Rinehart and Winston, 1971), pp. 231–233; William C. Gough and Bernard J. Eastlund, *Energy Wastes, and the Fusion Torch* (Washington, D.C.: U.S. Atomic Energy Commission, 1971), pp. 3–5. If solar energy is collected on Earth, it would be nonpolluting (except for possible local or regional thermal imbalances), since no new heat would be added to the Earth's surface. If solar energy is collected via artificial satellites, in outer space, it would contribute to thermal pollution.

41. According to EPA estimates, it cost the nation $3.5 billion in 1973 (at 1974 prices) to collect and dispose postconsumer solid waste alone (i.e., excluding production and similar waste). Only 7 percent of total postconsumer gross discards were recycled. Between 1973 and 1985, disposed (i.e., excluding recycled) postconsumer waste is expected to rise from 135 million tons to 167 million, and the cost will reach $8.3 billion in current prices. Short of recycling, sanitary landfills appear to present the most satisfactory method of waste disposal, but local opposition to them is very strong, and they also create environmental problems (underground leachate is serious, affecting the quality of ground and surface waters). U.S. Environmental Protection Agency, *Resource Recovery and Waste Reduction; Third Report to Congress* (Washington, D.C.: GPO, 1975), pp. 8–13. Japan is running out of land for garbage dumps and is creating artificial islands in Tokyo Bay for dumping garbage, along with such devices as a giant incinerator disguised as a 60-story skyscraper. Local residents, however, strenuously oppose these garbage disposal mechanisms in their neighborhoods. "Tokyo Plans Novel Projects for Garbage Disposal," *New York Times*, Dec. 20, 1973, p. 3.

42. See Helmut W. Schultz, "Cost/Benefit of Solid Waste Reuse," *Environmental Science and Technology* 9 (May 1975): 423–425; and James G. Abert, Harvey Alter, and J. Frank Bernheisel, "The Economics of Resources Recovery from Municipal Solid Waste," *Science* 183 (March 15, 1974): 1052–1058.

43. See Lester Brown, "Rich Countries and Poor in a Finite, Interdependent World," *Daedalus* 102 (Fall 1973): 157.

44. To limit the unfavorable effect of investments abroad on the balance of payments, restrictions have already begun to appear. For example, in January 1974, Japan was reported to have imposed a series of such restrictions. These restrictions, however, covered only "nonessential" investments such as real estate, hotels, restaurants, and stocks. "Japan Restricts Investments Abroad," *New York Times*, Jan. 19, 1974, p. 41.

45. It is assumed in this scenario that elimination of cardiovascular diseases progresses to the point of increasing life expectancy by 9 years, and advance in gerontology adds 15 years of vigorous life. However, they are partially overlapping and thus result in a total increase of life expectancy by 20 years. These assumptions are rather conservative, as compared with projections of such gerontologists as Comfort and Bjorksten (see chap. 3 and note 44).

46. Population growth has a considerable momentum that cannot be easily stopped. The age structure of growing societies is more favorable to future births. See Thomas Frejka, "Reflections

14. Advanced Societies   377

on the Demographic Conditions Needed to Establish a U.S. Stationary Population Growth," *Population Studies* 22 (Nov. 1968): 380, 382–383. In the year 2000, the population structure will be much less favorable to future births than in the 1960s, but it is not likely to decline sufficiently to prevent an increase in the rate of population growth because of the drop in mortality.

47. This scenario is based on the following assumptions. Advance in gerontology, which adds 15 years of vigorous life to senior population, at the same time decreases their vulnerability to diseases. This, in effect, would mean that nearly all people would stop dying from diseases for a period of time (up to 15 years in our scenario), provided that the new pharmaceutical agents are administered to all concerned. Series 2 projections (most generally used) of the Bureau of the Census give us a death rate of 10.4 per 1000 in the year 2005. In 1974, 90.6 percent of U.S. population died from diseases. Let us assume that about the year 2005, 75 percent of the population due to die from diseases do not actually die. Using Series 2 projection deathrate for the year 2005 (10.4 per 1000), we obtain an increase of population of 7.8 per 1000. Projections Series 2 give us a net population increase (including immigration) in the year 2005 as 6.0 per 1000. To be on the conservative side, let us assume it is 100 percent off; the actual increase is only half that much, 3.0 per 1000 (births are lower, the U.S. restricts immigration). This gives us a total population growth of 10.8 per 1000 (7.8 plus 3.0), or 1.08 percent for the year 2005. Sources: U.S. Bureau of the Census, *Current Population Reports*, Series P-25, No. 601, "Projections of the Population of the United States: 1975 to 2050" (Washington, D.C.: GPO, 1975), p. 35; U.S. Department of Health, Education, and Welfare, National Center for Health Statistics, *Monthly Vital Statistics Report* 24, No. 11 (Feb. 3, 1976), *Supplement,* pp. 2 and 14.

48. The increase in waste would not be limited to that of the household waste of the additional population but would include the extra commercial, municipal, industrial, mining, and agricultural wastes generated as a by-product of sustaining the consumption pattern of the added population and would run into many millions of tons. The present breakdown of solid wastes, though not necessarily typical for the 21st century, may give a general idea of how much waste is generated in various sectors of the economy. Of the estimated 4.3 billion tons of material thrown away annually, the largest categories are as follows: urban wastes, 250 million tons; industrial, 110 million; mining, 1.7 billion; agricultural, 2.3 billion. "Solid Waste: Disposal, Reuse Present Major Problems," *Congressional Quarterly Weekly Report* 31 (April 28, 1973): 1019. Statistics on solid waste are inadequate, and estimates vary.

49. See Frank A. Smith, "Technical Possibilities for Solid Waste Reduction and Resources Recovery: Prospects to 1985" (Resource Recovery Division, Office of Solid Waste Management Programs, U.S. Environmental Protection Agency, Dec. 10, 1974), p. 4. Effective recycling requires a series of policies with regard to source reduction, source separation, and mixed waste processing, which are time consuming to implement.

50. "Surprise-free" projections of trends in solid waste reduction in the United States indicate that, between 1973 and 1985, material recycling and energy conversion will increase by 25.4 million tons annually (from 9.4 to 35 million tons), while gross discards will increase by 57 million tons annually. Postconsumer residential and commercial solid waste only were analyzed. *Ibid.*, p. 2.

51. See "Statements of Dr. Talbot Page and Dr. James W. Sawyer, Research Associates, Quality of the Environment Program, Resources for the Future," in U.S. Senate, Committee on Public Works, Subcommittee on Environmental Pollution, Panel on Materials Policy, *The Need for a National Materials Policy; Hearings* (Washington: GPO, 1974). Part I, pp. 21–24, 29–30, 53–57. Statistics on recycling are usually either nonexistent, inaccurate, or misleading. E.g., in the case of the ferrous metals industry—an industry among the highest in recycling—about 50 percent of the total iron feed consists of scrap. But most of this is factory-produced scrap, and only about 8 percent is "obsolete" (i.e., post-consumer) scrap. *Ibid.*, pp. 53–54.

14. Advanced Societies

52. Even in the present state of the art, the costs of recycling, given a sufficiently large scale of application, are quite favorable in comparison with alternative, ecologically acceptable (though not ideal) ways of disposal of waste. A 1973 study concluded that, given a plant capacity of 2000 tons/day (a population base of 500,000–700,000 is required for this plant capacity), five recycling methods were cheaper than a resort to remote sanitary landfill, requiring transportation of waste 100 miles or more. See Midwest Research Institute, *Resource Recovery: The State of Technology*; Report Prepared for the Council on Environmental Quality (Washington, D.C.: GPO, 1973), especially pp. 55–58.

53. It is conceivable that the fusion torch could prove economical short of fusion power's reaching a commercial stage. In this latter event the necessary high temperature and the resultant plasma (extremely hot gas, not liquid as in medicine) could be obtained by conventional electric power. Alternatively a reactor torch could be constructed in the near future using a fusion-type unit, even though the reactor would not meet the so-called Lawson criterion for a self-sustaining system. Here the input of energy would be greater than the output. See George H. Miley, "Summary of a Meeting on Non-Electrical Energy Extraction from Fusion Reactors at AEC Germantown Headquarters, March 20, 1973" (Unpublished document, July 20, 1973), pp. 3–5.

54. For a more extensive discussion of the fusion torch, see William C. Gough and Bernard J. Eastlund, "The Prospects of Fusion Power," *Scientific American* 224 (Feb. 1971): 59–61; Gough and Eastlund, *Energy, Wastes, and the Fusion Torch*, and Zeinab A. Sabri, "A Study of the Feasibility of Fusion Torches" (University of Wisconsin Ph.D. Dissertation, 1972).

55. Even if a nation intended to keep the amount of materials completely stationary, some consumption of virgin raw materials would still be necessary because of losses in the solid waste collection and recycling process. Some of these losses may result from using waste for purposes other than their reconversion into materials; for example, certain wastes might be used to generate heat or electric power.

56. The definition of the semistationary society as used here is somewhat arbitrary. For example, it could be defined as a society in which the total stock of materials is stationary (this being a closed-materials economy), while both the population and the total stock of wealth are growing. To be a viable society, however, the growth of population should not exceed the growth of the total stock of wealth; that is, the per capita income of the population must be stationary or growing, and not declining. Since just about the only source of growth in material wealth would be technology (the total quantity of materials being fixed), the margin for the growth of the population would be rather narrow. Even a stationary per capita income (let alone a declining one) is likely to produce internal social and political problems jeopardizing society's viability. Thus, being guided by the criterion of the viability of the semistationary society, it would be prudent to define its "pure" model as one having a stationary population that would allow some growth of per capita income.

57. It might be helpful to point out for the benefit of those readers who are not familiar with the (largely economic) literature on the subject of growth and future societies that the existing literature deals with the stationary-state (no-growth) society and does not analyze stages of societal development short of the no-growth society as entities in their own right that need to be examined. In this context my introduction of the semistationary society is an innovation. The principal conceptual differences between the semistationary and the stationary-state society lie in the assumptions of the concept of the semistationary society to the effect that: (1) We do not have yet, and will not have for a long time, limits to growth from the raw materials end of the production cycle; the limitations are likely to come from the waste end. (2) The total stock of wealth will not be large enough for a long time so as to impose limits on growth in the sense that its sheer size combined with the rate of the flow of materials (throughout) might obstruct the natural ecological processes that form the biophysical foundation of wealth. Even if the stock of wealth begins to

reach a critical limit, the room for growth could be expanded further by lowering, through technological advance, the rate of throughput. (In the light of these two assumptions it is conceivable that a semistationary society, once it masters the problem of solid waste by converting to a closed-materials economy, might resume the consumption of fairly large quantities of virgin raw materials and thus increase its growth—a development not conceivable in the stationary-state society.) Apart from the conceptual distinction between the semistationary and the stationary-state society, there is also a practical one. The stationary-state society has limited political utility at the present, since it is a remote development in terms of time. The emergence of the semistationary society presents a much more real problem—it is much closer. And, as we shall see later, it is in the process of reaching the semistationary society and within it that strong pressures will emerge for societies to modify the way of governing themselves, and not in the far more distant stationary-state society.

For a conceptual analysis of the stationary-state society (which I found most useful), see John Stuart Mill, *Principles of Political Economy* (London: Longmans, Green and Co., 1929), pp. 746–751; Kenneth E. Boulding, "The Economics of the Coming Spaceship Earth," in *Environmental Quality in a Growing Economy*, Henry Jarrett, ed. (Baltimore: Johns Hopkins Press), pp. 3–14; and Daly in *Patient Earth*, Harte and Socolow, eds. pp. 226–244.

58. Stephen H. Schneider points out that adverse effects of waste heat on climate are likely to occur much sooner than generally realized. Global effects of heat-generating human activity will be more remote ("sometime after the year 2000"), but local, medium-scale, and regional climatic disruptions can be anticipated sooner. See his *The Genesis Strategy; Climate and Global Survival* (New York: Plenum Press, 1976), pp. 135–136, 180.

59. Harvey Brooks, "The Technology of Zero Growth," *Daedalus* 102 (Fall 1973): 144.

60. See the discussion of ideological and value differences between generations in chap. 3.

61. On the other hand it is entirely possible that further progress in gerontology and in eliminating diseases will take place in the first two decades of the 21st century and thus further sustain the momentum of population growth.

62. See Harvey Brooks, "Technology of Zero Growth," p. 149. Brooks estimates a possible growth due to productivity increase at 1.5 percent. He does not seem, however, to allow for the spread of leisure, which would further decrease economic growth.

One can hardly predict differences in productivity between the service and the manufacturing sector, but some recent figures are of interest. In 1967–1972, productivity per man hour rose 3.7 percent annually in the manufacturing sector; the corresponding figure for the service sector was 0.3 percent. Source: U.S. Bureau of Labor Statistics.

63. Harold D. Lasswell, *Politics: Who Gets What, When, How* (New York: Peter Smith, 1950).

64. This is not to suggest that income will be equally distributed; inequalities in income and wealth will remain. The point is that, in a society that uses material rewards as the single most important measure of success, a reasonably decent and comfortable living will not be enough for most people—to demonstrate success, they must move into unnecessary luxuries and thus widen and spread the inequalities in the standard of living. Once, however, material possessions are deemphasized as a value and as a measure of success, the opposite trend is likely—the discrepancies in the standard of living will decrease and the existing wealth of society is likely to be more than adequate to meet the needs of the population.

65. Some of these effects of the highly advanced societies were discussed in chap. 3 and chap. 12. The analysis here focuses on the effects not discussed in these chapters, but that discussion is certainly relevant to the present analysis.

66. Roland N. McKean, "Growth vs. No Growth: An Evaluation," *Daedalus* 102 (Fall 1973): 213.

67. *Ibid.*, pp. 213–214.
68. Staffan B. Linder, *The Harried Leisure Class* (New York: Columbia University Press, 1970), *passim*. The following discussion on the scarcity of time is largely based on Linder's analysis.
69. If, however, the "forerunner" trend of the late sixties reemerges in future generations and a large segment of population will not be interested in productive pursuits and will reject materialistic values, a less hectic and perhaps more agreeable society might evolve. But the extent to which society would discard the undesirable effects of advanced economies and become truly agreeable would depend on the specific nature of the values developed by future generations. It is conceivable that the "forerunner" values may result in social instability, too, but for a different reason.
70. A nation's position with regard to national security in the early 21st century has to be qualified by the following considerations:

 a. It is conceivable that prudent and forward-looking policies on the part of most early semistationary societies and LACs might succeed in avoiding strong manifestations of instability and thus allow the non-zero-sum game to penetrate the international political process more firmly. If so, even a high degree of instability on the part of some isolated societies will not jeopardize their national security—they simply will take a "free ride" on the nature of the international system.

 b. It is conceivable that the attitudes of the intellectual and cultural elites in a particular semistationary society might be largely antimilitary, as is the case now in the United States. Having greater power in a nonmaterialistic semistationary society, these elites might curtail allocation of resources to national security to the point of jeopardizing it. This development is, however, unlikely. Historically societies concerned with intellectual and cultural pursuits effectively combined these pursuits with sufficient safeguards of national security. This was true of ancient Greece, which combined high civilization with strength; at the height of her empire, Great Britain managed to combine her military power with a reasonable degree of intellectual and cultural achievement. It is thus entirely possible—though somewhat paradoxical in the light of present beliefs—that the U.S. military might find itself in a much happier position in some fifty years, leaning on the intellectual and cultural elites, than it is in now, when it leans largely on industry. This development, however, would require—and probably result in—greater intellectualization of the military.

71. By present standards these questions seem almost unreal. However, even present economies have reached a sufficient degree of complexity to show some symptoms of vulnerability suggested by these questions. For example, during the Vietnam war—which was a minor war, if viewed in the context of America's war potential—the U.S. government's policy in 1967–1969 of trying to have both guns and butter seriously overheated the economy; led to inflation, which the government proved to be incapable of controlling; and resulted in one of the most severe stock market declines. A half or three quarters of a century from now, closed-materials economies will not only be immensely larger in size but also much more complex, and they will have much narrower tolerances for sudden upheavals.

15
The National Purpose in an International Context

1. Herbert Croly makes a distinction between a "national promise" and a "national purpose." The former exists when the nature of the socioeconomic system and other conditions that society established for its functioning make the attainment of the "national promise" automatic, without its being explicitly promoted by the government. When the attainment of "the national promise" is impeded by some developments, and governmental intervention becomes necessary to pro-

mote its realization, then "the national promise" becomes "the national purpose." See Herbert Croly, *The Promise of American Life* (Cambridge, Mass.: The Belknap Press, 1965), pp. 16–21 and *passim*. (This is a reprint of the original 1909 edition of Croly's book.) No such distinction is made here; both are defined as "national purpose."

2. *Cf.* Daniel Bell's "One of the deepest human impulses is to *sanctify* their institutions and beliefs in order to find a meaningful purpose in their lives and to deny the meaninglessness of death." *The Coming of Post-Industrial Society*, p. 480.

3. See, for example, Pitirim A. Sorokin, *Modern Historical and Social Philosophies* (New York: Dover Publications, 1963), especially pp. 275–322. This is a comparative study of more than a dozen historians and philosophers of civilizations, in which Sorokin finds a substantial body of agreement among them on certain fundamental characteristics of civilizations and their development.

4. See, for example, Max Lerner, "America Agonistes," *Foreign Affairs* 52 (Jan. 1974): 287–300. Unlike some other analysts Lerner is more optimistic about America's future.

5. See Irving Kristol, " 'When Virtue Loses All Her Loveliness'—Some Reflections on Capitalism and 'The Free Society,' " *The Public Interest*, No. 21 (Fall 1970): 4–5.

6. The standard reference to Adam Smith's economic theory is, of course, *The Wealth of Nations* (1776) in its many editions. However, a much more digestible and brilliant account of his theory is provided in Robert L. Heilbroner, *The Worldly Philosophers*, 3rd ed., (New York: Simon and Schuster, 1967), pp. 49–67.

7. Some illustrations of the magnitude of the problems involved and of the great difference in this respect as we compare the past and the future can be given. In the early years of the introduction of the automobile, the latter was viewed as an epitome of individual freedom by making an individual highly mobile and flexible in his mobility. To make this freedom realizable, the federal government had to provide assistance in the construction of roads. Today an individual could not seriously claim that he is exercising his expanded freedom when he is crawling, at 8:30 in the morning, bumper-to-bumper, to work in the city. His freedom could probably be better served by creation of an efficient and comfortable mass transportation system. But the city, state, and federal government, facing the tug and pull of various bureaucratic interests in their own midst as well as pressures of vested interests from the outside, proved to be lacking in timely response. Still another example is the energy crisis. Certainly our national decision making with regard to both forestalling and coping with the energy crisis did not quite succeed in preventing conditions that played havoc with the individual's freedom.

8. Kristol, " 'When Virtue Loses All Her Loveliness,' " p. 5.

9. The above examples of how the economic purposes have tended to impede moral and social amelioration are not intended to be exhaustive. John Kenneth Galbraith, for example, discusses how "the industrial system" has subordinated esthetic considerations to the interest of its growth. See his *The New Industrial State*, 2nd ed. (New York: The New American Library, 1971), pp. 334–336.

10. Staffan B. Linder, *The Harried Leisure Class* (New York: Columbia University Press, 1970), p. 10. See also chap. 14.

11. There is no single document in which present national goals are specifically spelled out, but they appear in various pronouncements by the President and his advisers. The State of the Union messages provide perhaps the most nearly comprehensive documents addressed to national goals. The national goals stated below have been extracted from the State of the Union messages of 1972, 1973, 1974, and 1976. For the text of the messages, see "Text of Nixon's State of the Union Address," *Congressional Quarterly Weekly Report* 30 (Jan. 22, 1972): 116–129; "State of the Union Message," *Congressional Quarterly Weekly Report* 31 (Feb. 10, 1973):

382   15. The National Purpose

297–298; (Feb. 17): 363–366; (Feb. 24): 395–398; (March 3): 435–441; (March 10): 560–510; and (March 17): 604–608; "State of the Union Text," *Congressional Quarterly Weekly Report* 32 (Feb. 2, 1974): 239–242; "The State of the Union," *Weekly Compilation of Presidential Documents* 12 (Jan. 26, 1976): 43–52.

12. "Text of Nixon's State of the Union Address," *Congressional Quarterly Weekly Report* 30, pp. 126–127.

13. *Toward Balanced Growth: Quantity with Quality*. Report of the National Goals Research Staff (Washington, D.C.: GPO, 1970), pp. 133–147.

14. "State of the Union," *Congressional Quarterly Weekly Report* 31, p. 440.

15. Appropriations for FY 1976 were $172.4 million, and for FY 1977, $180.5 million. Source: The National Foundation on the Arts and the Humanities.

16. *Cf. Toward Balanced Growth*, pp. 23–24, 157–167, and *passim*.

17. Actually Galbraith's emphasis is on intellectual, esthetic, and cultural aspects; higher ethical standards and what can be broadly characterized as "a virtuous society" (including a generally shared concept of justice) escape Professor Galbraith's scrutiny. See Galbraith, *New Industrial State*, pp. 356–383.

18. Although Galbraith is discussed here as an example of a proposal for implementation of societal goals that is of doubtful practicality, he is considerably more practical than other analysts. Victor Ferkiss, for one, is much more radical in his conclusions and much less pragmatic in his solutions. According to Ferkiss, the momentum of our "bourgeois" civilization, with its "bourgeois values," carries us to killing nature and destroying ourselves in the process. The solution of this basically ecological crisis will come with the advent of "technological man"—almost a superman of the technological age, with a new philosophy, a new system of values, and a new culture of his own. The new philosophy, values, and culture will emphasize coexistence with nature and with other human beings. But it is not clear how a society of "technological men" could be created. See Victor C. Ferkiss, *Technological Man: The Myth and the Reality* (London: Heinemann, 1969), pp. 101–116, 257–259, 265, and *passim*.

19. See, for example, "Aimless America; Affluent, Poor, Old and Young All Turn Inward, Grow Moody," *The Wall Street Journal*, Oct. 16, 1972, p. 1.

20. U.S. Senate, Committee on Government Operations, Subcommittee on Intergovernmental Relations, *Confidence and Concern: Citizens View American Government; Hearing on a Survey of Public Attitudes* (Washington, D.C.: 1974), pp. 6–8 (hereafter cited as *Confidence and Concern, Citizens View American Government*).

21. The drop in confidence in key institutions between 1966 and 1976 was as follows: Executive branch, from 41 to 11 percent; Congress, from 42 to 9, major companies, from 55 to 16 percent; the military, from 62 percent to 23 percent; and U.S. Supreme Court, from 51 to 22 percent. Louis Harris, "Confidence in Leadership Down Again," *The Harris Survey*, March 22, 1976.

22. Peter Schrag, *The End of the American Future* (New York: Simon and Schuster, 1973). The title of Schrag's article in *The New York Times* (January 16, 1974, p. 39) aptly summarizes his point of view: "The Tunnel at the End of the Light."

23. *Cf.* Kenneth E. Boulding, *The Meaning of the 20th Century; The Great Transition* (New York: Harper and Row, 1965), p. 164. Boulding cites Fred Polak to the effect that "the ability of an ideology to organize society depends in large measure on the optimistic or pessimistic quality of its images of the future and on whether it holds that the future can be changed by human activity."

24. Clifford Geertz, "The Impact of the Concept of Culture on the Concept of Man," in *New Views on the Nature of Man*, John R. Platt, ed. (Chicago: The University of Chicago Press, 1965), pp. 109–113.

15. The National Purpose 383

25. For an incisive analysis of the relationship between biological aspects of man, civilization, and culture, see Caryl P. Haskins, *Of Societies and Men* (New York: Viking, 1960), pp. 200–215.
26. John Rawls, *A Theory of Justice* (Cambridge, Mass.: The Belknap Press, 1971).
27. The reader will note that this concept of justice, although it changed only a few words in Rawls' definition, significantly differs from that of Rawls. His definition (*ibid.*, p. 303) is as follows: "All social primary goods—liberty and opportunity, income and wealth, and the bases of self-respect—are to be distributed equally unless an unequal distribution of any or all of these goods is to the advantage of the least favored." Thus, Rawls would justify inequalities in rewards only by their being "to the advantage of the least favored." It is much too narrow a justification for the inequalities. This is especially so if "the least favored" are viewed mainly in material terms, as appears to be the case in Rawls' analysis.
28. For example, the possession of a TV set (a luxury in the 1950s) is no longer considered to be incompatible with poverty. The decrease in the price of TV sets is certainly a factor in this phenomenon, but not the only factor. The general affluence, aided by the advertising, changes the concept of "luxuries" and "necessities," artificial as the change might be in some cases.
29. Vietnam and Watergate stimulated interest in the issue of Presidential power. In the wake of these events, Arthur M. Schlesinger, Jr. wrote *The Imperial Presidency* (Boston: Houghton Mifflin, 1973), wherein he addresses himself to a "runaway Presidency," the "virtually unchecked" power of the executive, especially in foreign policy and in the employment of armed forces. Whereas the issue of abuses of Presidential power and responsibility is a serious one, an equally serious issue is the ability of the President to control his own government and special interests to carry out the responsibilities of his office and leadership. Watergate demonstrated, not only that Presidential power can be abused, but also that special interests have considerable influence over the President. In the last analysis the moral of Watergate appears to be, not the institutional limitations on the Presidency or the lack thereof, but a strong personal sense of responsibility and a high ethical standard, which the President must possess. Richard E. Neustadt's *Presidential Power; The Politics of Leadership* (New York: Wiley, 1960), provides a more nearly comprehensive and better balanced analysis of both the limitations and the potential of Presidential power.
30. Louis Harris, testifying before the Senate Subcommittee on Intergovernmental Relations on the study his organization conducted for the Subcommittee in the Fall of 1973, pointed out that "9 in every 10 people . . . expressed the cardinal article of faith that government can be made to work efficiently and effectively, and within the parameters of liberty a free people require." Not more than 5 percent of the public were willing to scrap the major institutions of American society. Harris further noted: "The great hope I find in this study is not that people are turned off and negative and cynical and say it is all gone. To the contrary, they are just waiting, just waiting for some kind of leadership that will communicate confidence in sharing with them the tremendous problems which the country faces. . . ." A majority of Americans disagreed with the statement that "That government is best which governs least." Preferences for stronger leadership were distinctly indicated. See *Confidence and Concern: Citizens View American Government*, pp. 9, 13–14, 17, and 26–27.
31. The founding fathers of the American republic, by establishing a popular government, expressed faith in common man. At the same time, they were well aware that common men are not infallible, and they thus introduced a system of checks and balances as a safeguard against possible mistakes that the electorate might make in the selection of its leaders. While safeguards against potential evils of democracy were provided, no corresponding provisions existed to ensure excellence of choice. But, as Irving Kristol somewhat wistfully remarks, the founding fathers "would have regarded as a fair test of their labors and degree to which common men in America could rise to the prospect of choosing uncommon men, speaking for uncommon ideals, as worthy

384   15. The National Purpose

of exercising authority over them." Irving Kristol, *On the Democratic Idea in America* (New York: Harper and Row, 1972), p. 54.

32. See *ibid.* pp. 28–29. Daniel Bell's views on the role of culture were discussed in chap. 14. See also Pitirim A. Sorokin, *Society, Culture, and Personality; Their Structure and Dynamics* (New York: Cooper Square Publishers, Inc., 1962), pp. 602–606, and his *The Crisis of Our Age* (New York: Dutton, 1957).

33. For an account of the state of affairs in the United Nations, see "Effectiveness of the U.N., 28, Is at an Ebb," *New York Times*, Sept. 10, 1973, p. 1; "Small Nations' Interests Dominate U.N.," *New York Times*, Sept. 11, 1973, p. 1; and "U.N.'s Bureaucracy Is Hobbled by Uncertain Skills and Loyalty," *New York Times*, Sept. 12, 1973, p. 2.

34. This phenomenon may, at least in part, explain why we do not seem to have major international movements to deal with such world problems as overpopulation, food, or the development of poor countries in general. To be successfully launched, a crusade must have a promise of success. In view of the complexity of the matter and the difficulty of making the entities and the people concerned to do their share, there does not seem to be enough promise of success to energize strong action. In part this phenomenon explains why a great deal of idealism is gone from international organization. To be viable, idealism must feed on a reasonable prospect of success.

35. See Max Lerner, *America as a Civilization* (New York: Simon and Schuster, 1957), pp. 883–884.

36. A benevolent visitor from another planet, observing the American policy aimed at developing a pluralistic Soviet society, might mutter: "Their vested interest aside, I cannot really blame the Soviet leadership for resisting the American effort. Look at the fix the Americans are in!"

37. Robert W. Cox, "On Thinking about a Further World Order," *World Politics* 28 (Jan. 1976): 195–196, makes a distinction between those conflicts in world politics that might lead to a change of the system' and those that merely effect changes within the system. The distinction is important and, as Cox points out, the focus of analysis should be on the former. The present study differs, however, from his point of view in the following respects: (1) It is not necessarily conflicts that change the system; in fact, it would be fruitful to focus on nonconflictual phenomena effecting systemic change and thus learn to steer it in a desired direction short of generating instability. (2) Individual changes in the elements of a particular international system may not change the system, but if their aggregate is sufficiently large, its impact may. It therefore might be important to focus the analysis on the implications of individual elements of the system (including behavior of major actors in world politics) and on the "critical mass" of them necessary to produce systemic change. Otherwise, certain major actors (e.g., the superpowers) may produce change in the international system in a mild spell of absentmindedness, as they appear to have done in the last 20 years.

38. Enlargement of our knowledge of biological mechanisms could well have in the future an important part in the further progressive evolution of man.

Index

Abegglen, James C., 353nn7, 8, 10, 354n31
Abert, James G., 376n42
Abu Dhabi, 20
Acton, Lord, 150
Advanced Research Projects Agency (ARPA), 369n14
Advertising industry, U.S.: expenditure of, 57
Adviser to the President for a Future International System (proposed), 254, 373-74n22; and Assistant to the President for National Security Affairs, 373n18
AEC, see Atomic Energy Commission
Affluence, 33, 263, 264, 299; as U.S. national purpose, 294, 296, 298; entrapping attributes of, 297, 302
Agency for International Development (AID), 373n17; concessional vs. reimbursable technical assistance, 255
Aging, see Life expectancy; Gerontology
Air cushion systems (in ground transportation), 334n16
Air cushion vehicles (marine), 137, 336n35; trends in, 28. See also Surface-effect ships
Airlines, and jet engine, 42
Alexander, Tom, 339n5
Alienation, social, 234; effect of institutions on, 34-35; in postindustrial society, 273; extent of in U.S., 303
Allison, Graham T., 364n3
Alter, Harvey, 376n42
Amalrik, Andrei, 360n25
America, see United States

"American challenge," 102, 111
Amman, R., 343n5
Amphibious capability, and USSR, 136
Anti-Ballistic Missile Defense (ABM), 155-56
Antitrust policy, need for reexamination, 229
Arab nations, surplus of petrodollars and industrial development, 50
Argentina, 12; and nuclear weapons, 136; and NPT, 158
Ariane (European satellite launcher), 101
Aristotle, 296; and high civilization, 306; and normative approach, 313-14
Arms Control and Disarmament Agency (ACDA), 373n17
Arts, 284
Arts and humanities, encouragement of as U.S. national goal, 298, 300; U.S. appropriations for, 382n15
Asai, Tsuneo, 354nn15, 19
Ashbrook, Arthur G., Jr., 357n13
Assistant to the President for National Security Affairs, and future international system, 373n18
Atlantic Ocean, potential resources of, and Western Europe, 249
Atomic Energy Commission (U.S.), 202, 373n12; nuclear vs. other energy programs, 223
Attwood, William, 367n22
Australia, 373n16
Autocratic regimes, and postindustrial society, 267, 268
Automotive industry, U.S., 43

386 Index

Bacot, Eugene, 350n20
Baier, Kurt, 338n52
Bailey, Norman A., 356n10
Baikal-Amur Mainline (BAM), 72; and strategic significance, 346n48
Balance of power, 167; decline of importance, 153
Baldwin, M. S., 338n47
Baltic Sea, convention to reduce pollution of, 362n22
Barfield, Claude E., 340n13
Barnett, A. Doak, 357n18
Basiuk, Victor, 340n8, 360n27, 365nn6, 7, 11, 366n16, 372n12, 373nn15, 21
Bauer, Raymond A., 371n27
Baylis, Thomas A., 348n71
Behrens, William W., III, 376n39
Bell, Daniel, 259, 301, 314, 374nn2-15, 375n27, 376n36, 381n2; on postindustrial society, defined, 260; university as central institution, 260, 271; theory vs. practical application of the concept, 261-62; governmental decision-making and struggle among interests, 262; and culture, 262; politicization of decision making, 262; communal ethic, 262-63; theoretical knowledge as base of power, 268-69; business corporation, 271
Benoit, Emile, 371n27
Bergman, Walter R., 334n14
Bering Strait and climatic amelioration, 69; political implications of, 244
Berman, Jack N., 362n28
Bernheisel, J. Frank, 376n42
Berry, M. J., 343n5
Bessemer process, 333n9
Biological clock, 30, 337n40
Biology, 29; and evolution of mankind, 384n38. *See also* Cloning
Bipolarity, 13; military aspect, 151; and power, 168; decline of, 168; in scenario, 171; and Soviet rise in power (scenario), 174
Bischoff, J. L., 363n34
Bjorksten, Johan, 337n44, 376n45
Bolotin, B., 347nn59, 60

Bombers, strategic, 156
Boretsky, Michael, 340n12
Borisov, P. M., 372n8
Boulding, Kenneth E., 34, 338n52, 379n57, 382n23
Braybrook, David, 364n3
Brazil: economic growth, 140; and rise to power, 144; and energy, 144; and NPT, 158; nuclear capability, 158; economic growth, compared with Japan, 356n5; oil and gas production (projections), 356n10; and agreement with West Germany on nuclear technology, 359n14
Breeder reactors: and U.S. energy R&D, 49; and USSR, 68
Brennan, D. G., 358n6
Brezhnev, L. I., 69, 156, 344n18
Brezhnev Doctrine: in scenario, 173
Brierly, J. L., 360n4
Britain, 25, 97, 100, 333n9; and steam propulsion, 6; and global projection of power, 24; and North Star project (Siberia), 72; science and technology policy, 94, 351n31; capability in science and technology, 94; and leisure, 147; North Sea oil and gas, 102, 104 (scenario), 352n38; in scenario, 107, 108
Bronson, David W., 345n29, 31
Brooks, David B., 363n33
Brooks, Harvey, 349n7, 365n5, 367n28, 379nn59, 62
Brown, Dail W., 362n23
Brown, Harold, 358n6
Brown, Lester R., 356n7, 376n43
Brown, William, 374n16
Bruce-Briggs, B., 374n16, 375n19, 376n37
Brzezinski, Zbigniew, 355n43, 360n23, 367n26, 372n3, 373n13, 374n1
Buehrig, Edward H., 331n1, 365n4, 373n19
Bundy, McGeorge, on "acceptable damage" as viewed by political leaders vs. strategic thinkers, 358n10
Bureau of International Scientific and

Index   387

Technological Affairs (SCI), see State, U.S. Department of
Bureau of Oceans and International Environmental and Scientific Affairs (OES), see State, U.S. Department of
Bureaucracy: in scenario, 273, 274; and productivity, 376n38
Bush, Keith, 347n52
Business: values of and control of technology, 209, 367n24; changing values of, 227; and government (U.S.), 228-30; and power in postindustrial society, 271-73; values and public responsibility, 312. See also Multinational corporations; Public-service-oriented corporations; Values
Bylinski, Gene, 338n44, 356n7

Calder, Nigel, 333n10
Cambel, Ali B., 369n8
Canada, and nuclear proliferation, 158
Cancer: effort to eradicate (U.S.), 30; research appropriations for (U.S.), 337n41; resolution of and life expectancy, 337n42
Candlin, A. H. S., 346n42
Cardiovascular diseases: effort to eradicate (U.S.), 30; research appropriations for (U.S.), 337n41; resolution of and life expectancy, 337n42; in scenario, 376n44
Careers: changes in, 228
Carey, William D., 230, 371n30
Carter, Luther J., 341n20
Caspian Sea, 69
Central Intelligence Agency (CIA), 76
CEQ, see Council on Environmental Quality
Chemical industry, U.S., 42; compared with Western Europe and Japan, 339n3
Chemistry, and resources, 17
Chevignard, D., 345n28
China, 15, 24, 81, 84, 357n18; and strategic nuclear weapons, 135; prospects for rise in power, 145-46; and oil, 145, 357n14; and food production, 145, 357n16; population growth, 145; imports of Western technology, 145-46, 357n16; and Japan, 146, 316; and nuclear proliferation, 158; "convergence" and international stability, 162; in scenario, 171; and United Nations, 253; and international system, 253; as world civilization (prospect), 316-17
Chung, William K., 362n26
Churchill, Winston, 314
Civil Aeronautics Board (U.S.), 229, 370n24
Civil rights movement: and technological impact, 231-32
Civilization, high, 296; grass-roots support for in U.S., 297; as a U.S. goal, 306-7; implementation of in U.S., 310-14; and balance between pluralism and centralization, 310-11; grass-roots support for and leadership, 312; and culture, 329
Civilizations, 294; rise and fall of, and U.S., 326. See also World civilization under China; Japan; Soviet Union; United States; Western Europe
Clark, Peter O., 333n8
Claude, Inis L., 361nn12, 16, 18
Climate and weather modification, see Weather and climate modification
Cloning, 30-31; prospects for, 31
Closed-materials economy, 276, 277; narrow tolerances of, 380n71
Coal–iron ore complexes, 17
Coffey, Joseph I., 359n13
Cohen, Richard E., 370n25
Cohn, Stanley H., 76, 347n62
Cold War, 177, 253, 320
COMECON (Council of Mutual Economic Assistance), 177; in scenario, 170, 172
COMEX, 21
Comfort, Alexander, 337n44, 376n45
Commerce, U.S. Department of, 182; and technology exports, 239; as "focal point" for policy development in industrial R&D, 340n15

388   Index

Commission on National Goals (1960), 303
Committee on Science and Technology, U.S. House of Representatives, 52-53
Committee on the Challenges of Modern Society (CCMS), NATO, 52
Common Cause, 370n20
Communal ethic: in postindustrial society, 262; as a national goal, 308; and virtuous society, 308
Communications, 25-26. See also Global TV network; Millimeter waveguide; Optical fibers; Satellite-based TV; Telecommunications
Communications Satellite Corporation (COMSAT), 29
Communism, 79, 237, 344n18
Communist bloc: pluralization of, 237
Compagnie Européenne Diffusion Gazeuse (Eurodif), 97; in scenario, 107
Competition: in Adam Smith's philosophy, 294; and ethics, 296-97; role in U.S. socioeconomic system, 296-97
Comprehensive Research and Development Corporation (Japan), 122
Computer industry (U.S.), 45
Computers, 15, 26, 29, 181, 210; as a public utility, 22; technological trends in, 22; and USSR, 66, 69-70; bit (defined), 335n19; U.S. share in Free World, 339n2
Computer technology: and priorities for R&D (U.S.), 224
Concert of Europe (1815), 186
Concorde (SST), 97; compared with Boeing 747, 351n26
Congress, U.S., 214; and decision-making, 200
Congress of Vienna (1815), 167
Consumers, protection of, 298, 299-300
Containment, military: overextension of resources, 323. See also U.S. foreign policy
Control of technology and of its impact, 53, 328, 368n2; defined, 3-4, 196; and technology as a dependent variable, 5; and national strategy, 195-97; and selectivity in R&D support, 221; and composition of leadership in science and technology policy, 221
Convention on the Continental Shelf (1958), 235; and technology as an independent variable, 165-66
Convention on the Prevention of Marine Pollution by Dumping Wastes ("Ocean Dumping Convention"), 52, 362n22
"Convergence" (U.S.-USSR): and international stability, 162
Cook, Paul K., 346n41
Coplin, William D., 360nn6, 28
Cost-benefit analysis, 41
Cost-effectiveness analysis, 208; utility of, 41; limitations of, 365n10
Council of Economic Advisers (CEA), 21
Council on Environmental Quality (CEQ), 29, 51; and Science Adviser to the President, 205
"Counterculture," 309
Cox, Robert W., 373n19, 384n37
Craven, John P., 369n12
Crime, 234, 298, 299; in semistationary society, 287; and human relations, 309
Croan, Melvin, 348n71
Croly, Herbert, 380n1
Cruise missile, 156
Cuba, 237
Cuban missile crisis, 76
Culture: and postindustrial society, 262; trends in and elevation of man, 313, 329; responsibility for, 313; and high civilization, 329

Dales, Sophie R., 342n38
Daly, Herman E., 376n40; 379n57
Davies, R. W., 343n5
Deane, John R., 342n2
Decision-making, 3, 58; "rational"

method (described), 197-98, 364n3; "constituency" method (described), 198-99, 364n3; constituency method and balance of power process, 198; constituency method and rapid technological change, 199; and executive agencies, 199-200; U.S. national, characteristics of, 200; new balance between "rational" and "constituency" approaches, 210-14; rational, dependence on forecasting and long-range planning, 222; constituency method and new technology, 222; constituency approach and effect on priorities in science and technology, 222-23; "incrementalist" model, 364n3; constituency and incrementalist models compared, 365n3
Decision-making process, international: 179; managerial vs. legal basis, 361n19
Deep ocean technology: as a high priority, 218
Defense, U.S. Department of, 24, 212, 213, 373n17; and military science and technology, 200-202; interest groups in, 365n8
Degens, Egon T., 363n34
Democracy, and postindustrial society, 266-67
Détente, 236-41; and Soviet imports of technology, 64; and Soviet military power, 243-44
Deterrence, 179, 189, 237; nonmilitary elements in, 240-41; economic and technological factors in, 369n5
Deutsch, Karl W., 375n32
Development assistance, 359n19. See also Agency for International Development; Foreign aid
Development Assistance Committee (DAC), OECD, 359n19
Dewhurst, J. Frederick, 349n3
Diebold, John, 349n6, 369n13
Diplomats: values and control of technology, 208, 366n22

Direct Broadcast Satellites (DBS), 333n6
Director, Defense Research and Engineering (DDR&E), Office of, 201
Director, Office of Science and Technology Policy (OSTP), 206-7; and international security, 206-7
Diseases, elimination of: and effect on economy, 30, 337n43
Djerassi, Carl, 357n17
Dobrov, G. M., 344n19
Donnelly, Warren H., 351n27
Dougherty, James E., 331n2
Dror, Yehezkel, 364n3
Drucker, Peter F., 37, 148, 339n61, 358n24, 367n25, 370n26
Drug addiction, 309
Dunn, Frederick S., 360n5

Earthwatch, 180-81
Eastern Europe: pluralization of (scenario), 170; and cooperation in science and technology, 248; societal pluralization, 323
Eastlund, Bernard, J., 279, 376n40, 378n54
Easton, David, 375n34
EC, see European Community
Eckstein, Harry, 331n44
Ecology, global: 189; and international regulation, 178; future maintenance of, 319, 320, 329. See also Environment; Environmental protection; U.N. Environmental Program; Waste heat
Economics, 270
Economy: "linear" (defined), 275, 277; "closed-materials" (defined), 276, 277; closed-materials, 280; conversion from linear to closed-materials, 280-81
Economy, U.S.: orchestration of technological composition of, 230
Education: as national goal, 298, 299
Edwards, Imogene U., 347n56
Egypt: and nuclear weapons, 136, 158; as candidate for international power, 144-45; and "Arab Marshall Plan," 357n12

Electro-optical weapons, 13-14, 137, 332n8; and international stability, 159-60; and stability on strategic nuclear level, 160; as high priority area for R&D, 218. *See also* Laser weapons
Emmett, John L., 336n33
Energy: and national power, 15; advanced technology and costs of, 26; potential reduction of costs, 31-32; as a belated priority (U.S. case), 221-23. *See also* Breeder reactors; Fusion energy; Magnetohydrodynamic (MHD) power; Oil; Solar energy; Superconductivity; and energy-related entries under individual nations or regions
Energy crisis: effect on fundamentals of U.S. science and technology policy, 48-49; potential benefits to U.S., 50-51; and Western Europe, 95; in scenarios, 103-6; 107-8, 172; and Japan 114; and USSR, 317
Energy Research and Development Administration (ERDA), 27-28, 212, 373n12; reorientation of R&D policy, 49; as a constituency for energy, 223; and cooperation in R&D with International Energy Agency, 249
Engineers ("technologists"): values of and control of technology, 208
Environment: and USSR, 73; and LDCs, 142; future global problems, 188; vs. energy in constituency decision-making, 223; and IIASA, 240; in scenario, 278
Environmental protection: cost of (U.S.), 29; and technological progress, 56. *See also* Ecology; Environment; Recycling
Environmental Protection Agency (EPA), 51, 212
Environment Committee of OECD, 52
Epstein, William, 359n12
ERDA, *see* Energy Research and Development Administration
Ethics: and competition, 296-97
Etzioni, Amitai, 364n3

European Atomic Energy Community (Euratom), 97; in scenario, 107, 350n17
European Center for Nuclear Research (CERN), 96; in scenario, 107
European Coal and Steel Community (ECSC), 350n17
European Communication Satellite (ECS) system, 101
European Communities (EC), 350n17. *See also* European Community
European Community: foreign aid of, 25; and energy, 95; and the environment, 95-96; science and technology policy, 101; Committee for Scientific and Technological Research (CREST), 101; and MNCs, 183
European Economic Community (EEC), 350n17
European Launcher Development Organization (ELDO), 100
European Nuclear Authority (ENA): in scenario, 107, 108
European Science Foundation (ESF), 101
European Space Agency (ESA), 100-101; cooperation with NASA on Spacelab, 101; in scenario, 107
European Space Research Organization (ESRO), 100
Evolution, human, 329; biological and cultural aspects of, 307; and U.S., 326; and culture, 329; and governing of societies, 329; and advance in knowledge of biological mechanisms, 384n38
Ewing, David W., 367n24
Export Administration Act of 1969 (U.S.), 239

Fainsod, Merle, 348n70, 370n24
Faltermayer, Edmund, 338n56
"Favorable determinism," 302; defined, 6; and U.S., 44
Federal Coordinating Council for Sci-

ence, Engineering, and Technology (FCCSET), 366n17
Federal Council for Science and Technology (FCST), 366n13
Federal Maritime Commission (FMC), 370n24
Federal R&D contract: concept of (U.S.), 41
Fedorenko, R. P., 349n74
Feraru, Ann Thompson, 364n41
Ferkiss, Victor, 382n18
Feshbach, Murray, 345n26, 347nn57, 58
Findley, Paul, 339n57
Fisher, George, 343n13
Foch, René, 351n25
Ford, Gerald R., 156
Forecasting, 222, 230, 261
Foreign aid: Soviet compared with U.S., 335n26; U.S. and EC nations compared, 335n27. *See also* Development assistance
Foreign aid, U.S.: and molding international system through science and technology, 255
Foreign policy: and domestic policy, 195
Foster, William C., 358n6
Founding fathers, U.S.: and common man, 383n31
Fox, William T. R., 333n1, 361n17, 364n39
France, 25, 27, 45, 100, 373n14; and North Star project (Siberia), 72; science and technology policy, 90, 92-93, 351n31; and nuclear proliferation, 158
Franco-Prussian War (1870–1871), 168
Freedom, conditions for its enjoyment: and decision-making, 295, 381n7
Freedom, individual: as U.S. national purpose, 294, 295
Freedom vs. control: as an issue, 310
"Free rider" problem, 285; in semi-stationary society, 286, 287
French-American Mid-Ocean Undersea Study (FAMOUS), 373n14
Fujitsu-Hitachi Group, 121
Full employment, 298, 299

Functionalism, 176-78, 186; as independent variable, 176-77; and technology, 176, 361n13; as dependent variable, 177; and Western Europe, 177-78; prospects for, 320
Furash, Edward E., 371n27
Fusion energy, 26; magnetic confinement, economies of scale of, 27; magnetic confinement program (U.S.), costs of, 27; laser fusion program (U.S.), costs of, 28; fusion-fission hybrid reactor, 28; dual-purpose fusion reactor, 28, 336n34; and cost of energy, 32; and USSR, 68; and Western Europe, 91, 352n34; magnetic confinement fusion, described, 336n33; laser fusion, described, 336n33; hybrid reactor, advantages of, 336n34
Fusion torch, 279; economies of, 378n53
Future: as a criterion of U.S. national purpose, 305
"Future shock," 233
"Futurism," 42

Galbraith, John Kenneth, 381n9, 382nn17, 18; and societal goals, 301-2
Gallagher, Charles F., 355n34
Gallik, Dimitri M., 343n14
Gardner, Richard N., 361n15
Gas centrifuge technology, 337n36; development in Western Europe, 97; in scenario, 108
Geertz, Clifford, 307, 382n24
Geiger, Theodore, 351n30
General purpose forces, 11; and allocation of resources, 217-18. *See also* Subnuclear umbrella military technology
Gerasimov, I., 69, 344n20, 346n37, 347n49
Gergen, Kenneth J., 371n27
Germany, 45; and railroads, 6; and Thomas-Gilchrist process, 333n9
Germany, East: and Economic System of Socialism, 81; and New Economic System, 82

Germany, West: 54, 96, 97, 100, 110, 352n46, 359n22; and North Star project (Siberia), 72; and European regional cooperation, 93; science and technology policy of, 93, 351n31; in scenario, 105; nonmilitary technology and growth, 139; and market ideology, compared with U.S., 349n4
Gerontology, 30, 278, 379n61; in scenario, 326n44, 377n47; advance in and vulnerability to disease, 337n42; cross-linkage theory of aging, 337n44; advance in and effect on human vigor and personal appearance, 337n44
Gillette, Robert, 341n18, 359n13
Gilpin, Robert, 341n16, 349nn5, 6, 350n11, 366nn14, 19, 22, 368n1
Global functional institutions, 178-81, 186
Global TV network, 20
Goldie, L. F. E., 361n19
Goldman, Marshall I., 345n27
Gordon, Lincoln, 370n24
Gough, William C., 279, 376n40, 378n54
Graham, Loren R., 343nn13, 15
Gravity-vacuum transit concept, described, 334n16
Great Britain, *see* Britain
Grechko, A. A., 77
Greenberg, Daniel S., 340n7, 350n14
Green Revolution, 141, 143, 187
Greensdale, Rush V., 343n12
Gregory, William H., 343n4
Griffiths, Franklyn, 348n65
"Group of 77" (LDCs), 162
Gushchenkov, Ye., 372n8

Haas, Ernst B., 335n23, 360n1, 361n13
Haas, Michael, 154, 358n2
Haldi, John, 357n20
Haldi Associates, Inc., 357n20
Halstead, Thomas A., 359n11
Hamilton, Alexander, 229
Hamrin, Robert, 342n39
Handler, Bruce, 356n5

Hardt, John P., 343nn13, 14, 344n21, 346nn44, 47, 347n63, 348n65
Harris, Louis, 383n30
Harris, Louis, and Associates, Inc., on alienation and drop in confidence in U.S. institutions, 303
Harte, John, 376n40, 379n57
Haskins, Caryl P., 351n24, 383n25
Hayes, Robert H., 351n28
Hayflick, Leonard, 337n40
Hazard, John N., 361n9
Heilbroner, Robert L., 381n6
Hein, R. A., 334n16
Hemy, G. W., 343n10
Hiestand, Dale L., 370n22
Hirasawa, Kazushige, 355n38
Hirsch, Robert L., 336n31
Holland, Wade B., 346n39
Holliday, George D., 344n21
Holmfeld, John D., 346n36
Hong Kong, 19
Horisaka, Kotaro, 353n12
Hough, Jerry F., 344n23, 348n71
Housing business, U.S., technological shortcomings of, 225-27
Housing and Urban Development, U.S. Department of, 226-27
Howard, Michael, 359n17
Howe, Richard, 352n37
Hsiao, Gene T., 357n19
Huddle, Franklin P., 365n11, 369n9
Humanities, 284
Human relations: in advanced societies, 284-85
Hungary, 79
Hunsberger, Warren S., 353n5
Hunter, Holland, 343n9
Huntington, Samuel P., 360n23
Hydrofoils, 137, 369n4

IBM Japan, Ltd., 121
Ike, Nobutaka, 355n32
Iklé, Fred Charles, 358n9
India: and nuclear proliferation, 158; and satellite community TV network, 333n5

Index   393

Indochina: and decision-making, 234
Indonesia, 373n16
Inglehart, Ronald, 37
Instability, international: 58, 156; and military power, 150; and technological change, 154, 155, 165-66; and relationship to internal stability, 154, 161, 163-64; and change in power, 155; and nuclear proliferation, 157-59; and transnational violence, 161; and marine resources, 165; and international law, 165-66. See also Stability, international
Instability, social: 34, 58, 232, 234, 290; and potential increase in human longevity, 36-38, 148; and Western Europe (scenario), 107; and technological impact, 147; and LDCs, 163-64; and autocratic regimes, 274; and post-industrial society, 274; and intergenerational conflicts, 282; in LACs and LDCs, 289
Institute of Mathematical Statistics (Japan), 126
Intelsat, 25
Intercontinental Ballistic Missile (ICBM), 201
Interdependence, 23-24, 151, 154, 195; and USSR, 84; and Western Europe, 95-96; among semistationary societies, 288; and U.S. national purpose, 305
Intergenerational changes, 36-38; in Western Europe, 37; in Japan, 126
Interim Agreement on Limitations of Strategic Offensive Arms (1972), 156
Intermodality, 137
International Atomic Energy Agency (IAEA), 179
International Business Machines (IBM) Corp., 98
International cooperation, and international political process, 187-88
International cooperation in science and technology, requirements for, 29; U.S.-USSR, 238-40; U.S.-Western European, 247-49; U.S.-Japanese, 252

International Energy Agency (IEA), 49, 95; and oil companies, (scenario), 104; in scenario, 104, 105, 107, 108; and large R&D programs, 249; and fluidized combustion coal facility, 372n11
International Institute of Applied Systems Analysis (IIASA), 362n21; and U.S.-USSR cooperation, 239-40
International law: conventional vs. customary, and technology, 165; and technological change, 165-66; "socialization" of international society and technological change, 166; as instrument of power, 168-69; "socialist," 171; "socialist" (in scenario), 174
International power, see Power, international
International relations: as evolving field, 1-2; and theory of, 2; and steering of society, 314; and world politics, 331n1; and political science, 331n4. See also World politics
International Seabed Authority (ISA), 184
International system, 236, 240; and power, 167-69; and MNCs, 182, 183; and peace, 187; and satisfaction of human wants, 187; and planet's ecological system, 187; directions of evolution, 186-92; pluralization and diffusion of power in, 190-91; U.S. universities and comprehensive conceptual framework for, 253; U.S. government organization for, 254; Adviser to U.S. President for Future International System (proposed), 254; and constituency for, 254-55; and allocation of resources for forming of, 255, 373-74n22; and societal change, 327; dialectical characteristics of change in, 327-28; future goals of, 328-29; and world order, usage of the terms, 371n2; and U.S. organizations concerned with, 373n17; conflict and change, 384n37. See also World order

Interstate Commerce Commission (ICC), 229, 370n24
Israel, 12; and nuclear weapons, 136, 158
Italy: science and technology policy of, 94; in scenario, 104, 106; organization for science and technology policy, 350n15

Jacquency, Theodore, 370n20
Janowitz, Morris, 367n23
Japan, 4, 45, 54, 110, 293, 352n46, 359n22; steel industry, 17-18; as U.S. competitor, 48, 90; in scenario, 104, 108; GNP compared with U.S., 114; energy crisis, impact of, 114, 129; conditions favorable for economic growth, 114-16; quality controls, 115; and cheap labor, 115; defense expenditures as percentage of GNP, compared with NATO and U.S., 115; and marine transportation, 115-16; Economic Planning Agency, 116; relations between government and business, 116-17; Ministry of International Trade and Industry (MITI), 116-17; assimilation of Western technology, 117, 118-19; labor force, 117-18; rate of savings, 118; literacy rate, 118; trading companies, 118; employee loyalty, 119, 126; Council for Science and Technology (STC), 119; Science and Technology Agency (STA), 119, 120, 354n15; Science Council of Japan, 119; organization for science and technology policy, 119-20; science and technology policy of, 119-20; shipbuilding, 120; chemical industry, 120; photographic equipment industry, 120-21; computer industry, 121, 354n18; "big science" programs, 120, 121-22; allocation of resources to science and technology, 121; nuclear program, 121; outer space program, 121-22; imports of technology, 122; and offshore oil resources, 122, 354n24; compartmentalization in R&D, 122, 124; university system, 122; "think tanks," 122; promise of future technology for, 122-23; science and technology policy, requirements of, 123-24; R&D, allocation of resources to, 124; and infrastructure, 125; social security, 125; foreign aid, 125, 128-29; change in values, 126-28; and allocation of resources to defense, 125; and women, 126-27; and consensual decision-making, 127, 131, 316; pluralization of society, 127; and life expectancy, 127; and national strategy, 128-30; and Southeast Asia, 128, 131; and China, 129, 130, 131, 132, 146, 251, 316; foreign investment (projections), 129; and USSR, 129, 131, 251; and world order, 130, 132; cultural characteristics and foreign affairs, 130-31, 355n43; and nonmilitary technology and economic growth, 139; and leisure, 143; and differential impact of technology, 149-50; and shaping West European future, 249; and future international system, 250-51; a stake in future stability of Western Europe, 251; technological and economic cooperation with USSR and China, objectives of, 251; and postindustrial society, 268; as world civilization (prospect), 315-16; technology and rise to power before World War II, 333n9; and robots, 354n26; confrontation politics, 355n36; technology imports vs. exports, 355n40, and moratorium on whaling, 355n42; and restrictions on investment abroad, 376n44. *See also* Pluralizing effect of technology; Nixon, Richard M.
Japan Desert Development Institute, 20
Japan Economic Research Center: projections of Japan's growth, 353n1
Japan External Trade Organization (JETRO), 118
"Japan, Inc.," 117
Jarrett, Henry, 379n57

Index 395

Jensen, Clayton E., 362n23
Jews: and U.S. Soviet policy, 242
Johnson, D. Gale, 347n50
Joint European Torus (JET) project, 352n34
Jotischky, Laszlo, 348n69
Justice: in virtuous society, 308, 383n27
Juviler, Peter H., 343n15

Kahn, Herman, 259, 301, 355n45, 359n16, 374-75nn16-25, 376n37; on post-industrial society: open ended, 263; affluence in, 263; and hedonism, 264; not a learning or purposeful society, 264; synthesis of values and interests vs. deadlock, 264-65; "Westernistic" society, 264; and multinational corporations, 271-72
Kalachek, Edward D., 369n14, 370n26
Kanamori, Hisao, 353n1, 354n29, 355n39
Kaplan, Max, 357n21
Kaplan, Morton A., on stability of bipolar and multipolar systems, 258n1
Katz, Milton, 342n36
Katz, Zev, 348n73
Kennan, George, 372n3
Keohane, Robert O., 361n17
Khachaturov, T. S., 345nn24, 26, 34
Khrushchev, N. S., 62; and diversion of resources from military sector, 348n66
Kindleberger, Charles P., 340n6
Kintner, Edwin E., 336n32
Kirkpatrick, Evron M., 271, 375n33
Kissinger, Henry A., 332n5, 359n15
Komiya, Ryutaro, 353nn2, 7
Korean War, 61, 115
Kosygin, A. N.: on science and economic competition between Communism and "capitalism," 344n18; on all-European power grid, 361n10
Kozhevnikov, F. I., 361n9
Kraar, Louis, 354n28, 356n11
Kristol, Irving, 381nn5, 8, 383-84nn31, 32

LACs, see Less advanced countries
Ladd, Everett Carll, 37, 339n60, 366n21

Laissez-faire, 208, 209, 295, 297
Land Satellites, *see* Landsats
Landsats (Land Satellites), 20-21, 334n9, 362n24; and LDCs, 143; and U.N., 181
Lasers, 22, 91
Laser weapons, 14, 332n8; and long-range guidance, 332n8
Lasswell, Harold D., 331n1, 372n3, 373n19, 379n63
Latin America, 19, 144; in scenario, 105
Latvians, 348n73
Law of the Sea Conferences, 183
Lawrie, James A., 334n14
Layton, Christopher, 349n6, 350n16, 352n44
LDCs, *see* Less developed countries
Leadership, 314; importance of, 311; and U.S. President, 311-12; emanating from society, 312; U.S. and the world, 324; and U.S. public, 383n30
League of Nations, 186
Lebanon: U.S. military landing in, 13
Lebedev, V., 345n24
Lee, J. Richard, 343n6, 343n7, 346n44
Leisure, 357n21; and national power, 147; as a value, 233; and semistationary society, 284; in advanced societies, 285; projections of, 357n20
Leitz, Ernst, GMBH, 121
Lend-lease: and USSR, 60, 342n3
Lenin, V. I., 375n29
Lepawsky, Albert, 331n1, 373n19
Lerner, Max, 381n4, 384n35
Less advanced countries (LACs), 317, 328; defined, 288
Less developed countries (LDCs), 13, 147, 164, 165, 181, 317, 373n21; ability to mold their future through technology, 18-19; and international power, 139-144; advanced nations and economic gap, 139-40; and economic growth, 139-40; growth compared with advanced nations, 140; and international regime of the seabed, 141, 184; and narrowing power gap with ad-

Less developed countries (*Cont.*)
vanced nations, 142-43; and differential effect of technology, 143; "Group of 77," 162; and internal vs. international instability, 164; and "Socialist" system of international law (scenario), 174; and Earthwatch, 180; and civilization, 324. *See also* Foreign aid; Green Revolution; Power, international; and entries of individual LDCs
Liberman, Ye. G., 85
Life expectancy, 29-30; and eradication of cardiovascular diseases and cancer, 30; projections, 30, 337-38n44; potential increase of and effect on society, 36-38; potential impact on social trends, 36-38; breakthroughs in (scenario), 278; assumptions, in scenario, 282, 376n45, 377n47. *See also* Gerontology; Longevity, human
Life magazine, on national purpose, 303
Lifespan, human: *see* Gerontology; Life expectancy; Longevity, human
Lijphart, Arend, 331n3
The Limits to Growth (book), 275
Lindblom, Charles E., 364n3
Linder, Staffan B., 285, 297, 380n68, 381n10
Lipset, Seymour Martin, 37, 339n60, 366n21
Lissitzyn, Oliver J., 361n9, 373n20
Long, T. Dixon, 353n14
Longevity, human, and social instability, 148, 163; and LDCs and LACs, 289
Long-range logistic capability, and USSR, 136
Long-range planning, and USSR, 68
Loving, Rush, Jr., 339n5
Loyter, M., 347n53
Luxembourg, 96
Lyashchenko, P. I., 342n2

McArthur, John H., 371n28
McDonald, Gordon J. F., 333n10
McElroy, William D., 341n28
McKean, Roland N., 379n66, 380n67

Mackintosh, Malcolm, 347n64
Maddison, Angus, 349n2
Maene, J. Rey, 355n36
Magnetic levitation systems (in transportation), 334n16
Magnetohydrodynamic (MHD) power, 26; economies of scale of, 27; described, 32, 338n50; and cost of energy, 32; and USSR, 68
Magruder, William M., 47
"Magruder exercise," 205
Main, Jeremy, 338n51
Malaysia, 355n37
Malone, Thomas F., 336n30
Manganese nodules, economic potential of, 184
Manheim, F. T., 363n34
Mannheim, Karl, 36-37, 339n58, 368n30
Manyushis, I., 344n19
Marine resources, development of: 15, 21, 26, 28, 183-86; and international stability, 165; and international regulation, 178; and priorities for R&D, 224; advantages of U.S.-Japanese cooperation on, 252. *See also* Maganese nodules; Mid-Atlantic Ridge deposits; Oil; Red Sea hot brine deposits
Marshall Plan, 89, 90, 117
Martel, Leon, 374n16
Marubeni Corporation, 118
Marx, Karl, 375n29
Marxism, 317
Marzouk, M. S., 356n5
Material rewards: and semistationary and stationary-state societies, 283-84; as a value, 370n64
Materials, 369n9; and priorities for R&D, 223-24
Matsumoto, Yoshio, 354n30
Matthias, Berndt T., 338n48
Meadows, Dennis L., 376n39
Meadows, Donella H., 376n39
Meissner, Paul, 335n21
Melman, Seymour, 339n6
Merchant marine, 25; U.S., EC and USSR compared, 335n28

Meritocracy, and postindustrial society, 35, 260-61
Mero, John L., 363n33
Merriam, Charles, 361n17
Mesthene, Emmanuel G., 342n36, 357n22, 371n31
Metz, William D., 336n34, 350n20
MHD power, *see* Magnetohydrodynamic power
Mid-Atlantic Ridge deposits: economic potential of, 185; and U.S.-European cooperation, 249
Middle East, 13, 89, 95, 142, 251
Miley, George H., 378n53
Military establishment: and attitudes of intellectual and cultural elites, 20th and 21st centuries compared, 380n70
Military officers: values of and control of technology, 209
Military power, *see* Power, military
Military technology: stalemate in, 11-15; compared with nonmilitary technology with respect to international stability, 166; and priorities in R&D, 216-19. *See also* Electro-optical weapons; General purpose forces; Nuclear weapons, strategic; Nuclear weapons, tactical; Nuclear umbrella; Subnuclear umbrella military technology
Mill, John Stuart, 379n57
Millimeter waveguide, 22; described, 334n17
Minolta Camera Corp., 121
Mirabito, John A., 362n23
MIRV, *see* Multiple independently targeted reentry vehicles
MITI, *see* Japan, Ministry of International Trade and Industry
Mitrany, David, 176, 361n11, 361n13, 361n14
Mitsubishi Corporation, 118
Mitsui and Co., 118
MNCs, *see* Multinational corporations
Modelski, George, 362n27
Mole, C. J., 338n47
Monroe Doctrine, 173

Morton, Henry W., 344n15
Morton, Rogers, 372n5
Multinational corporations (MNCs), 25, 181-83, 252, 362n28; Soviet, 66, 85; American, and Western Europe, 101; and LDCs, 182; in postindustrial society, 271–72; as actors in global arena, 320; regulation of, 363n30; and U.S. foreign policy, 373n21
Multiple independently targeted reentry vehicles (MIRV), 155-56; described, 358n5
Mutsu, 121
Mutual and Balanced Force Reductions (MBFR), 77

Nagy, Tamas, 348n69
Nakamura, Takafusa, 355n39
Nakane, Chie, 131, 132, 353n9, 355nn41, 43
Nasmith, Augustus, 372n6
National Aeronautics and Space Administration (NASA), 23, 100, 101, 213; and Japan, 122, 354n20
National Association of Building Manufacturers, 226
National Bureau of Standards (NBS), Experimental Technology Incentives Program, 340n15
National goals, 57; defined, 294
National Goals Research Staff, Executive Office of the President, 300-301, 303, 368n31
National goals, U.S.: criteria for, 304-6; delineation and implementation of, 304-5; as variables, 312-13; questions about, 313. *See also* National purpose, U.S.
National Institute of Technology (proposed), 369n14; and identification of priorities for R&D, 224
Nationalism: changing form of, 98-99, 351n29
National Oceanic and Atmospheric Administration (NOAA), 51, 184

National purpose: and ability to move society, 293; functions of, 293-94; defined, 294. *See also* National goals

National purpose, U.S.: 1; extent of achievement, 294-98; crisis of, 303-4; delineation of as an intellectual process, 304; implementation of as intellectual and political process, 304-5; criteria for, 304-6; and requirements of the future, 305; proposed, 306-7; and poverty, 308-9; and discrimination, 308-9; as a variable, 312-13; and culture, 313; and international organization, 326

National Science and Technology Policy, Organization, and Priorities Act of 1976 (P.L. 94-282), 206, 366n17

National Science Foundation (NSF), 43, 101, 205-6, 212-13; Experimental R&D Incentives Program, 340n15; Research Applied to National Needs (RANN) Program, 340n15

National security: and nonmilitary technology, 16; in 21st century, 289-90, 380n70

National security, U.S.: as a force in U.S. technological growth, 45; vs. ideological preferences, 323

National Security Council (NSC), Executive Office of the President: Military Technical Consultant Mechanism (MTCM), establishment of, 44; and decision making in science and technology, 200; and Science Adviser to the President, 205; and science and technology, 206

National Space Development Agency (NASDA), Japanese: 121-22

National Space Transportation System ("Space Shuttle"), 248

National strategy: defined, 197

Nation-state: as an actor in world politics, 179; functions of and technological impact, 189-90; as a variable, 191

Nation-state system, and effect of science and technology, 189-90

National resources, decline in deterministic effect of, 137

Naval power, Soviet, 160; expansion of, 76, 136, 217; in scenario, 173

Naval power, U.S.: 136

Nelson, Richard R., 369nn6, 14, 370n26

Netherlands, 96, 97

New Technological Opportunities (NTO) Program, 47

New Zealand, 373n16

Nippon Electric Corporation, 121

Nixon, Richard M., 48; and the President's science and technology advisory machinery, 43; trip to China and effect on Japan, 129; and Japan, 130

Noda, Kazuo, 353n2

"No-growth" society: *see* Stationary-state society

Nongovernmental organizations (NGOs), 191, 364n41

Nonmilitary technology, 15-16; and national security, 16; and nuclear umbrella, 137; and power, 137-39; and international stability, 161-66; criteria for priorities, 202; priorities in, 219-25

Non-Proliferation Treaty (NPT), 158, 179, 253; and promotion of nuclear technology for peaceful uses, 158

Non-Proliferation Treaty Review Conference, 158, 359n12

North Sea oil and gas, 90; and Britain, 102, 104 (scenario); and Norway, 102, 352n39

NPT, *see* Non-Proliferation Treaty

Nuckols, John, 336n33

Nuclear missile program, U.S., impact on economy, 42

Nuclear proliferation, 157-59; and India, 359n14; and Argentina, 359n14; and Brazil, 359n14

Nuclear umbrella, 11-12; effect on change in international power, 12. *See also* Nuclear weapons, strategic

Nuclear weapons, strategic: 135, 238; and third states, 12, 157-59; and international stability, 155-57; and electro-

Index 399

optical guidance, 157; candidates for acquisition of, 159; and electro-optical technology, 160; and allocation of resources, 216-217
Nuclear weapons, tactical: 14; and electro-optical weapons, 160
Nye, Joseph S., 361n17

Occidental Petroleum Corporation, 344n22
Ocean Dumping Convention, see Convention on the Prevention of Marine Pollution by Dumping Wastes
Oceanology: and priorities for R&D, 224. See also Marine resources, development of
O'Connor, J. J., 343n8
OECD, see Organization for Economic Cooperation and Development
Office of Management and Budget (OMB), Executive Office of the President: and Science Adviser to the President, 205
Office of Science and Technology (OST), Executive Office of the President, 47, 204, 366n12; abolition of, 44
Office of Science and Technology Policy (OSTP), Executive Office of the President, 206-7
Office of Technology Assessment (OTA), U.S. Congress, 52
Office of Telecommunications Policy (OTP—Executive Office of the President), and Science Adviser to the President, 205
Offshore oil, depth of exploitation, 21
Oil, 143; and USSR, 72; and OPEC, 144; and Brazil, 144; and China, 145, 357n14; and deep ocean areas, 364n38
Oil industry, U.S., and alternative sources of energy, 222-23
Oligopoly, 266, 267
OPEC, see Organization of Petroleum Exporting Countries
OPEC nations, and international power, 144

Operating agencies in science and technology, U.S., need for rationalization of, 213-14
Operation Breakthrough (U.S. housing), 226
Optical fibers, 22; described, 335n17
Organization for Economic Cooperation and Development (OECD), 162; report on U.S. science and technology, 44; as example of functionalism, 178; and MNCs, 183
Organization of American States (OAS), and MNCs, 183
Organization of Petroleum Exporting Countries (OPEC): in scenario, 104-5, 108; and foreign aid to LDCs, 356n9
Osgood, Robert E., 332n5
OST, see Office of Science and Technology
OSUMI (Japanese satellite), 121
Outer space, 137; U.S. program in, impact on economy, 42; technology of, U.S.-USSR compared, 68; activity in and international regulation, 178
Outer Space Treaty, 165, 253

Page, Talbot, 377n51
Pakistan, and nuclear weapons, 158
Palamountain, Joseph C., 370n24
Pardo, Arvid, 363n31
Pearson Report, 359n19
Peck, Morton J., 369n14, 370n26
Pentagon, see Defense, U.S. Department of
Perkins, Dwight H., 357nn16, 18
Perry, Matthew C., Commodore, 117
Peterson, Peter G., 340n12, 345n25
Petrodollars, 50, 143, 255
Pfaltzgraff, Robert L., Jr., 331n2
Philippines, 355n37
Piret, Edgar L., 350n13
Planning for science and technology policy, 222; "substantive" vs. "instrumental" planning, 212; instrumental planning: methodological vs. administrative, 212; and communications with

Planning (*Continued*)
professional societies, universities, and industrial groups, 213; and OES, 213
Platig, E. Raymond, 331*n*2
Plato, 296; and high civilization, 306; and normative approach, 313-14
Platt, John R., 382*n*24
Pluralism, 44, 59, 211; vs. centralization in U.S. science and technology; and freedom, 310-11
Pluralizing effect of technology, 2-3; and Britain, 55; and Western Europe, 98-99, 351*n*29; and Japan, 127-28; and world society, 191, 321; on subnational level, 192; and "Socialist" camp, 272; and steering of society, 302; and U.S., 302, 310-11; and competing interests, 328
Polak, Fred, 382*n*23
Poland, in scenario, 108
Polaris (submarine force), 201
Policy planning for science and technology, *see* Planning for science and technology policy
Political science, 270-71; and "relevance," 271; and combination of empiricism with normative approach, 313; and steering of society, 313-14; and international relations, 331*n*4
Politics, domestic vs. international and technology, 331*n*4
Pollack, Herman, 366*n*18
Pollution, 275, 276, 279; cost of (U.S.), 29. *See also* Ecology; Environment; Waste heat
Pollution-intensive industries, and Japan, 277
Population growth, 376*n*46, 379*n*61; decline of in LDCs, 141; and international stability, 161; and stationary-state society, 276; in semistationary society, 276, 280, 281-282; and human longevity (scenario), 278; and current world projections for year 2000, 356*n*7
Port Authority of New York and New Jersey, 370*n*21

Positive-sum game, in world politics: *see* World politics
"Postachievement ethic," 303
"Post-Apollo" U.S. cooperative program in space with Western Europe, 100-101; *see also* National Space Transportation System
Postindustrial society, 259; and American youth, 34-35; defined (Daniel Bell), 260; deterministic aspects of, 265, 267; questions concerning, 265-66; effect on various forms of government, 266-68; loci of power in, 268-74; intellectual and scientific communities as locus of power, 268-269; theoretical vs. applied knowledge as related to power, 269; government as locus of power, 269; and multinational corporations, 271-72; and communal ethic, 273; and universities, 273; and instability, 274; ability to mold, 375*n*27. *See also* Bell, Daniel; Kahn, Herman
Poverty: eradication of, 232, 298, 299; changing concept of, 383*n*28
Powell, David E., 347*n*52
Powell, Raymond P., 343*n*11, 345*n*26
Power, international: new elements in, 55-56; and law of diminishing marginal returns, 142, 151; new candidates for, 144-46; and pressures for allocating resources to other pursuits in advanced societies, 146-47; change of importance of various factors in, 150, 151-52; decline of importance of, 151-52; and rise of political and economic elements in, 151; pluralization of, 152; "power over" vs. "power toward," 152, 153; pluralization of, and MNCs, 183
Power, military: decline of importance of, 56, 150-51; and semistationary society, 291; and geographic considerations, 368-69*n*4
Power, presidential (U.S.), 311, 383*n*29
Power transmission grids, global, 21, 288
President, U.S., 214, 373*n*18, 383*n*29; and decision-making, 199; and interna-

Index 401

tional system, 253-54; and effective control over the Federal Executive, 311; and leadership, 311-12; and academic community, 314
President's Committee on Science and Technology, functions of, 207
President's Science Advisory Committee (PSAC), 204, 365n12, 366n18, 367n28; abolition of, 44
Price, Don K., 339n1, 366nn19, 20, 367n25
Price, Harry B., 349n1
Priorities in science and technology, 216-25; and the military sector, 201-2, 216-18; convergence of military and civilian technologies as high-priority area, 218; nonmilitary technology, 219-25; and energy, 221-23; and lack of mechanism for determination of, 225
Private enterprise, U.S., 42, 227. See also Business; United States: government and business
Production associations, Soviet: 70, 83, 84; significance of, 67, 345n33
Productive pursuits, decline as a value, 233
Productivity, manufacturing and service sector compared, 379n62
Project Independence (U.S.), 45, 48; in scenario, 105
Protestant ethic, 148, 260, 262, 264
Public-service-oriented corporations, 272-73, 291-92, 312
Pueblo incident, 13

Racz, Barnabas, 348n69
Radice, Jonatan, 352n40
Radio broadcasting, 297
Radio Corporation of America (RCA), 45
R & D, *see* Research and development
Raiffa, Howard, 372n6
Railroads, and rise of U.S. and Russia, 333n9
Randers, Jorgen, 376n39

Rapawy, Stephen, 345n26, 347nn57, 58
Rationality and intellectual pursuits as a value, need for upgrading of, 312-13
Rawls, John, 308, 383n26
Real-time detection and surveillance, global, 137; as high-priority area, 218
Recycling, 377n49; of waste, and energy, 276; technology of, 279; comparative costs of, 378n52. See also Fusion torch
Redman, Christopher, 352n41
Red Sea hot brine sediments, 363n35; economic potential of, 184-85
Reed, Laurance, 353n49
Regulatory commissions, 229
Reorganization Act No. 1 of 1973, 205-6
Rescher, Nicholas, 388n52
Research and development (R & D), 57; U.S. and USSR compared, 68, 70; tax treatment of (U.S.), 220; and U.S., 220-21; in semistationary society, 281; USSR: industrial vs. research institutes, 345n31; and economic objectives, U.S. and France compared, 349n7
Resource location, decline in deterministic effect of, 17-19
Riccardi, Ferdinando, 352n47
Ringwald, George B., 355n33
Ritterband, Paul, 366n21
Roosevelt, Franklin D., 314
Roosevelt, Theodore, 295
Rosecrance, Richard N., 358nn3, 4, 359n21, 360n24
Rosenau, James N., 360nn6, 28
Ross, David A., 363nn34, 36
Ross-Skinner, Jean, 345n27
Rostow, Walt W., 360n23, 372n3
Rothstein, Robert L., 358n6, 367n29, 368n30, 375n30
Roush, J. E., 371n27
Rumsfeld, Donald H., 358n8
Rumyantsev, A., 345n24
Russia: and railroads, 6. See also Soviet Union
Russians: as prospective minority in USSR, 317

Salisbury, Harrison E., 346n43
Sanitary landfills, 376n4
Satellite-based TV, 19-20
Saturated diving, 21
Sauvant, Karl P., 362n25
Sawyer, James W., 377n51
Scenarios, future: Western Europe and world, 103-13; world and decline of Soviet power, 170-71; world and rise of Soviet power, 172-74; emergence of semistationary society, 277-78; increase in life expectancy, 278; increase of life expectancy in U.S., assumptions of, 376n45; increase of life expectancy and population growth in U.S., assumptions of, 377n47
Schilling, Warner R., 332n4, 366n22
Schneider, Ronald M., 356n10
Schneider, Stephen H., 364n40, 379n58
Schrag, Peter, 382n22
Schroeder, Gertrude E., 346n40
Schultz, Helmut W., 376n42
Science: and Britain, 94; and postindustrial society, 260, 269; international cooperation in, 362n20. *See also* Technology
Science Adviser to the President (Special Assistant to the President for Science and Technology), 204; in National Science Foundation, 206. *See also* Director, Office of Science and Technology Policy (OSTP)
Science and technology: as factor in non-zero-sum international political process, 187-89. *See also* Technology; World politics
Science and Technology Council (proposed), 368n32; rationale for, 211; functioning of, 211-13; and priorities in R&D, 224
Science and technology policy: differentiated from "technological strategy" and "control of technology and of its impact," 368n2; in U.S., 219-20. *See also* this entry under European Community; Japan; Soviet Union

Science and technology, excellence in as professed U.S. goal, 219
"Science of science," in Soviet Union, 64
Scientific and technological community, U.S., as a constituency, 204-5
Scientific and technological cooperation, international, demand for, 29, 139
Scientists, 367n28; values of, and control of technology, 208; and power, 270
Scott, Andrew M., 366n22
Scott, Bruce R., 371n28
Seabed, internationalization of, 183-84, 363n31; political impact, 184-86; and oil, 185. *See also* Marine resources
Seaborg, Glenn T., 288, 334n15, 342n28
Seamans, Robert C., 372n11
Self-interest: role in U.S. society, 296; utility to society, 308
Seligman, Daniel, 338n51, 338n55
Semistationary society, 275-80; definition, 276, 378n56; sociopolitical implications of, 280-87; and economic growth, 280-81; technology as factor in economic growth of, 282; virgin raw materials as factor in economic growth of, 282; and rate of economic growth, 283; and system of rewards and incentives, 283-84; and social status, 284; and career security, 284; and social stability, 286; and leisure, 286; and human relations, 286-87; reorientations of values in, and human relations, 287; international implications of, 287-92; and international system, 327; conceptual differences from stationary-state society, 378n57
Senate, U.S.: 206
Senkaku Islands: and offshore oil deposits, 122
Servan-Schreiber, Jean-Jacques, 102, 349n6
Shabad, Theodore, 342n7, 343n8, 346n43
Sheldon, Charles S., II, 346n36
Sheren, Andrew, 347n63
Shimizu, Masami, 354n16

Shin Nippon Steel Co., 18
Shipbuilding industry, U.S., 43
Shiraishi, Fumiaki, 355*n*33
Shonfield, Andrew, 370*n*23
Shulman, Marshall D., 360*n*3, 372*n*7
Siberia: and exploitation of resources of, 71-72
Siegal, Jacob S., 337*n*42
Siemens AG, 27
Simonet, Henri, 350*n*21
Skilling, H. Gordon, 348*n*65
Skolimowski, Henryk, 331*n*6
Skolnick, Jerome, 357*n*23
Skolnik, Alfred M., 342*n*38
Skolnikoff, Eugene B., 366*n*19; 367*n*28
Smead, Elmer E., 370*n*21
Smith, Adam, 294-95, 296, 381*n*6; and competition, 294-95; and self-interest, 295; and social goals, 306
Smith, Frank A., 377*n*49
Smith, Holand, 350*n*16
Smolinski, Leon, 345*n*33
Snowden, Donald P., 338*n*49
Societal growth: and material growth, 329
Sociopolitical discontinuities: and technological impact, 33-38, 231-34. See also Semistationary society
Socolow, Robert H., 376*n*40, 379*n*57
Solar energy: and ERDA, 49; and Western Europe, 91; and thermal pollution, 376*n*40
Solid waste: as a problem, 276; reuse of, 276, 278-79; and USSR, 289; and Japan, 376*n*41; cost of disposal, 376*n*41; estimates of (U.S.), 377*n*48; trends in reduction, (U.S.), 377*n*50. See also Fusion torch; Recycling; Sanitary landfills
Sooy, Walter R., 332*n*8
Sorokin, Pitirim A., 294, 374*n*12, 381*n*3, 384*n*32
South Africa, and nuclear weapons, 158
Southeast Asia, 19; in scenario, 105
South Korea: and nuclear weapons, 158
Soviet bloc, 237

Soviet society, pluralization of, 237; scenario, 80; and technological impact, 317; and "open society," 323
Soviet Union, 4, 15, 17, 24, 45, 142, 155, 168; and global projection of power, 13, 25; and fusion energy, 43; and weather and climate modification, 43; and MHD power, 43; effect of World War II on technological capability of, 60; and technological composition of the economy, 61, 65, 68; economic policy, 61-62; and factors in economic growth (historical), 62; complexity of modern economy, 62; productivity, 62; economic reforms, 62-63, 65, 84; State Committee for Science and Technology, 63-64, 239, 344*n*17; and science, 63, 64; Academy of Sciences, 63; imports of technology, 64; Ninth Five-Year Plan (1971-1975), 65, 71, 72, 74, 75, 77; and consumer sector, 65, 73-74; economic vs. technological inefficiency, 65; productivity, compared with U.S., 66; All-Union Automated System for Planning and Management (OGAS), 66, 87; Tenth Five-Year Plan (1976-1980), 68, 73, 74, 75; Fifteen-Year Plan (1976-1990), 68; advanced technologies, compared with U.S., 68; Siberian rivers, 69; "economic levers" vs. "administrative methods" in economic management, 70; bureaucracy, 70, 85; consumer goods vs. heavy industry and defense, 71, 83; pressure on resources, 71; and China, 71; agricultural productivity, compared with U.S., 72; energy exports and balance of trade, 72, 346*n*45; North Star (Urengoy) project, 72; agriculture, 72-73; and manpower, 74-75; Shchekino system, 75, 83, 84; Ninth and Tenth Five-Year Plans and labor productivity, 75; and extent of allocation of resources to defense, 76, 347*n*61; military burden, 76-77, 86, 87; and military-industrial complex, 77;

Soviet Union (*Continued*)
  Defense Council, 77, 347-348n64; Politburo, 77; and internal competition for resources, 77; imports of Western technology and productivity, 77-78, 85; potential models of future society, 78-83; and the principle of relative economic advantage, 84; imports of technology and dependence on the West, 84; productivity as a problem, 85-86; and leadership, 87, 88; evolution of, and viability of the West, 88; in scenario, 105-9; political implications of energy exports (scenario), 105-6; projection of power by, and U.S. alliances, 135; differential advantages in stalemating U.S. power, 136; compared with U.S. in taking advantage of future technology, 138; and differential impact of technology, 150; pluralization of, and international stability, 162-63; *samizdat*, 164; "mellowing" of (scenario), 171; and problems of postindustrial society (scenario), 173; and industrial agreements with U.S. firms, 239; "linkages" with the West, 241; allocation of resources to military power (outlook), 243-44; as potential commercial competitor of U.S., 245; as semistationary society (prospective), 289, 317; as world civilization (prospect), 317-18; and zero-sum strategy, 317-18; and foreign aid, 335n26; science and ideology, 344n15; labor productivity and auxiliary workers, compared with U.S., 345n26; energy exports and trade surplus, 346n45; rate of economic growth, 347n59; change in energy base, 1960-1972 (statistics), 343n7; and export of uranium, 352n43; and difficulties with nationalities, 360n8; and international law, 360n9. *See also* Computers; Environment; Fusion energy; Long-range planning; Production associations; Research and development; Siberia

Space Shuttle, 248
Spain, and nuclear weapons, 158
Special Assistant to the President for Science and Technology, *see* Science Adviser to the President
Spencer, Daniel L., 353n4
Spengler, Oswald, 294
Sprout, Harold, 55, 233, 342n34, 352n48, 371n32
Sprout, Margaret, 55, 233, 342n34, 352n48, 371n32
Stability, international: 154-66; and internal stability, 154, 163-64; and USSR, 160; and nonmilitary technology, 161-66
Stability, social, and economic growth, 281
Stalin, J. V., 65, 375n29
Stans, Maurice H., 340n12
State, U.S. Department of, 212; Bureau of Oceans and International Environmental and Scientific Affairs (OES), lack of Congressional oversight over, 52; and science and technology in foreign policy, 202-4; as a designated constituency in national decision-making, 203; and control over resources, 203; regional vs. functional approach, 203-4; Bureau of International Scientific and Technological Affairs (SCI), 204; OES, problems of, 204; OES, potential role in science and technology policy, 213; and coordination of cooperation with USSR, 239; Bureau of International Organization Affairs (IO), 373n17
State, U.S. Secretary of: and international system, 253
Stationary-state ("no-growth") society, 280; prospect of, 275-76; definition, 276; emergence of, 283
Status, social: and semistationary society, 281, 284
Steinberg, Eleanor B., 341n23
Sterrett, C. C., 338n47
Stockholm Conference, *see* U.N. Con-

ference on the Human Environment (Stockholm, 1972)
Strategic Arms Limitation Talks (SALT), 77, 156
Strategic limited war, 332n1
Student unrest, U.S., 164
Subnuclear umbrella military level: and world stability, 159-60
Subnuclear umbrella military technology: defined, 11; growing stalemate in, 13-14, 135, 332n6
Subsystems, regional: rise of, 168
"Sufficiency" (military), 217; and general purpose forces, 217-18
Sunshine Program (Japan), 121
Superconductivity, 26; and cost of energy, 31-32, 338n47; and USSR, 68
"Superculture": defined, 34; rise of, 34; and international stability, 163, 164; and loyalty, 190
Surface-effect ships, 369n4; technological trends in, 28; and priorities in R&D, 218; work on in U.S., 337n35
Sutton, Francis X., 367n24
Sweden, 255
Switzerland, 96, 255
Systems analysis, 23, 41

Tanaka, Kakuei, 354n28, 355n42; and "remodeling Japanese archipelago," 125; demonstrations against in Southeast Asia, 355n37
Technetronic society, 374n1
Technical assistance, foreign, 323; reimbursable, 143; concessional vs. reimbursable, 255. See also Agency for International Development
Technological advance, and national power, 146-47
Technological change, rapidity of, and national power, 138
Technological composition of economy, orchestration of: in USSR, 68; in U.S., 230, 371n29
Technological impact: forms of, 2-3; and proliferation of interests, 2-3; and complexity of societal problems, 3; levels of, 11-16; and sociopolitical discontinuities, 33-38, 231-34, 275-80; and social stability, 147; and international vs. domestic politics, 331n4; "elite ethos" and international stability, 359n21. See also Control of technology and of its impact; Pluralizing effect of technology
Technological strategy, 57, 215-30; defined, 215; differentiated from "control of technology" and "science and technology policy," 368n2
Technological superstructure, 138; "determinism" of, 18; and Western Europe, 99; and LDCs, 142; modification of (U.S.), 215, 225-28, 230; defined, 333n3
Technology, civilian, see Nonmilitary technology
Technology: control of (defined), 3-4; as dependent variable, 4; as independent variable, 4; "deterministic" effect of (defined), 6; "volitional" effect of (defined), 6; "favorable determinism" (defined), 6; differential impact of, 6-7, 135; definition of, 8, 331n6; trends in and projections of, 17-38; factors in differential benefits from, 18; integrative effect of, 19-24; as an independent variable, divisive effect of, 23-24; global projection of influence through, 24-27; high cost of, 27; economies of scale in, 27; rapid change in, 29-33; differential effect of large scale and high cost, 138; sociopolitical impact on advanced societies of, 146-50; and changing values, 147-48; as dependent and independent variable and international stability, 154-55; and regional integration, 174-75; and functionalism, 176, 361n13; as independent variable and MNC's, 182; control of, and technology as dependent variable, 195-97, 364n1; interdependence of with institutions and administrative mech-

Technology (*Continued*)
anisms, 227; destabilizing potential of and U.S. policy, 234; as dependent vs. independent variable and foreign policy, 235-36; related to crises and leadership, 314. *See also* Control of technology and of its impact; Military technology; Nonmilitary technology; Science and technology; Technological impact

"Technology gap" (Atlantic Community), 90-92; limitations of as measure of Western Europe's viability, 92

Technology-intensive products, defined, 340n12

Technology, military, *see* Military technology

Technology transfer: from U.S. to Western Europe, 91; to Japan, 115, 122; U.S.-USSR, 239; U.S. internal, 371n27

Telecommunications, 15, 22; and Western Europe, 98-99. *See also* Communications

Telefunken AG, 27

Television, 232; global, 25; effect on student protest (U.S.), 34; quality of and culture (U.S.), 297; public, 297

Thailand, 355n37

Thermal pollution, and solar energy, 376n40. *See also* Waste heat

Thermionic converters, and USSR, 68

Thermonuclear power, *see* Fusion energy

Third World: and nuclear stability, 157-59. *See also* Less developed countries

Thomas-Gilchrist process, 333n9

Toffler, Alvin, 233, 371n33

Tominaga, Ken'ichi, 375n26

Toshiba International Corporation, 121

Totalitarianism and postindustrial society, 267

Toynbee, Arnold J., 294

Transgovernmentalism, 361n17

Transnationalism, 179, 361n17

Transnational violence, 160-61, 166; and social stability, 161

Transportation, 26, 181; and national power, 15; and mobility of raw materials, 17; advanced technology of, 22

Transportation, U.S. Department of, 370n24

Trans-Siberian railroad, 72

Treaty on Anti-Ballistic Missile Systems (1972), 156

Trotsky, Leon, 375n29

Tsuratani, Taketsugu, 355n46

Turkey, and nuclear weapons, 158

Twenty-First century, 259, 286; and human longevity, 338n44; and national security, 380n70. *See also* Semistationary society

Tyumen region (USSR), 71

Ukrainians, 348n73

U.N. Commission on Transnational Corporations, 363n30

U.N. Conference on the Human Environment (Stockholm, 1972), 52, 180; and NGOs, 364n41

Underwater technological capability of operations: projections of, 21

Unified commands, 335n22

United Arab Emirates, irrigation experiments in, 20

United Kingdom, *see* Britain

United Nations, 178, 190, 252, 253; specialized agencies of as actors in world politics, 179; and MNCs, 183; origins of, 186; prospect for, 318; relative ineffectiveness, 320

U.N. Environment Program (UNEP), 52, 180-81, 362n23

U.N. Industrial Development Organization, 359n19

U.N. Security Council, 251

United Nations specialized agencies, *see* Global functional institutions

U.N. World Science Conference, 141

United States, 4, 15, 17, 55, 142, 168, 373n14, and global projection of power, 24-25; strong points in science and technology, 41-42; television industry, 42, 45; computer industry, 42; chemical industry, 42; compared with Western Europe and Japan in selected industries, 43; weaknesses in science and technology, 43-45; and forces in U.S. technological growth, 44-45; and science and technology policy, 46-53, 219-220; technology-intensive products and balance of trade, 46-47; agricultural products and balance of trade, 49-50; and danger of economic decay, 51; and environmental policy, 52; problems and issues to be resolved, 53-58; and government-business relations, 54-55; productivity, need for increase, 57; environmental regulations, effect on economy, 58; pluralism of institutions as an asset, 58; in scenario, 104-6, 108; and Western Europe, 111, 113; and differential advantages over USSR in certain technologies, 136-37; compared with USSR in taking advantage of future technology, 138; and Soviet military power, 217; vitality of, and Soviet societal change, 245; rise and decline of as world civilization, 294; and lack of public confidence in U.S. institutions, 303, 382*n*21; Executive Branch, need for reorganization of, 311; and tendency toward internal stalemate, 323; and ideological appeal, 323, 324; and world responsibility as trailblazer for uncertain future, 325; and international system, 326-27; and materialistic values, 328; and investment abroad, 335*n*25, 339*n*4; and effectiveness of organization for developing marine resources, 369*n*12. *See also* Civilization, high; Decision-making; Energy; National purpose, U.S.; President, U.S.; Stationary-state society; U.S. foreign policy; United States as world civilization; and various U.S. government agencies

United States as world civilization, 315; factors in its rise, 321-22; decline in assets, 322-24; recoverable assets, 324-25; and LDCs, 324-25; U.S. values and world order, 326

U.S.-European cooperation in science and technology: desirability of, 247-48; and major multilateral vs. bilateral programs, 248-49; and marine R&D, 249, 373*n*15

U.S. foreign policy: and Soviet Union, 236-245; and "containment" of USSR, 236-237; and pluralization of USSR, 237; détente, 236-41; objectives with regard to USSR, 240-41; and outlook for pluralization of Soviet society, 241-44; U.S.-Soviet, weaknesses of, 245; and Western Europe, options of, 245-49; and Japan, options of, 250-52; and future international system, 252-55, 326-29; and promotion of "open society," 323; trilateral cooperation with Western Europe and Japan, 373*n*13. *See also* Foreign aid, U.S.; International system; State, U.S. Department of; U.S.-European cooperation in science and technology; U.S.-Japanese cooperation in science and technology; U.S.-Soviet cooperation in science and technology; Technical assistance, foreign

U.S. Navy, 13; diving experiments, 21; and surface effect ships, 337*n*35. *See also* Naval power, U.S.

U.S. President, *see* President, U.S.

U.S.-Soviet cooperation in science and technology, 238-41; and Soviet societal change, 241-44; economic benefits from, 244-45

U.S.-USSR Agreement on Cooperation in the Field of Science and Technology (May 1972), 238-39

U.S.-USSR Joint Commission on Scientific and Technical Cooperation, 239

U.S. Steel Corporation, 18

408　Index

Universities, U.S.: and mid-career changes, 228; in postindustrial society, 271, 273; and intellectual leadership, 375n35
Uranium Enrichment Co. (Urenco), 97; in scenario, 107
Urengoy (USSR), 71, 72
USSR State Committee for Science and Technology, *see under* Soviet Union
USSR, *see* Soviet Union
Ustinov, D. F., 77

Values: technological impact upon, 26; and U.S. youth, 33-35; consensus on and technology, 53-54; and priorities in science and technology, 56-57; and productive pursuits in advanced societies, 148-49; embedded in institutions, 207; and control of technology, 207, 209-10; social, and priorities in science and technology, 224-25; technology, and changing priorities in, 232; and reorientation of, 284, 292; materialistic and social instability, 286; and national purpose, 293; reorientation of, and improvement of social ills, 309-10; and material rewards, 370n64; and agreeable society, 380n69. *See also* Business; Diplomats; Engineers; Scientists
Van de Walle, Werner, 351n28
Van Zandt, Howard F., 353nn3, 8, 10, 354n31
Venezuela, 365n29
Vernon, Raymond, 350n10, 362n27
Verrier, Anthony, 332n6
Videophones, 22
Vietnam war, 237, 323; stalemate in, 13; and effect on U.S. youth, 34-35
Virgin raw materials, 276, 280; in semistationary society, 288, 378n55
Virtuous society: as national purpose, 294; and *laissez-faire* environment, 296; criteria for, 307-10; and justice, 308; implementation of, 310-14; and balance between pluralism and centralization, 310-11. *See also* Civilization, high
Vladivostok accord, 156
Voice of America, 25
"Volitional" effect of technology, defined, 6
Volkov, G., 344n18

Wakaizumi, Kei, 131, 132, 355n44
Walsh, John, 350n16, 351n27, 352n34
Waltz, Kenneth N., 358n1, 360n2
War: and semistationary societies, 290-91; prospect for incidence of, 318; future technological impacts on, 319-20
Ward, Barbara, 359n18
Ware, Willis H., 335n21
Waste heat: as a pollutant, 276; global control of, 319; and local and regional climatic disruptions, 379n58
Watergate, 297, 303, 310, 383n29
Weather and climate modification, 26, 28, 333n10; and national power, 15; and USSR, 68; and Western Europe, 91, 96; and international regulation, 178; and priorities for R&D, 229; future capabilities for, 336n30
Welfare system, U.S., 57, 342n38
Wells, J. Morgan, 334n14
Western Europe, 4, 138; and global projection of power, 25; as U.S. competitor, 48, 90; reemergence of vitality of after World War II, 89-90; technology transfer from U.S., 91; gain in productivity, vs. U.S., 91; and lag in advanced technology, 91; weaknesses in development and utilization of technology; 92-100; and energy, 94-95; and energy crisis, 95, 104-6 (scenario); and environment, 95-96, 106-7 (scenario); and computer technology, 98; and timely development of technology, 99-100; and utilization of technology, 100; patent system, 102; and venture capital, 102; and industrial standardization, 102; aircraft industry, 102, 107 (scenario), 352n44; potential

future models of, 103-10; merchant marine, 106; and "American challenge," 111; cooperation vs. competition in science and technology with U.S., 111-13; and differential impact of technology upon, 149; and U.S. policy options, 245-49; as world civilization (prospect), 315; labor productivity, compared with U.S., 349n9; changing form of nationalism, 351n29; Oceanic Development Commission (proposed), 353n49. *See also* European Community; Fusion energy; Gas centrifuge technology; Instability, social; Lasers; Pluralizing effect of technology; Weather and climate modification; and individual West European countries

"Westernistic" society, 264-65, 273

Wetter, F., 343n6

White House (Executive Office of the President): and science and technology policy, 204-5; politicization of science and technology policy in, 205

Wiener, Anthony J., 359n16, 374n16

Wienert, Helgart, 344n15

Wolfe, Thomas W., 332n7, 347n64, 349n75

Women: impact of technological change on, 231; equal rights for, 298, 299

Woods Hole Oceanographic Institution, 373n14

World Bank, 359n19

World government: and technologico-economic forces, 186-87

World order, 327, 371n2

World politics, 185; as positive (non-zero) sum game, 7, 187, 189, 288, 289; as zero-sum game, 7-8, 187; as mix of zero-sum and positive sum game, 188-89; and international relations, 331n1. *See also* Functionalism; International relations; International system; Political science; Politics; Transnationalism; World society

World society: and changing pattern of power in, 191; and future outlook, 318-21; pluralizing effect of technology on, 321; governing of, 321

"World system," 327, 328

Wright, Christopher, 366n14, 19, 22

Yager, Joseph A., 341n23

Yashica Corporation, 121

Yergin, Daniel, 350n21

Young, Oran R., 360n3

Yugoslavia, 80

Zaibatsu, 114

Zaleski, Eugene, 344n16

Zeiss, Carl (Co.), 121

Zero-sum game: defined, 7; and semi-stationary society, 281; and Japan, 316; and USSR, 317-18. *See also* World politics